Lecture Notes in Physics

The Lecture Notes in Physics

The series Lecture Notes in Physics (LNP), founded in 1969, reports new developments in physics research and teaching – quickly and informally, but with a high quality and the explicit aim to summarize and communicate current knowledge in an accessible way. Books published in this series are conceived as bridging material between advanced graduate textbooks and the forefront of research to serve the following purposes:

- to be a compact and modern up-to-date source of reference on a well-defined topic;
- to serve as an accessible introduction to the field to postgraduate students and nonspecialist researchers from related areas;
- to be a source of advanced teaching material for specialized seminars, courses and schools.

Both monographs and multi-author volumes will be considered for publication. Edited volumes should, however, consist of a very limited number of contributions only. Proceedings will not be considered for LNP.

Volumes published in LNP are disseminated both in print and in electronic formats, the electronic archive is available at springerlink.com. The series content is indexed, abstracted and referenced by many abstracting and information services, bibliographic networks, subscription agencies, library networks, and consortia.

Proposals should be sent to a member of the Editorial Board, or directly to the managing editor at Springer:

Dr. Christian Caron
Springer Heidelberg
Physics Editorial Department I
Tiergartenstrasse 17
69121 Heidelberg/Germany
christian.caron@springer-sbm.com

Richard Wielebinski   Rainer Beck (Eds.)

Cosmic Magnetic Fields

Springer

Editors

Richard Wielebinski
Rainer Beck
Max-Planck-Institut für Radioastronomie
Auf dem Hügel 69
53121 Bonn
Germany
E-mail: rwielebinski@mpifr-bonn.mpg.de
E-mail: rbeck@mpifr-bonn.mpg.de

Richard Wielebinski and Rainer Beck, *Cosmic Magnetic Fields*,
Lect. Notes Phys. 664 (Springer, Berlin Heidelberg 2005), DOI 10.1007/b104621

Library of Congress Control Number: 2005930474

ISSN 0075-8450
ISBN-10 3-540-24175-2 Springer Berlin Heidelberg New York
ISBN-13 3-540-24175-1 Springer Berlin Heidelberg New York

Springer is a part of Springer Science+Business Media
springeronline.com

Printed in The Netherlands

Typesetting: by the author using a Springer LaTeX macro package

Printed on acid-free paper SPIN: 11369875 55/Techbooks 5 4 3 2 1 0

Preface

Recent years have seen great progress in the field of cosmic magnetic fields, both in the available observational data and in the theoretical interpretations. Now we have extensive data on magnetic fields in virtually all types of cosmic objects: the Earth, the Sun, the Milky Way, out to magnetic fields in the most distant reaches of the Universe.

The first detection of a magnetic field outside the Earth was made by G.E. Hale, who made optical Zeeman effect observations of magnetic fields in the Sun, nearly 100 years ago. The detection of the Zeeman effect in stars by H.W. Babcock followed some 40 years later. Optical starlight polarization observations were made in 1949 and, when combined with the Davis–Greenstein effect interpretation, suggested that magnetic fields were present in our Milky Way. Radio polarization measurements confirmed this deduction in 1962. With developing sensitivity of radio and optical observations magnetic fields have been shown to be present practically everywhere. We know that the Sun is driven by magnetic fields. Supernova remnants show us the evolution of the magnetic field in shock fronts that follow a stellar explosion. Pulsars and X-ray sources have been shown to possess extremely intense magnetic fields. The Milky Way is a magnetic laboratory, with complex magnetic field structures, worthy of exploration. Regular patterns of large-scale magnetic fields are observed in nearby galaxies and radio galaxies. Also clusters of galaxies were shown to be permeated with detectable magnetic fields. In spite of this mounting evidence about the presence of magnetic fields in the cosmic Universe only a few attempts have been made to interpret the situations taking into account all the relevant parameters, in particular those resulting from the presence of magnetic fields. In many publications elaborate numerical investigations are carried out but without the consideration of the action of a magnetic field. Only recently did we get magneto-hydrodynamic codes. These developments show us that the inclusion of magnetic fields is indeed necessary. Possibly the whole approach to the interpretation of observational data will change as a result of magnetic effects. The next decade in astrophysics will shed more light on the role of magnetic fields in the Universe. We should soon find out if the magnetic fields are only a consequence of gas motion, or are they at the heart of the matter?

The scheme of this book is to start with the most distant magnetic fields in the Universe and then make our way back to our Galaxy. Contributions based on observational data will describe magnetic fields in clusters of galaxies, in radio galaxies, nearby galaxies and in the Milky Way. Contributions on various theoretical effects and considerations of the magnetic fields in the Universe are interspersed between the observational chapters where appropriate. The review articles in this book do not cover all aspects of cosmic magnetic fields. In particular no contribution on solar magnetic fields has been included since in recent years numerous reviews and books have covered this subject adequately.

In the first contribution Martin Rees considers the magnetic fields in the early Universe. A substantial magnetic field could have been generated in the early Universe, however no relevant physics to this early stage of our development has been proposed do far. It is argued that the first significant magnetic fields must have been formed in the course of formation of the early non-linear structures giving the needed seed fields needed for the dynamo action. The build-up of the magnetic fields is an important aspect of the cosmogonic process.

The second contribution by Philipp Kronberg deals with the magnetic fields in galaxy systems, clusters of galaxies and beyond. The history of the realization that cosmic rays and magnetic fields are essential for the observed radio emission is given. This chapter focuses on the possible role of stars, black holes and supernovae, in injecting magnetic energy into the intergalactic medium. These two broad categories of energy output from galaxies have been recognized for some time. The discussion begins with the environment of galaxies out to a distance of the local Supercluster of galaxies and proceeds all the way back to the formation of first stars and galaxies. The magnetic effect in radio galaxies are also considered.

Rainer Beck's contribution describes the magnetic fields in nearby galaxies. There has been a tremendous progress in this field in recent years. This progress in observations reveals a wide range of large-scale magnetic phenomena. Spiral fields exist in grand-design and flocculent galaxies, and even some dwarf galaxies host ordered fields. Regular magnetic fields trace the gas flow in barred galaxies. Vertical magnetic fields observed above disks of edge-on galaxies indicate strong galactic winds into the halos. Magnetic fields possibly help to feed the active galactic nuclei, which may solve a long-standing problem.

A contribution by Russel Kulsrud on the origin of Galactic magnetic field follows. From considerations of the origin of cosmic rays the existence of magnetic field became obvious. Assuming that a magnetic field did exist there is no problem in sustaining it. On the other hand there are problems to create such a magnetic field. The discussion touches on the well known alpha–omega disc dynamo as well as evolution of primordial magnetic fields. Arguments for and against either interpretation are clearly given, pointing out that the question of the origin of the galactic magnetic fields remains open.

In the fifth contribution Richard Wielebinski describes the present knowledge about the magnetic fields of the Milky Way, derived from radio continuum and Faraday effect observations. The basics of the synchrotron emission theory are sketched showing their application to the observations of magnetic fields. The development of radio observations is given and the latest results described. A combination of radio polarization surveys and Faraday Rotation Measure studies of pulsars and extragalactic radio sources are expected to lead to a 'tomography', a three-dimensional description of the magnetic field of the Milky Way.

The following contribution by Anvar Shukurov on the mesoscale magnetic structure in spiral galaxies attempts to give interpretation for the observations described in Chapters 3 and 5. Various observed phenomena, like regular magnetic fields in the inter-arm regions of spiral galaxies or the observed field reversal in the Milky Way are considered. It is shown that the dynamo theory has been impressively successful in explaining the gross features of galactic magnetic fields. Systematic studies of structures on intermediate scales should advance our understanding on the nature of cosmic magnetism.

The seventh contribution by Carl Heiles and Richard Crutcher deals with the details of magnetic fields in diffuse H I and molecular clouds. The Zeeman effect allows direct measurement of the magnetic fields in these objects and the recent increase in sensitivity gave us a large number of new results. Polarization of starlight is discussed as a basic phenomenon that delineates the local magnetic fields. Polarization of thermal grains is sketched as well as spectral-line polarization. The history of H I Zeeman observations is discussed in detail as well as the development of gathering the data on molecular lines. Although a large volume of reliable data has been collected the future developments, like the new large radio telescopes, are described since they hold a key for future new results.

In the contribution that follows Leon Mestel and John Landstreet give us an overview on the state of stellar polarization observations. Stellar magnetic fields have been detected across the whole Hertzsprung–Russell diagram. This contribution concentrates on pre-main sequence and on late- and early-type stars, with some discussion of red giants and white dwarfs. The current observational situation is reviewed and some consequences on star formation, stellar structure and evolution are summarized.

The two final contributions deal with details of some theoretical aspects of magnetic fields. Axel Brandenburg considers the importance of helicity in dynamo theory. Magnetic helicity is conserved and its evolution provides a dynamical feedback on the alpha effect that is distinct from alpha quenching. The explicit connection with catastrophic alpha quenching is reviewed and the alleviating effects of magnetic and current helicity fluxes are discussed.

The final chapter by Max Camenzind deals with numerical magnetohydodynamics (MHD) in astrophysics. In the past 10 years powerful numerical algorithms and computational methods have been developed for simulating the time evolution of magnetic fields in astrophysical environments. The most recent trends go to fully conservative schemes and adaptive mesh refinement for large-scale supercomputing. Examples of such computer simulations are given. A discussion of General Relativistic MHD codes, still in development, is given.

The collection of contributions for this book needed more than one year. Some authors used this time to update their manuscripts in view of the quick changes in this subject. In this time also concrete plans for new instruments (ALMA, LOFAR, SKA) have been crystallizing. 'Cosmic Magnetism' is one of the Key Science Projects for the Square Kilometer Array (SKA). Great steps forward in our understanding of magnetic fields will be possible.

The editors wish to thank Gabi Breuer for adapting all the contributions to the Springer macro system. We wish to thank Anton Zensus for the support that made this book a reality.

Bonn
February, 2005

Richard Wielebinski
Rainer Beck

Contents

List of Contributors

Martin J. Rees
Institute of Astronomy
Madingley Road
Cambridge, CB3 0HA, UK
mjr@ast.cam.ac.uk

Philipp P. Kronberg
Institute of Geophysics
and Planetary Physics
Los Alamos National Laboratory
NM 87501, USA
kronberg@lanl.gov

Rainer Beck
Max-Planck-Institut
für Radioastronomie
Auf dem Hügel 69
53121 Bonn, Germany
rbeck@mpifr-bonn.mpg.de

Russel M. Kulsrud
Princeton University
Plasma Physics Laboratory
Princeton NJ 08544, USA
rkulsrud@astro.princeton.edu

Richard Wielebinski
Max-Planck-Institut
für Radioastronomie
Auf dem Hügel 69
53121 Bonn, Germany
rwielebinski@mpifr-bonn.mpg.de

Anvar Shukurov
School of Mathematics
and Statistics
University of Newcastle
Newcastle upon Tyne
NE1 7RU, U.K.
anvar.shukurov@ncl.ac.uk

Carl Heiles
Astronomy Department
University of California
Berkeley, CA 94720, USA
cheiles@astron.berkeley.edu

Richard Crutcher
Astronomy Department
University of Illinois
Urbana, IL 61801, USA
crutcher@uiuc.edu

Leon Mestel
Astronomy Centre
University of Sussex
Falmer, Brighton
BN1 9QH, England
lmestel@sussex.ac.uk

John D. Landstreet
Department of Physics
and Astronomy
University of Western Ontario
London, ON N6A 3K7, Canada
jlandstr@astro.uwo.ca

Axel Brandenburg
Nordita, Blegdamsvej 17
2100 Copenhagen Ø Denmark
brandenb@nordita.dk

Max Camenzind
Landessternwarte Königstuhl
69117 Heidelberg, Germany
M.Camenzind@lsw.uni-heidelberg.de

Magnetic Fields in the Early Universe

Martin J. Rees

Institute of Astronomy, Madingley Road, Cambridge, CB3 0HA, UK
mjr@ast.cam.ac.uk

Abstract. A substantial magnetic field could conceivably have been generated in the ultra-early Universe. However, the relevant physics at those eras is very uncertain and no very plausible mechanism has been proposed. It seems more likely that the first significant cosmic fields, and the seed field for galactic dynamos, had to await the formation of the first non-linear structures. The history of star formation is controlled, at least in part, by how and when galaxies and their precursors acquired their fields. The build-up of magnetic fields is an important aspect of the overall cosmogonic process.

1 Introduction

Magnetic fields pervade stars, galaxies, and even perhaps the intergalactic medium. They probably owe their present strength to dynamo amplification. But there must then have been an initial seed field – otherwise the dynamo process would have had nothing to feed on. It seems to be generally 'taken for granted' that the requisite seed field will be there. In many astrophysical contexts this confidence may be justifiable: if the dynamical (and amplification) timescale is short enough, there can be a huge number of e-foldings; a merely infinitesimal statistical fluctuation might then suffice. But the large-scale fields in disc galaxies seem to pose a less trivial problem. The amplification timescale may be 2×10^8 years; even by the present epoch there would then have been time for only 50 e-foldings. The galactic field could not, therefore, have built up to its observed strength by the present day, unless the seed were of order 10^{-20} G – very weak, but not infinitesimal. Moreover, if substantial fields exist even in high-z galaxies whose discs may have only recently formed, the seed would need to have been correspondingly higher.

Magnetic fields are crucial in cosmic radio sources, and for the physics of cosmic-ray production and propagation. Moreover, star formation would proceed differently (with regard both to its rate, and the shape of the initial mass function) if there were no magnetic field: the field modifies the Jeans mass and contributes to transfer of angular momentum in protostars. So we cannot hope to model galactic evolution adequately without knowing when the field builds up to a significant strength. If several galactic rotation periods elapsed before a dynamically-significant field built up, then the oldest stars may well, for this reason alone, have a different luminosity function (Rees, 1987). (Moreover, a field can be significant even when it is not, overall, of dynamical importance: a far weaker and dynamically negligible field – one that is merely strong enough to render the gyroradius smaller than the collisional mean free path and thereby inhibit transverse conductivity – may be significant.) There is as much reason to believe that the absence of a magnetic field affects the IMF as to believe that a lack of heavy elements does so (though the quantitative nature of the effect is as uncertain in the one case as the other).

The question of how quickly the interstellar magnetic field built up is germane to several other aspects of galactic evolution. Obviously the behaviour of supernova remnants in high-z galaxies is sensitive to this (as is the trapping of cosmic rays, and the possibility of inflating loops into the halo – a process which may itself affect the efficiency of galactic dynamos). So where did the 'magnetic history' of the cosmos actually start?

2 Magnetic Fields from the Ultra-Early Universe?

Could a field have been created in the exotic ultra-dense stages of the big bang? According to current theories, the ultra-early Universe possibly underwent various phase transitions; and maybe one of these transitions could (as in a cooling ferromagnetic material) spontaneously create a field. Because the relevant physics is exotic and poorly understood, we plainly cannot rule this possibility out (see, e.g., the comprehensive review by Widrow (2003). Various authors have considered what might have happened at the electroweak phase transition (at around 10^{-12} s) and the later quark-hadron transition (at around 10^{-5} s) where it has been speculated that there might be 'bubble' formation in a first-order phase transition. Even if the transition that created the field occurred as late as the quark-hadron transition era, when the horizon encompassed only 10^{-6} solar masses of baryons, the resultant field on galactic scales (Hogan, 1983) would be only 10^{-30} G. Even worse is the situation at the GUT (10^{15} GeV) era. A phase transition back then might be more efficient in converting ambient energy into magnetic form, but if the correlation scale is limited to the scale of the horizon, the smallness of this scale imposes a very severe constraint on the net field strength at astrophysically-interesting scales. Suppose that, at a very early time t, some physical process generates an ordered field on the scale of the horizon at t, whose strength is such that $B^2/8\pi \simeq F\left(aT^4\right)$, with $F < 1$.

Suppose also that the Universe subsequently expands according to the ordinary (decelerating) Friedmann equations. Then on a galactic scale we would expect an ordered field with energy density.

$$F \times \left\{\frac{\text{horizon mass where field is created}}{\text{mass of a galaxy}}\right\} \times \left(aT^4\right) \tag{1}$$

A seed field of 10^{-20} G has energy density 10^{-29} (aT^4). At the GUT era (when the horizon was only large enough to encompass about 10^4 baryons) the ratio in the brackets in (1) is of order 10^{-65}. So, even if the field had a high local energy density (and F was not very small), it would be primarily on such small scales that it would quickly decay, and there would seem no chance of getting even 10^{-20} G on the scale of a protogalaxy.

The limit set by the horizon scale of course only straightforwardly applies if the field, along with the background radiation, is created after any inflationary phase was completed. This is a reasonable assumption because any field produced before or during inflation would have been exponentially diluted. It assumes also that there is not an effective 'inverse cascade' mechanism, as invoked by Field and Carroll (2000), that can transfer energy from small to large scales; this requires a helicity in the primordial field that does not seem likely to be present in such models. The gulf between the small scales of these exotic processes and the large scale of galaxies poses a generic problem with attempts to attribute an exotic origin to galactic-scale fields (even merely of 'seed' strength); certainly no convincing microphysical mechanism has yet been found. (Of course, this problem would be surmounted if there were an overall cosmic anisotropy.)

A magnetic field already present at the recombination era, might affect the cosmogonic process (cf. Rees, 1971; Wasserman, 1978; Subramanian and Barrow, 1998). The constraints are summarised in Fig. 1, in terms of the field's characteristic lengthscale. A field whose comoving strength were now $> 4 \times 10^{-10}\Omega_b$, where Ω_b is the fraction of the critical cosmological density in baryons, would, at recombination (and at all later epochs until reheating occurred), have contributed more pressure than the baryons and electrons; it would therefore have affected the Jeans mass, and raised the minimum mass of the first generation of bound systems that would be expected in all 'hierarchical' models for the build-up of cosmic structure. Moreover, even a field too weak to affect the Jeans mass could still be cosmogonically important in two ways: (i) A field with characteristic scale ℓ would (because of the inhomogeneous stresses) induce motions at about the Alfvén speed on those same scales. Any resultant density fluctuations whose amplitude, at t_{rec}, exceeded 10^{-3} would have become non-linear, via the ordinary gravitational instability, by the present epoch. Thus, even a present-day intergalactic field as low as 10^{-13} G could have been cosmogonically significant if it dated from the pre-recombination

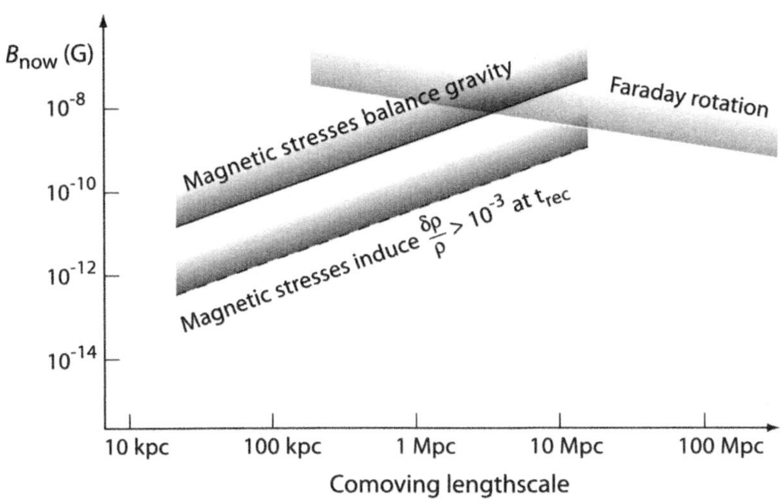

Fig. 1. Constraints on the magnetic field on various length scales at the time of recombination t_{rec}. A field is cosmogonically important if it can generate density perturbations of amplitude $\sim 10^{-3}$ at t_{rec}, since these would have developed into gravitationally bound systems by the present time. See text for further explanation

era. (ii) The very first stars, 'Population III', are believed to have formed at redshifts of order 20 from gas that underwent radiative cooling via molecular hydrogen and formed stars in the cores of minihalos of dark matter. The final masses of these stars depend on details of cooling and pressure. The relevant gas compresses almost isothermally to more than 10^{10} its turnaround density before becoming a genuine protostar that thereafter collapses quasistatically. During the near-isothermal collapse, a frozen-in field would become dynamically significant even if it were 1000 times below the equipartition pressure in the uncollapsed intergalactic medium. This field is similar in strength to that inferred in (i) above, but to be significant it need not be of such large scale: indeed it can on scales encompassing a sub-stellar baryonic mass – all that is necessary is that it remains 'frozen in' to a collapsing Population III protostar without suffering dissipative decay.

The present constraints on an intergalactic field come from upper limits to intergalactic Faraday rotation. These depend on the field's correlation length, ℓ and are sketched in Fig. 1. These Faraday limits are not stringent enough to rule out *tangled* fields of interesting strengths. (And the strength and topology of such fields is of course important for conductivity in clusters of galaxies, intergalactic propagation of ultra-high energy energy cosmic rays and so forth).

3 Protogalactic Batteries

A battery can start to operate whenever there is some large-scale vorticity. If the primordial fluctuations were irrotational (as they are in most cosmological models), then this would have to await nonlinearities that lead to shock waves, ionization fronts, or the formation of bound systems that exert tidal torques on each other. Compton drag can then (cf. Zeldovich et al., 1983) gradually build up a current in a rotating protogalaxy. If plasma moves at speed V relative to the frame in which the microwave background is isotropic, its motion would be damped out on a timescale $(m_\mathrm{p}/m_\mathrm{e})\, t_\mathrm{comp}$, where $t_\mathrm{comp} = m_\mathrm{e}c/\sigma_\mathrm{T}(aT^4)$ is the usual Compton cooling timescale for electrons. To couple electrons and ions, an E-field of strength $m_\mathrm{e}V/et_\mathrm{comp}$ must maintain itself in the plasma. A protogalaxy of radius R rotating with speed V would be gradually braked by Compton drag, and the E-field within it (with, of course, non-zero curl) would build up a B-field at a rate $\left(m_\mathrm{e}c^2/et_\mathrm{comp}\right)(V/R)$. For a protogalaxy at redshift $z \simeq 5$, this process yields a field of order 10^{-21} G. If, contrary to most cosmologists' expectations, vorticity were present in the original fireball (i.e. before density perturbations became non-linear), then an ingenious variant of this mechanism, proposed by Harrison (1970), could start even earlier than t_rec. But not even this mechanism could generate more than 10^{-19} G on a galactic scale. We should therefore explore other possibilities; and we would indeed be impelled to do so if even high-z galaxies turned out to have strong magnetic fields.

The battery effects due to Compton drag would be important at very high redshifts (and for very supersonic motions at lower redshifts). However, there is then the more generic possibility of a thermal battery behind oblique shocks, or indeed whenever there is a discontinuity or gradient in plasma temperature (Lazarian, 1992). This has been discussed in the context of protogalactic collapse by various authors (see, for instance, Kulsrud et al., 1997, Chap. 4), but generically the field on galactic scales due to this process is below 10^{-20} G. A variant on this process, though even less efficient than shocks on the same scale, is the battery that occurs at an ionization front (Subramanian et al., 1994; Gnedin et al., 2000).

4 Magnetic Fields from the First Stars

Protostars condensing in the present-day interstellar medium, as is well known, start off with too much magnetic flux rather than too little and shed the excess via (for instance) ambipolar diffusion. But the field in a star at the end of its life may be insensitive to the conditions at its birth: even if a star initially had zero field, a stellar-scale Biermann battery could generate a seed field, on which dynamo amplification (by a huge number of factors of e if necessary) could operate. If such a star exploded as a supernova, then a wind spun off the remnant pulsar could pervade several cubic parsecs with a field of order 10^{-4} G (just as in the Crab Nebula). The ejecta from the first Population III stars would expand more widely – either individually or as part of a wind from a starburst in a small 'pregalaxy' forming in a halo with a shallow potential well (Kronberg et al., 1999). Such fields could become widely dispersed in intergalactic space. By the time gas came to form galactic discs like ours, it could have been contaminated by Population III remnants. These could have created a stronger field throughout the galactic disc than a larger-scale battery could have done, unless there were turbulent amplification (cf. Kulsrud et al., 1997, Chap. 4) during the collapse. The scale would still be that characteristic of a supernova remnant (albeit sheared and partially mixed), so to end up with a galactic-scale field the large-scale modes would need to be preferentially amplified by the dynamo mechanism. As a quantitative estimate of the large-scale 'seed', note that each hemisphere of the Crab Nebula contains an (equal and opposite) flux of order 10^{34} G cm^2. If N remnants of early (perhaps even Population III) supernovae ended up as part of the gas in a young galactic disc, the net flux would then be larger by a factor N^x. The appropriate value for x isn't obvious. The net effect depends on the two hemispheres evolving differently – otherwise the net flux cancels out. To assume that $x = 1/2$ may therefore be over-optimistic. A better guess might be $x = 1/3$. This is appropriate if the remnants are randomly oriented, and the galactic disc can be modelled as the interior of a surface which slices a fraction N of the remnants. As an example, if $N = 10^6$, the large-scale component of the field in a protogalactic disc of 10 kpc radius would be $3 \times 10^{-8} - 3 \times 10^{-9}$ G, for x in the range 1/3–1/2.

One key issue is the degree of turbulent mixing and diffusion of fields from (for instance) supernova remnants. We know these processes are very effective in a galactic disc, where there are active churning and shearing motions. However, at eras when star formation has just begun, and in quiescent locations like the intergalactic medium, it is not at all obvious that such homogenisation occurs. The flux may remain restricted to the supernova ejecta, and may not penetrate pristine intergalactic matter. There is an interesting connection here to the diffusion of the heavy elements that result from early stellar nucleogenesis. These are known from QSO absorption line studies to pervade the intergalactic medium widely, even at redshifts as large as 3–5. But it is unclear whether they are fully mixed, or confined to clumps or filaments with a small volume-filling factor (but nonetheless

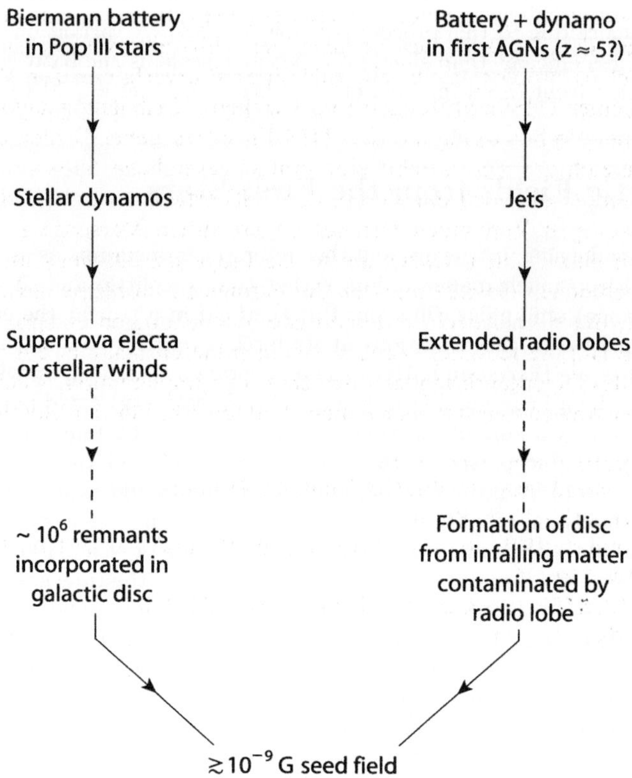

Fig. 2.

with a large enough surface-covering factor to account for the ubiquity of heavy elements along lines of sight to QSOs). The dispersal of the first heavy elements and of the first significant magnetic fields are probably linked, and how these processes proceed between redshifts of $\gtrsim 10$ and the present is a fascinating study that is now becoming observationally feasible.

5 AGNs and Radio Lobes

Even some of the highest-redshift radio galaxies have radio lobes up to 50 kpc in size, containing ordered fields of 10^{-5} G, implying a flux of order 10^{41} G cm^2. We are observing them at a cosmic epoch when the Universe had only a tenth of its present age; radio galaxies like 4C 41.17, with a redshift $z = 3.8$ (Chambers et al., 1990) may well have formed when the formation of typical galaxies (especially those with discs) still lay in the future.

The fields in the lobes of radio galaxies could have been generated in the active nucleus of the associated galaxy and expelled along collimated jets (resembling a scaled-up and directional version of the relativistic pulsar wind that generates the Crab Nebula's field) (Chap. 2). In the nucleus itself, the dynamical timescale may be as short as a year, or even a few hours if the relevant processes occur close to a black hole. So we need not worry about what seeded the AGN itself: just as in a star, there is time for millions of e-foldings – more than enough time for a battery process to operate, or for a dynamo to be seeded by an infinitesimal field. Thus, a radio galaxy's field, like that in a supernova remnant, can be accounted for even if the progenitor central object had zero field when it formed.

Galaxies may acquire their discs at $z \lesssim 2$ via collapse of a slowly-rotating cloud with turn-around radius > 50 kpc. The diffusion of the huge amounts of magnetic flux from radio lobes into the wider medium is as uncertain as the similar case on supernova scales discussed in the last section. If the infalling material had been 'contaminated' by a fraction f of a radio source lobe, the large-scale component of the seed field would be $3\times10^{-8}(f/10^{-4})$ G. So only a small value of f might suffice. However the seed fields in discs could only be attributed to early radio sources if the lobe material were subsequently mixed into a larger volume. This is because radio galaxies are relatively thinly spread through the Universe, being far less common than disc galaxies.

6 Summary

The seed field for the galactic dynamo poses a more challenging problem than the seeding of smaller-scale cosmic dynamos because the galactic timescale is so long, and the amplification correspondingly slow (and, of course, the problem is far worse if the galactic dynamo mechanism is less efficient – cf. Chap. 4 by R. Kulsrud). There are as yet no firm grounds for expecting significant fields in the ultra-early Universe – indeed there are good reasons for expecting the large-scale components of any such field to be uninterestingly small. And the galactic-scale batteries where Compton drag or temperature gradients provide the emf would be barely enough to yield an adequate seed. More promising, in my view are the two in Fig. 2, either of which could yield $\sim10^{-9}$ G.

These mechanisms are not mutually exclusive; and there are clearly strong inter-relations between fields in stars, in AGNs or radio galaxies, and in galactic discs. The build-up of a galactic magnetic field depends on how strong the seed field is and when it was generated. Because of the field's importance in star formation, we have little chance of really understanding what a high-redshift galaxy should look like until issues of magnetic field amplification and field diffusion have received a good deal more attention.

References

Chambers, K. C., Miley, G. K., van Breugel, W. J. M.: 1990, ApJ 363, 32
Field, G. B., Carroll, S. M.: 2000, Phys. Rev. D 62, 10308
Gnedin, N. Y., Ferrara, A., Zweibel, E. G.: 2000, ApJ 539, 505
Harrison, E. R.: 1970, MNRAS 147, 279
Hogan, C. J.: 1983, Phys. Rev. Lett. 51, 1488
Kronberg, P. P., Lesch, H., Hopp, Y.: 1999, ApJ 511, 56
Kulsrud, R. M., Cen, R., Ostriker, J. P., Ryu, D.: 1997, ApJ 480, 481

Lazarian, A.: 1992, A&A 264, 326
Rees, M. J.: 1971, in *General Relativity and Cosmology*, ed. R. K. Sachs (New York: Academic Press), p. 315
Rees, M. J.: 1987, QJRAS 28, 197
Subramanian, K., Barrow, J. D.: 1998, Phys. Rev. Lett. 81, 3575
Subramanian, K., Narashima, D., Chitre, S. M.: 1994, MNRAS 271, L15
Wasserman, I.: 1978, ApJ 224, 337
Widrow, L. M.: 2003, Rev. Mod. Phys. 74, 775
Zeldovich, Y. B., Ruzmaikin, A. A., Sokolov, D. D.: 1983, *Magnetic Fields in Astrophysics* (New York: Gordon and Breach)

Magnetic Fields in Galaxy Systems, Clusters and Beyond

Philipp P. Kronberg

Institute of Geophysics and Planetary Physics,
Los Alamos National Laboratory, NM 87501, USA
kronberg@lanl.gov

1 Introduction and Background

The discovery of significant magnetic fields beyond our Milky Way was unwittingly made in the early 1950's when new 'discrete' cosmic radio sources were identified with distant galaxies. This also revised the prevailing view that these cosmic radio emitters were associated with stars in our Milky Way. The non-thermal nature of these sources' emission was recognized as synchrotron radiation, and as the cause of the Crab Nebula supernova remnant's radiation by I.S. Shklovskii (1953) and others.

Synchrotron radiation uniquely traces a magnetic field, since its volume emissivity is $\epsilon(\nu) \propto n_{\mathrm{e}}^{\mathrm{r}} \, B^{(\mathrm{s}+1)/2} \, \nu^{(1-\mathrm{s})/2}$ ($\mathrm{erg\,s^{-1}\,cm^{-3}\,Hz^{-1}}$), where B is the magnetic field strength and the frequency dependence ($\alpha = (1-s)/2$) is tied to s. The latter defines the distribution in energy space of the radiating relativistic electrons $n_{\mathrm{e}}^{\mathrm{r}} \propto n_0 E^{-\mathrm{s}}$ and s is usually $\approx$2.7 (see Chap. 5). The relativistic, cosmic-ray particles are also important to mention here, since synchrotron radiation also requires the presence of cosmic-ray (CR) electrons $n_{\mathrm{e}}^{\mathrm{r}}$, and by implication, protons having the same density, $n_{\mathrm{p}}^{\mathrm{r}}$. Synchrotron radiation from the accompanying protons is virtually invisible because of their 1836 times lower charge/mass ratio. This is a further interesting fact, in that in distant extragalactic systems so far no direct estimate has been made of the energy content of the protons, $E_{\mathrm{p}}^{\mathrm{r}}/E_{\mathrm{e}}^{\mathrm{r}}$, commonly parameterized as k (e.g. Pacholczyk, 1970). The only direct measurement of k comes from the CR particles that arrive on Earth, where k is $\approx$100. In extragalactic systems k has been variously estimated between 1 and 100.

I include CR's in the context of extragalactic magnetic fields because magnetic fields are required to accelerate the CR's to their relativistic energies. This fact was recognized early in the development of cosmic-ray physics by Fermi (1949), who proposed a way of accelerating CR's in a magnetized interstellar shock zone. The important point is that, wherever we detect magnetic fields in diffuse astrophysical plasmas they are accompanied by cosmic-ray electrons and nuclei. Unverified as yet is whether some extragalactic regions, e.g. cosmic voids that may contain significant magnetic energy, have a much smaller CR energy density. Some primordial (i.e. pre-recombination) mechanisms for magnetic field generation might generate 'CR-less' intergalactic fields, but these possibilities are not yet experimentally verifiable. Discussion of some of these early **B**-generation possibilities during the primordial inflation and plasma epochs, and that are largely linked to fundamental particle physics issues can be found in Kronberg (1994), Enqvist (1998), and in Chap. 1 of this book by Martin Rees.

This chapter will focus on the possible role of (i) stars and supernovae, and (ii) galactic black holes in injecting magnetic energy into the intergalactic medium (IGM). These two broad categories of energy *output* from galaxies have been recognized for some time. We begin with the environment of galaxies out to the distance of the local Supercluster of galaxies and proceed all the way back to the formation of the first stars and galaxies.

2 Stellar Sources of Extragalactic Magnetic Fields

An important question to answer is whether galaxies and stars can eject enough magnetized gas into the intergalactic medium to seed a significant fraction of the IGM with magnetic fields. Any IGM field seeding of this type will be especially effective if it also happened at an early phase of the Universe when the average inter-galaxy separation was small. Many of the clues and parameters required to answer this question can be gleaned from the nearby, $z \approx 0$ Universe where stars and galactic systems can be observed in great detail and can be physically characterized. The information gained can then be built into model calculations that are extrapolated backwards in proper time and embedded into cosmological evolution scenarios.

In a typical spiral galaxy disk like the Milky Way the interstellar medium (ISM) energy density is a few times $10^{-12}\,\mathrm{erg\,cm^{-3}}$, and it is approximately equally shared by the ISM magnetic field, the ionized gas density, and the galactic CR particle density (see Chap. 3). But in local regions with elevated star-formation rates, such as in H II regions with an overabundance of young hot stars, the local energy density is increased by one to two orders of magnitude, $\approx 10^{-11}$ to $10^{-10}\,\mathrm{erg\,cm^{-3}}$. It is easy to see that such overpressures will cause outflow from the galactic disks. Indeed galactic outflow halos are seen in X-rays, neutral H I, Hα emission, synchrotron emission and X-ray bremsstrahlung emission. That is, all the products of hot star outflow, novae and supernovae can be traced in an outflow halo whose velocity, spectroscopically verifiable, is in the range 500–2000 $\mathrm{km\,s^{-1}}$. Figure 1 illustrates the optically visible outflow for the edge-on–viewed nearby galaxy NGC 891.

Examination of our local Galactic environment on smaller scales reveals that, apart from H II regions, individual stars and protostellar systems can have outflow velocities that are close to or exceed the escape velocity of the outer Galactic disk. Even the polar coronal zones of the quiescent Sun have outflow velocities up to $\approx 800\,\mathrm{km\,s^{-1}}$. Collimated outflow jets from individual stars and protostars have recently been observed with outflow velocities up to several thousands of $\mathrm{km\,s^{-1}}$ (Feigelson and Montmerle, 1999).

Given that star-forming regions in Milky Way-like galaxies create an order of magnitude or more of local ISM overpressure, it is not surprising that regions of extreme starbursts in other galaxies can produce an outflow wind of the kind that we see in M 82 – see Fig. 2. Although M 82, the prototypical nearby starburst galaxy, has less than 1/10th the mass of the Milky Way, its nuclear region has a star-formation rate of order $1\,\mathrm{M_\odot\,yr^{-1}}$ within the inner 600 pc, causing an extreme overpressure and outflow. Smaller 'dwarf' star-forming galaxies with much lower escape velocities are also observed to generate prominent outflow halos (e.g. Chyży et al., 2000).

Fig. 1. Edge-on view of the galaxy NCG 891 showing the optically visible outflow gas and dust. (Source: C. Howk (UCSD), B. Savage (U Wisconsin), N. A. Sharp (NOAO/WIYN/NOAO/NSF)) See also Howk and Savage (1997)

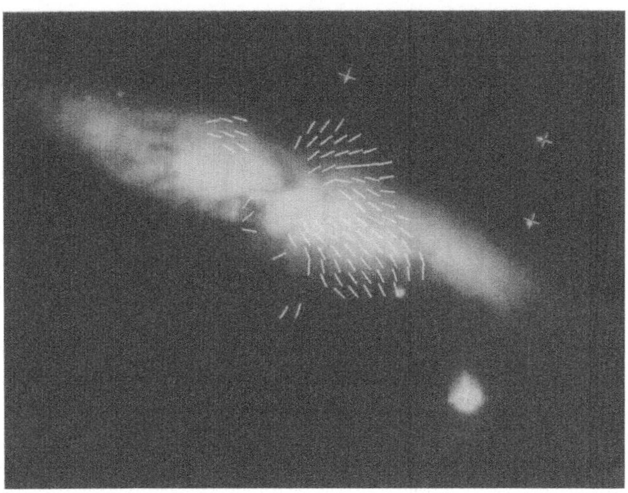

Fig. 2. Superposition of the Faraday RM-corrected projected magnetic field orientation within the outflow halo of M 82. The combination of the X-ray bremsstrahlung halo emission (not shown) and the Faraday rotation of the polarized emission shown reveals magnetic field strengths in the range of 10–50 μG, and having coherent zones of $\gtrsim 1$ kpc. (Source: Reuter et al., 1994)

The surprisingly strong magnetic fields in local Universe galaxy outflow halos suggests that some of the kinetic outflow energy is converted into magnetic fields in the course of the outflow, probably through shearing instabilities as some recent simulations have shown (e.g. Birk et al., 2000). In addition to magnetic fields, we expect the outflow material to contain cosmic rays and metal-enriched gas, products of the supernovae, as well as dust and some neutral gas. This means that evidence for early starburst outflow can be searched for at high redshift by looking for all these components (Lesch and Hanasz, 2003). For this reason, dust emission and metal-enriched gas at large redshifts are examples of proxy indicators, discussed further below, of the ejected magnetic fields which are less easy to detect directly at the earliest galactic epochs.

The outflow volume ΔV_F (Fig. 3) for a given starburst 'event' is the product of an assumed constant outflow velocity (v), chimney cross-section (σ) and outflow time (τ). At $z \approx 10$, the average (mostly dwarf) galaxy–galaxy separation, $\langle|\mathbf{r}_{ij}|\rangle$, is only 30–40 kpc (comparable to the size of a present-epoch galaxy), and the IGM density and pressure becomes comparable to that in the core of a galaxy cluster. Furthermore the ambient photon energy density ($\propto (1+z)^4$) will, at very early times exert a significant counter-pressure to the ejection and outward diffusion of magnetized gas. All of these conditions become more 'extreme' as we proceed into the interesting and little-observed epochs beyond $z \approx 10$. Partially compensating the higher ambient pressure at high z's is the fact that starburst-driven outflow winds seem to contain an unresolved 'forest' of very high velocity 'galactic spicules' having v of several thousand km s^{-1}, each with small σ, by which they can more effectively penetrate the higher density environment than the simplified global assumptions suggest.

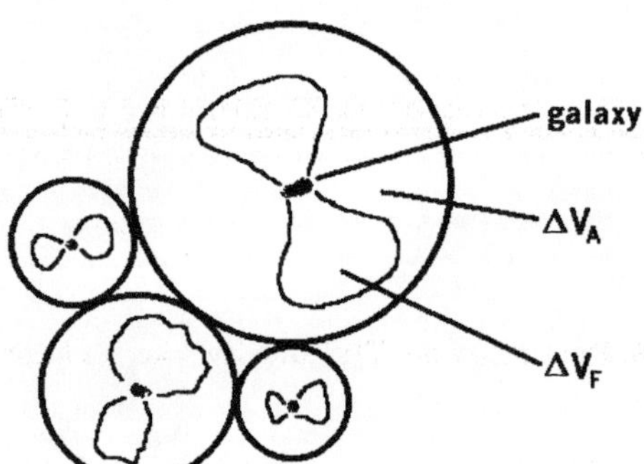

Fig. 3. Schematic illustration the filling of a fiducial volume ΔV_A which is 'available' to be filled by starburst-driven outflow halo gas that fills ΔV_F of this volume

At small intergalaxy spacings $\langle|\mathbf{r}_{ij}|\rangle$ at $z \gtrsim 10$, galaxy–galaxy tidal interactions will be very strong. Tidal removal of interstellar matter, not as yet well simulated, will most probably increase at these early epochs, and will enhance the star/supernovae-driven outflows' ability to spread magnetized (and metal-enriched) gas into the IGM. One indirect observational check on these effects is the excitation state of the primeval gas, which is becoming accessible to measurement. Tidally dispersed gas may contain cooler components than direct star-driven outflow material, hence the IGM gas temperature at these early times may provide good proxy information on the physics of both metal- and magnetic field-seeding of the early IGM by primeval galaxies.

As the Universe expands, the star+SN outflow-produced fields and associated gas will expand and dilute with time. An important result, illustrated in Fig. 4, is that the dwarf galaxy starburst outflow of magnetized gas is most effective in seeding the IGM at $z \gtrsim 7$. At lower (later epoch) redshifts, $z \lesssim 6$, the co-moving volume growth sharply reduces the ability of starburst outflows to increase the volume fraction of B-polluted IGM gas in the Universe. This 'fiducial' redshift of ≈ 7 is relatively insensitive to galaxy merging scenarios, outflow parameters, and to the cosmological parameters.

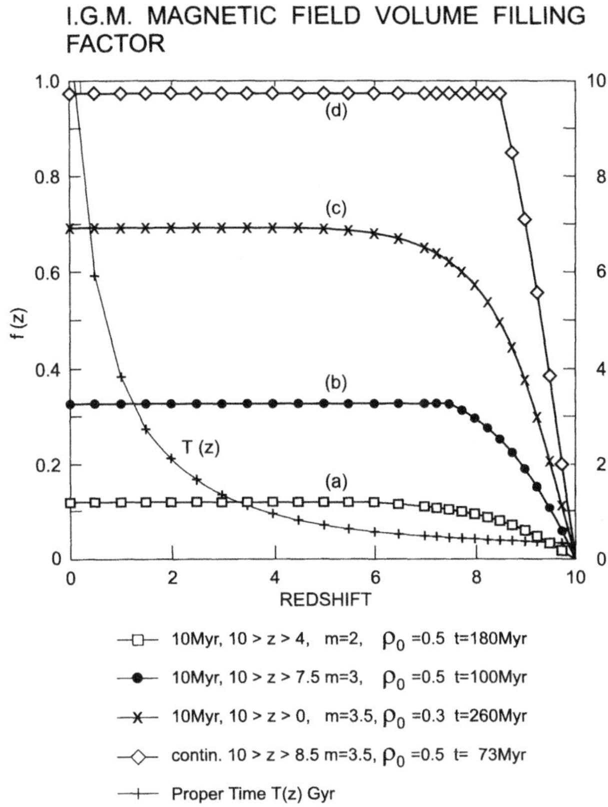

Fig. 4. Plot showing how the fraction of 'available' intergalactic space, $f(z) = V_F(z)/V : A(z)$, grows with proper time ($T(z)$, r.h. scale) for 4 different parameter combinations of starburst-driven outflow. The parameter m is a galaxy merging rate index, t is the total starburst 'on'-time, and ρ is the galaxy density at $z = 0$. The void fraction was adopted as 80% and was assumed independent of z for $z \lesssim 10$ (Kronberg et al., 1999)

The foregoing enables us to extrapolate the well-studied outflow properties of galaxies in the local Universe, where the galaxy count census is most complete, to the earlier Universe in which the fraction of dwarf galaxies (and globular clusters) within a co-moving volume element was much higher because mergers into larger galaxies had largely not yet taken place. Because the co-moving volume, hence the average inter-galaxy separation was small, the IGM volume available to be filled with outflow gas and magnetic fields, V_{A}, was much smaller, at say $z = 10$, than at the present.

As cosmic time proceeds (r.h. scale of Fig. 4), magnetized halos of the kind seen in Fig. 3 grow, and diffuse outwards at the same time as the total volume of intergalactic space grows, and as the galaxies undergo progressive merging. Völk and Atoyan (2000) have calculated the contribution of galactic outflow winds to the magnetization of the IGM within galaxy clusters, also based on observed galactic wind outflow properties, and arrive at similarly high magnetic field filling factors for the IGM within clusters.

These results seem robust to different scenarios of galaxy mergers and galaxy evolution. However the current limited observations of individual galaxies at large z, our currently incomplete knowledge of the beginnings of gravitational collapse, star formation, tidal interactions among primeval galaxies, and the first stellar luminosity functions (when the average metallicity was still lower) etc., indicate that they may be considerably refined in the future.

3 Early Plasma-driven Seeding Scenarios

Another early field-seeding mechanism in which the magnetic field is seeded and/or amplified in a pre-galactic gaseous medium is discussed by Harrison (1970). Other situations favourable to early field seeding have been proposed by Gnedin et al. (2000) to occur in the breakout of ionization fronts from protogalaxies or field amplification later, up to $z \approx 5$ due to the propagation of ionization fronts through hypothesized dense neutral filaments in the evolving web of large-scale structure. Via the Biermann battery effect and the magnetic induction equation, $< \partial|\mathbf{B}|/\partial t >$ is positive, thereby seeding and amplifying weak pre-stellar intergalactic fields by effectively converting kinetic into magnetic energy. Readers are referred to Rees (1987) and references therein for an earlier discussion and summary of these ideas.

4 The Galactic Large-scale $\alpha - \Omega$ Dynamo Theory

Over the past few decades, the origin of current μG-level magnetic fields in galaxies, excluding AGN-fed radio lobes which we discuss below, was sought in large-scale galactic dynamo theory. The galactic $\alpha - \Omega$ dynamo is an elegantly constructed theory to explain how the current magnetic energy density ($\approx 10^{-12}\,\mathrm{erg\,cm^{-3}}$) in a spiral galaxy's disk can be obtained by tapping into the energy reservoir of galactic rotation, assisted by a combination of large-scale shearing ($d\Omega/dR$) and turbulent outflow (α) from locally over-pressured star-forming regions. The galactic $\alpha - \Omega$ dynamo assumes that galactic magnetic fields were orders of magnitude weaker at early epochs, and derives an exponential growth rate (the solution to the magnetic induction equation) over a galactic dynamical lifetime. For example our Galaxy could amplify the field strength by many orders of magnitude to the present few-μG level in the galactic disk over its $\approx 10^{10}$ yr lifetime. At some point, $|\mathbf{B}|$ saturates when energy equilibrium is reached with other components of the interstellar medium, such as cosmic rays and ISM turbulence.

Later in this chapter we will encounter strong evidence that microgauss-level magnetic fields have existed since the earliest galaxies formed, and were a natural consequence of galaxy evolution (e.g. starburst outflows) as discussed above, and also of black hole formation as I discuss later. An instructive piece of evidence against large-scale galactic dynamo amplification from initially small seed fields is the discovery (Chyży et al., 2000; Chyży et al., 2003, see also Chap. 3) that some nearby compact dwarf galaxies (e.g. the dwarf spirals NGC 1569, NGC 4449, IC 10) have little ordered rotation or velocity shear ($d\Omega/dR \approx 0$ in their central regions). Yet they have similar, or even greater ISM and halo magnetic strengths when compared to the Milky Way.

It seems more likely that a galactic $\alpha - \Omega$ dynamo serves to re-organize the large-scale galactic field over a galaxy evolution time, rather than as a global field strength-amplifying mechanism in galaxy disks. This does not mean, however, that the coherent $\alpha - \Omega$ dynamo theory is obsolete in all astrophysical situations. Essentially the same mechanism has been proposed to operate on sub-parsec scales to explain the process of electromagnetic energy extraction from a galactic black hole-accretion disk system (Colgate et al., 2001). The $\alpha - \Omega$ dynamo may be a very important mechanism for the launching of jets. Further, the semi-infinite scalability of this important fundamental MHD mechanism means that it is one of few astrophysical magnetic phenonena that can be tested in the laboratory (Colgate et al., 2002; Beckley et al., 2003). Several laboratory astrophysical $\alpha - \Omega$ dynamo experiments have been designed to verify the MHD characteristics of the $\alpha - \Omega$ dynamo; at Riga (Latvia), Los Alamos/New Mexico Tech (USA), the Universities of Maryland and Wisconsin (USA) and Karlsruhe (Germany).

5 Proxy Measurements for Cosmologically Early, Star-ejected Magnetic Fields

Originally predicted by Harwit and Pacini (1975), a surprisingly robust correlation has emerged between the far infrared dust-generated emission in star-forming galaxies and the co-extensive synchrotron emission (Dickey and Salpeter, 1984; Kronberg et al., 1985; Völk, 1989; Helou and Bicay, 1993). This suggests that cosmologically early ($z \approx$ 10–50) starbursting activity, which will produce intense thermal dust emission will be accessible to observation, redshifted to submillimeter wavelengths. Emission from dust, produced and heated by early massive stars and supernovae (SN), produces these galaxies' highest emitted spectral density (W Hz^{-1}) at $\lambda \approx$ 40–400 μm. At the otherwise inaccessible redshifts of interest, say $z \approx$ 4 to 15, this relatively intense, and little-absorbed radiation is redshifted to wavelengths in the range ≈0.2 mm to 6 mm. Angular sizes of early dwarf and starbursting galaxies at these redshifts are of order a few arcseconds. This means that they can be resolved and spectroscopically studied with sub-mm telescopes. Examples are the SCUBA detectors of the JCMT 15-m telescope, and the coming generation of submillimeter interferometers such as the ESO–NRAO ALMA array, which will have the sensitivity, resolution, and spectroscopic ability to image individual galaxies in detail back to these epochs.

Since SN will metal-enrich star/SN-driven outflow winds at early epochs, spectroscopic evidence for the production of B, Be and elements upward in atomic number also provide meaningful observational constraints on cosmic chemical evolution.

Spectroscopic observations of age-sensitive species in stars in the local Universe, e.g. in the Galactic halo (e.g. Duncan et al., 1992), globular clusters, or low-z elliptical galaxies can tell the age of stellar giant branch turnoffs, or when star formation ceased. For dwarf spheroidal galaxies it tells us when the outflow-generating SN were formed. By these and related observational means we have cosmic time-markers for major galaxy outflow events in the past.

Another possible proxy indicator of galactic magnetic field strengths at earlier epochs might come from analyses of B and Be (spallation products of CR nuclei). Prantzos and Aubert (1995) have suggested that these were present in the early Galaxy, and that their (magnetic) confinement to the Galaxy is required to be stronger in the past to explain the observed B/Be ratio. This would suggest an early galactic field that was possibly even stronger than the present-day galactic magnetic field. This is independently consistent with the claim by Parizot and Drury (2000) that B and Be production was confined to the early Milky Way, which in turn is consistent with a large supernova rate very early in our galaxy's history. This lends further support to the existence of cosmologically early magnetized plasma outflows as discussed above. Observations of metal absorber species over a range of redshifts in the spectra of high-z quasars give firm, if indirect, evidence of early SN production. This is another kind of proxy measurement for magnetized gas and CR outflow at earlier cosmological epochs.

6 Post-amplification of Initially Weak Intergalactic Seed Fields of All Kinds

The dilution to $z \approx 0$ of early starburst-seeded, intergalactic magnetic fields would produce only a quite weak intergalactic field $\lesssim 10^{-9}$ G within the general IGM in galaxy filaments and voids, assuming the fields were passively frozen in to the co-expanding gas. Calculating another way, the total thermonuclear energy released from all the galaxies up to $z \approx 7$ is not energetically sufficient to magnetize the Universe to a level that I will describe below. However the gravitational collapse of the baryonic and dark matter into the 'filaments' of large-scale structure could naturally and passively amplify an original all space-filling weak field, close to microgauss levels, as simulated by Ryu et al. (1998). In a further development of these simulations Ryu et al. (2003) explored the development and consequences of shock structure on various scales due to infall into cosmic sheets, filaments and knots, and at a resolution now extending down to subclump structures in a ΛCDM Universe. The large-scale gravitational infall energy is converted to thermal heat, CR acceleration and IGM magnetic field amplification in the process. In these models, the field amplification is 'passive' in that the magnetic energy does not feed back into the gravitational dynamics. The next simulation challenge, somewhat more difficult, is to incorporate feedback of magnetic fields into the calculations. The magnetic, or back-reaction is of basically two types: The first is the effect of magnetic fields that have been amplified by compression, shocks and shearing instabilities. The second is due to magnetic and CR pressure that emanates from the central black holes of the galaxies which we discuss next. The magnitude of this latter feedback energy (Kronberg et al., 2001) has only been more recently appreciated in the context of intergalactic magnetic fields and structure evolution on the scale of galaxy filaments. This leads us to the effect of galactic black holes on the magnetic properties of the intergalactic medium.

7 Massive Black Hole Seeding of Intergalactic Magnetic Fields

Much of what I have discussed thus far concerned a mix of observations and theory to explain intergalactic magnetic fields that derive from the thermonuclear energy of stars and supernovae, combined with some as yet incompletely quantified conversion of virial energy of early galaxy–galaxy interactions, and from the kinetic energy of large-scale motions.

In addition to all of these processes the radio-emitting lobes of radio galaxies and quasars, which are energized by a central black hole (BH), rather than stars and SN, represent an additional form of magnetic energy injection into the IGM. This happens through extragalactic radio lobes which can form in a very short cosmological time ($\approx 10^7 - 10^8$ yrs), also in the 'mature', low-z Universe. They are over-pressured relative to the surrounding IGM by a large factor unless they are inside of galaxy clusters. Thus, deposition of magnetic fields into the IGM by galactic BH-driven energy flows is inevitable. An AGN-jet radio galaxy system 3C 303, one of many morphological variants of such systems, is shown in Fig. 5. The locations of the central BH, jet, and expanding lobes are clearly seen at cm radio bands.

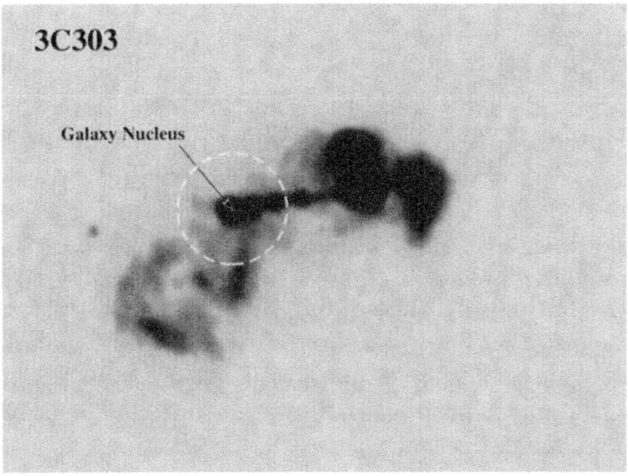

Fig. 5. The radio galaxy 3C 303, associated with an elliptical galaxy at $z = 0.141$ (Kronberg, 1986)

Whereas the relativistic electrons' radiative lifetimes are $\approx 10^8$ yrs, limited by either synchrotron or inverse Compton radiation losses, the relativistic protons and the magnetic fields in the radio lobes dissipate their energy on a very much longer time scale – 10^9 yrs or more. Thus the energy in the protons and magnetic fields outlasts that of the synchrotron-visible CR electrons by a factor of 10 or more, i.e. much of a Hubble time, during which black-hole generated intergalactic magnetic fields will fill a much larger volume that we see in currently visible radio sources. So it is appropriate to ask if an IGM magnetic field filling factor analogous to that for starbursting galaxies in Fig. 4 can be calculated for radio galaxies. Such model calculations have been done by Furlanetto and Loeb (2001), who calculate that the magnetic fields from QSOs can fill up 5–20% of the IGM volume. It is interesting that this *volume* fraction is comparable to the estimates described above for primeval starburst galaxies.

From the question of what IGM *volumes* can be filled with magnetic field, we now consider the amount of galactic energy that is released into the IGM: It has been known for many years (Burbidge, 1856) that the energy content of some extragalactic radio source lobes is $\gtrsim 10^{60}$ erg, and subsequently realized that this enormous energy can only come from the gravitational infall energy of a supermassive black hole. The importance of a $\gtrsim 10^8\,\mathrm{M}_\odot$ rotating central black hole/accretion disk arises from its enormous formation energy reservoir, $E_\mathrm{G} \approx M_\mathrm{BH}\, c^2$, where we use the Schwarzschild radius, r_s, as a convenient fiducial 'end point' radius. For $M_\mathrm{BH} = 10^8\,\mathrm{M}_\odot$, $r_\mathrm{s} = 2GM/c^2 = 2.96 \times 10^{13}$ cm, $\approx$2 AU. For the gravitational collapse of this mass to the Schwarzschild radius, $E_\mathrm{G} = 9 \times 10^{61}$ erg. Comparison of this energy with that of the total stellar thermonuclear energy reservoir is instructive: If we take a galaxy with mass $M_\mathrm{G} = 10^{11}\,\mathrm{M}_\odot$ of baryonic matter and assume generously that 0.1% of all the initial galactic baryonic matter is processed by thermonuclear fusion with a mass conversion efficiency of 0.7%, then $0.001 \times 0.007\, M_\mathrm{G}\, c^2$ is 1.4×10^{60} erg. Even this generous estimate is nearly a factor of 100 below the gravitational infall energy of a $10^8\,\mathrm{M}_\odot$ galactic supermassive black hole. The energy content of a starburst outflow halo would be smaller still.

A remarkable puzzle of modern astrophysics is how, as we discuss later, a substantial fraction of the BH gravitational infall energy is converted to magnetic fields and relativistic particles that we see as powerful FRI and FRII radio sources (Thorne, 1974). Relevant analyses of radio source observations can be found in Rawlings and Saunders (1991), Falcke and Biermann (1995), Kronberg et al. (2001), and Gopal-Krishna et al. (2001). It has recently become clear from observational estimates of central black hole masses that most, if not all large galaxies harbour a central black hole (Richstone et al., 1998; Tremaine et al., 2002). Those in larger elliptical galaxies range from 10^7 to $10^9\,\mathrm{M}_\odot$.

We can use the released energy combined with global galaxy density statistics to now ask how much *energy* and energy density, as distinct from volume filling factor, could have been recycled into the IGM from galactic black holes, given that they represent an enormous energy reservoir. To relate the global energy output of AGNs to the global space density of massive black holes, it has been assumed that galaxies in their QSO phase produce a photon output that scales to the mass of the central BH engine. Calculations applied to the QSO epoch at $z \approx 2$ have been laid out by Soltan (1982), Chokshi and Turner (1992) and Small and Blandford (1992).

These authors used the QSO photon output to estimate the average space density of BH's.

Since extended RLQSOs have globally similar radio properties to radio galaxies at the same cosmological epoch, the QSO energy to BH mass scaling applied to the photon-intense QSO epoch will also apply over a wider range of cosmological epochs. Specifically, the co-moving BH density at later epochs toward $z = 0$ will certainly not be less than at the 'quasar epoch'. Observation-based estimates of the integrated QSO photon output, and an assumed conversion efficiency from BH mass to light led Chokshi and Turner (1992) to estimate the comoving BH density for the QSO epoch

$$\rho_{\rm BH} \approx \frac{2.2 \times 10^5}{\eta_{\rm RAD}} \, {\rm M_\odot \, Mpc^{-3}} \; , \tag{1}$$

where $\eta_{\rm RAD}$ is an efficiency factor for generating radiation, taken as 0.1. Depending on the QSO luminosity function evolution, about half of this mass density is already accumulated by $z \approx 2$ (refer to Fig. 1 of Chokshi and Turner, 1992). If it is also assumed that only a fraction $\eta_{\rm RL} \lesssim 10\%$ of all QSOs are radio-loud, (i.e. are powerful jet-lobe sources), and that for those, about 10% of the black hole accretion energy is converted into magnetic fields, $\eta_{\rm BH}$, then by $z \approx 2$ the mean IGM magnetic field energy density can be calculated as

$$\varepsilon_{\rm B} = 5 \times 10^{-3} \left(\frac{\eta_{\rm RL}}{0.1}\right) \left(\frac{\eta_{\rm B}}{0.1}\right) \rho_{\rm BH} \approx 7.3 \times 10^{-17} \, {\rm erg \, cm^{-3}} \tag{2}$$

(Kronberg et al., 2001). It is interesting to compare this energy density with a global estimate of the IGM *thermal* energy density, also at $z \approx 2$.

$$\varepsilon_{\rm TH} = 5 \times 10^{-16} n_{-4} T_4 \, {\rm erg \, cm^{-3}} \; , \tag{3}$$

where n_{-4} is the IGM gas density in units of $10^{-4}\,{\rm cm^{-3}}$, and T_4 is the temperature normalized to 10^4 K. It is salutary to note that if both the magnetic field energy and the intergalactic thermal plasma were uniformly spread throughout a co-moving volume element their energy densities are comparable, subject to some uncertainty in the various normalization factors.

If, instead of using the quasar density and luminosity function statistics at $z \approx 2$, we calculate the average magnetic energy injected by all radio galaxies close to our epoch, we arrive at similar cosmologically scaled smoothed-out $\varepsilon_{\rm B}$ at $z = 0$ in the galaxy-overdense filaments of large-scale structure. This is of order 10^{-6}–10^{-7}G when we apply a minimum total energy criterion where the energy is approximately equally divided between the particles and magnetic field.

Finally, using a recent estimate of the mass density of galactic black holes, $\approx 4 \times 10^5 \, {\rm M_\odot \, Mpc^{-3}} \times \eta_{\rm BH} M_{\rm BH} \, c^2$ (average BH energy fed back into the IGM, where $\eta_{\rm BH} \sim 0.1$–0.3), this gives an energy density-equivalent IGM magnetic field of order $1\,\mu$G ($B_{\rm minE}$). It assumes that energy densities of the energized CR's and magnetic fields are comparable. What is interesting is that all of the above calculations more or less agree. They indicate that galactic BH feedback energy into the IGM in galaxy filaments is at least as important as that due to large-scale gravitational infall. And these numbers did not require elaborate simulations requiring various assumptions about parameters of the primeval IGM that are still difficult to measure.

If, as is likely, $B_{\rm minE}$ is comparable with the co-spatial CR energy density and both are similarly distributed in space, then it might be observationally tested as weak, diffuse emission at very low radio frequencies. The advantage of an observational search at low radio frequencies is (1) that the spectral density ($\mathrm{W\,Hz^{-1}}$) of synchrotron radiation increases towards low frequencies (assuming the same total bandwidth), and (2) the radiating relativistic (CR) electrons have the longest energy loss times against I.C. and synchrotron radiation. As the observing frequency lowers to 30 MHz, their radiative lifetime $\tau_{\mathrm{l}}^{\mathrm{e}}$ approaches 10^9 yrs (at $z = 0$), meaning that the radiating electrons have the longest possible time do diffuse from their sites of acceleration. Intergalactic field strengths at near μG levels are sufficiently strong that they might well have had an impact at earlier phases of galaxy formation.

A significantly magnetized IGM appears tentatively consistent with the discovery of diffuse, 326 MHz synchrotron emission well beyond the boundaries of the Coma cluster of galaxies by Kim et al. (1989), shown in Fig. 6. The low-level synchrotron 'glow' that they found extending beyond the Coma cluster gave supracluster intergalactic B_{minE} values between 10^{-7} and 10^{-6} G over linear dimensions a few times the core size of the Coma cluster itself. If the IGM is directly energized by the relativistic particles and magnetic fields from AGN-powered 'clouds', then the magnetic flux generated could cause a widespread synchrotron glow to be seen be over larger IGM volumes, especially at the lowest radio frequencies. This can be better tested in future when much more sensitive low-frequency radio images become available.

Adapted from: Kim, Kronberg, Giovannini and Venturi, NATURE, 1989

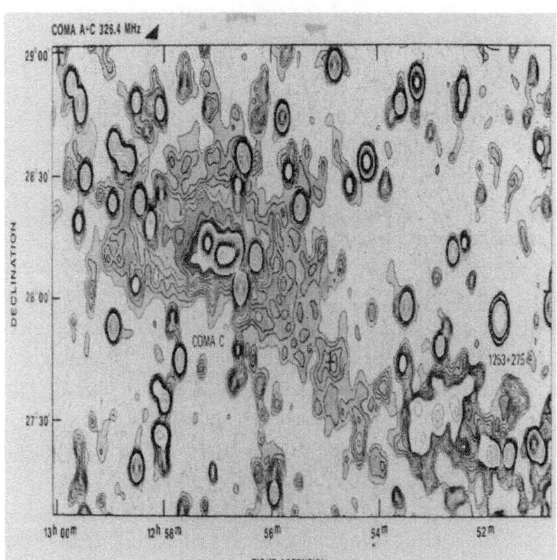

Fig. 6. A deep $1^{\circ}\!.5\times1^{\circ}\!.5$ (2.6×2.6 Mpc) radio continuum image of intergalactic space around the Coma Cluster of galaxies at 326 MHz made with the Westerbork Synthesis Radio Telescope in the Netherlands. Weak, intergalactic diffuse emission is seen to the east and SW of the Coma cluster itself, which is at $12^{\mathrm{h}}57^{\mathrm{m}}$, $+28^{\circ}20'$(B1950) (Adapted from Kim et al., 1989)

8 Gravitational Collapse and Black Hole Electromagnetic Energy Generators

It is not yet completely understood (1) how the angular momentum of the material spiraling into the black hole is transferred outward. An explanation of this process presents a major agenda for current astrophysical research. This process is the prerequisite for (2) the subsequent conversion of energy into a form that can be collimated, as is observed to happen, and then (3) transferred to $\approx 10^{11}$ times larger scales (A.U. to megaparsecs!). Somewhere, probably early in this chain of processes, the energy is probably (4) converted into magnetic fields – a process that is likely closely coupled with (2). Finally (5) how are the relativistic (CR) particles energized that produce the synchrotron-emitting radio lobes? Some answers are beginning to emerge for some, though not yet for all of these questions. Even the current most ambitions models and simulations – briefly discussed below – may be at least partly superceded in future as the complex plasma and MHD physics phenomena they contain become better understood.

Energy outflow models involving coherent, electromagnetic Poynting flux energy flow have been discussed since the 1970's by Richard Lovelace, Martin Rees, Stirling Colgate, Roger Blandford, Max Camenzind, Donald Lynden-Bell, Harald Lesch, David Meier and others. Particle beam jets have also been widely discussed over this period, but I will focus on electro-magnetic BH-jet models here, since they have made the most impressive recent progress, especially with the help of supercomputer simulations. An example is the recent BH-jet simulation been undertaken by Koide et al. (2002), in which they model extraction of the rotational energy 'deep in' within the complex, rotationally distorted Kerr metric space-time very close to the massive galactic black hole (Fig. 7 below). In their simulations the torsional Alfvén wave generated in the field by the rotating ergosphere carries pure Poynting flux and angular momentum outward from the black hole.

Another, quite different combination of an $\alpha - \Omega$ dynamo, hydrodynamic and electromagnetic model has been proposed by Colgate et al. (2001). Figure 8 illustrates in cartoon form some of the proposed physics to explain the formation of a supermassive BH accretion disk. A model of the outer accretion disk around the galactic black hole, shown in false color, illustrates a non-linear hydrodynamic calculation of angular momentum transfer outward in the BH accretion disk via a Rossby vortex mechanism that was recently proposed by Li et al. (2001). A partial model of stage (2), by Stirling Colgate and colleagues at Los Alamos Laboratory is based on the physically plausible assumption of a naturally magnetized accretion disk around the black hole at a galaxy's nucleus that is regularly punctured by a densely packed 'beeswarm' of thousands of stars that we expect to orbit close to the black hole. The stellar speeds are of order 20,000 $\mathrm{km\,s^{-1}}$. The shock from each puncture of the 30,000 Gauss-magnetized accretion disk drags a loop of magnetic field vertically out of the rapidly rotating disk. This sets up an $\alpha - \Omega$ dynamo analogous to the large-scale galactic dynamo that was originally proposed to explain large-scale fields of differentially rotating spiral galaxy disks. This powerful dynamo generates magnetic loops that are proposed to combine into a rapidly rotating helical field that is tied to the rotating accretion disk. A cartoon illustration of the $\alpha - \Omega$ dynamo appears in the inset in the figure. Under certain plausible

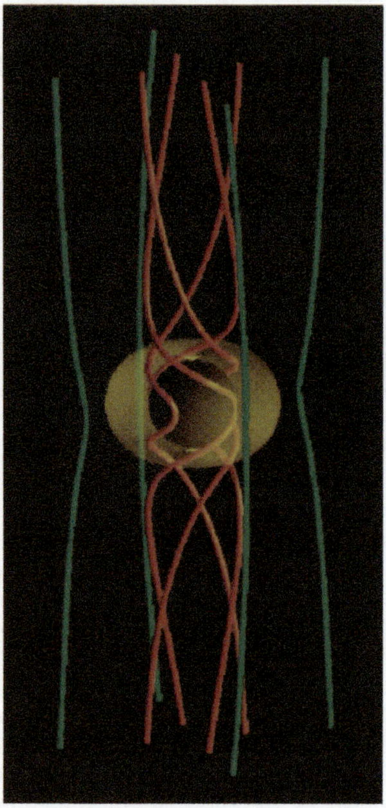

Fig. 7. Three-dimensional illustration of magnetic field lines around a Kerr black hole. The black sphere at the center depicts the black hole, and the yellow zone is that of the ergosphere, inside of which any material, information, or energy must rotate in the same direction as the black hole. In this case of high rotation, the shape of the surface is like that of an apple. The red tubes show the magnetic field lines that cross into the ergosphere and the green lines show those that do not (Source: Koide et al., 2002)

assumptions the field is self-collimating, and it propagates away, initially carrying most of the energy as illustrated in Fig. 8.

Such a system can be aptly called Nature's ultimate electricity generator. In the initial phases illustrated here, the energy is carried away almost entirely by magnetic fields, and electric currents. The current along the inner jet axis in Fig. 8 is of order 10^{19} ampères (Lovelace, 1976), and close to the jet axis it flows mostly parallel to the local magnetic field. In the model conceptually illustrated in Fig. 8, the coupling out of the rotational energy occurs in the BH accretion disk at radii that are beyond the inner zone where Kerr metric space-time distortions dominate. Details of phase (5) – conversion of the magnetic into cosmic ray energy – have yet to be well understood and are not included in Fig. 8. It probably involves a combination of magnetic reconnection and shock acceleration, that will occur much

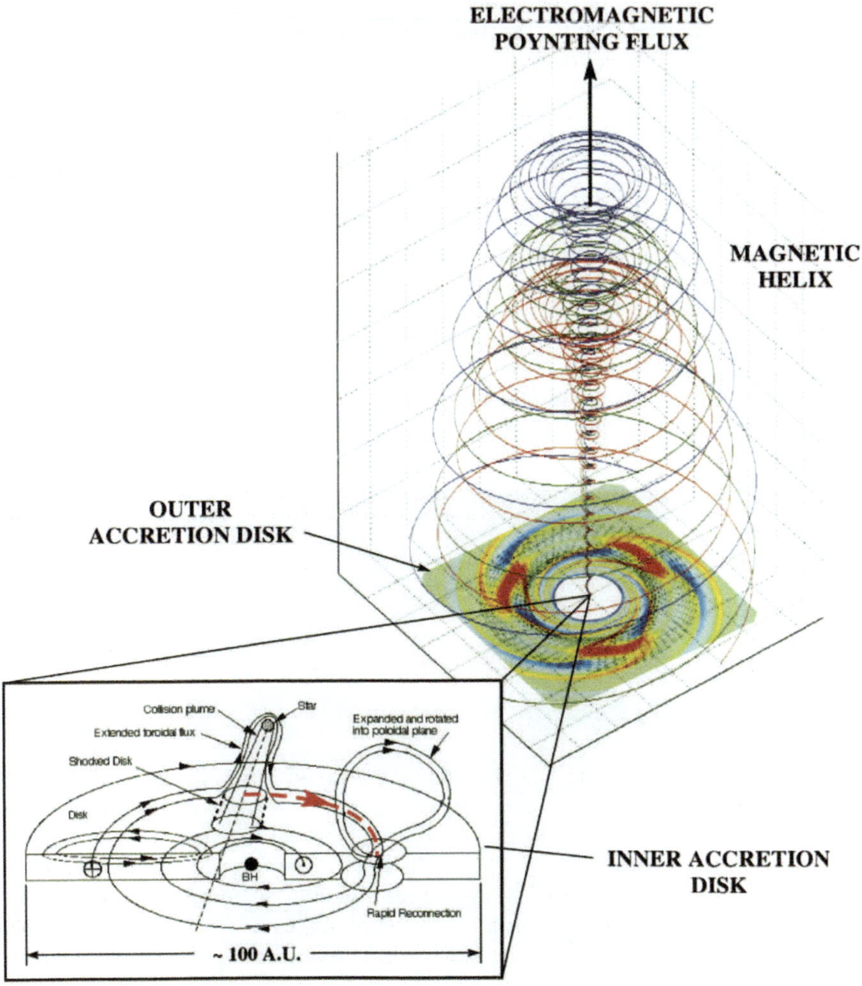

Fig. 8. Simplified illustration of three interconnected models to explain how the gravitational infall energy to a massive galactic black hole is converted to a highly collimated electromagnetic Poynting flux energy flows along the rotation axis of the black hole's accretion disk. (1) angular momentum is transferred outward by hydrodynamic vortices – a non-linear hydrodynamical instability. (2) The shock zones of stars colliding with the BH accretion disk drag magnetic field lines out of the disk. These are distorted, stretched and reconnect to form merging flux loops that act as a dynamo, thereby converting the disk's rotational energy to magnetic field energy. (3) a helical magnetic field structure carries about 10^{19} ampères along the rotation axis. This forms the base of the jet which, in this early phase is almost entirely electromagnetic. Subsequently, and on much larger scales, the magnetic energy is partially converted to particle energy as the energy flow proceeds out of the parent galaxy and into intergalactic space. This latter phase occurs beyond the scale shown in the figure (Sources: Colgate et al., 2001; Li et al., 2001)

further up the jet axis, off the scale of our figure. In this phase magnetic energy is gradually converted into CR particle energy as the system gradually relaxes to a minimum energy configuration in the mature radio lobes (see e.g. Kronberg et al., 2004). Magnetic reconnection (which also operates at the base of the helix in Fig. 8) is currently an active area of research in plasma physics.

9 Radio Galaxies as 'Calorimeters' of BH-injected Magnetic Fields and CR's

A recent analysis of measured radio source energies by Kronberg et al. (2001) showed that the best calorimeters of BH energy release into the IGM are the physically largest, 'giant' radio galaxies. These, for which 2147+816 (Fig. 9) is an example, are typically found in rarified IGM environments, extending well beyond the small galaxy groups to which their host galaxies typically belong.

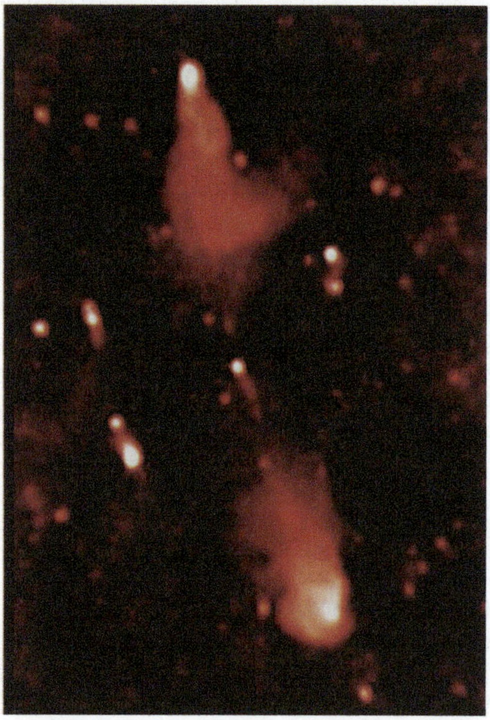

Fig. 9. 1.4 GHz radio emission from the 'giant' radio galaxy 2147+816 at $z = 0.146$. The nucleus of its host galaxy (not shown) is presumed to contain a super-massive BH. The radio-emitting lobes, whose overall projected size is $\approx$2.5 Mpc, are fed by highly collimated, oppositely directed jets that have transported at least 5×10^{60} erg into the radio lobes (Source: NRAO/AUI)

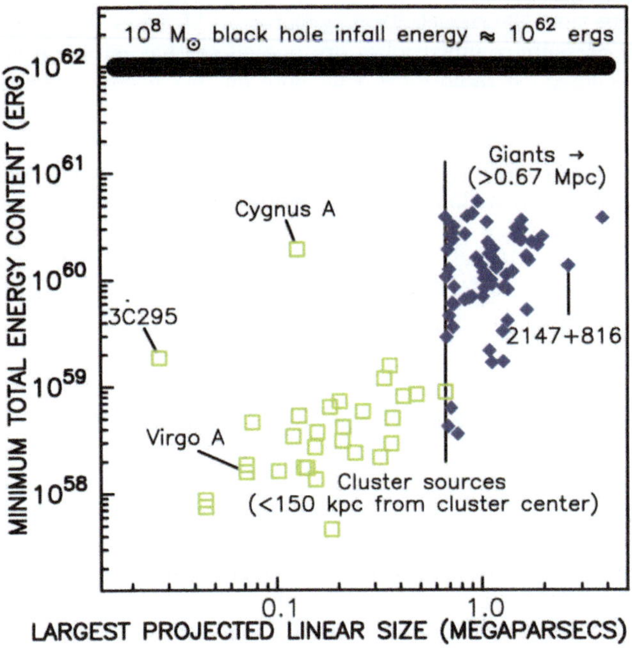

Fig. 10. Comparison of the total energy content in CR's and relativistic electrons with projected linear size for 2 classes of radio source; those with the highest average ambient density near galaxy cluster cores (open squares) and sources with overall projected dimension $\geq 0.7\,h_{75}^{-1}$ Mpc (solid diamonds) which are in lowest density environments (Adapted from Kronberg et al., 2001)

Figure 10 compares the total radio lobe energy content and maximum projected radio size for (1) the giant radio sources ($\geq 0.7\,h_{75}$ Mpc projected), and (2) radio sources whose host galaxy is within 0.15 Mpc of the cluster center. The noteworthy facts that emerge from Fig. 10 are: (a) the asymptotic upper limit to the energy content of the giants is close to 10^{61} erg, which is greater than 1%, more probably 10% or possibly more, of the supermassive BH's gravitational infall energy. (b) Sources near galaxy cluster cores have a total CR+B energy upper envelope that is typically $\approx$30 times lower – in absolute terms a deficit of $\approx 10^{60}$ erg relative to the giants.

It appears that the giant radio sources have lost the least amount of their BH reservoir energy, when compared with other classes of radio source. In this sense they are the best calorimeters of BH-released energy into intergalactic space. In particular, the PdV work done as they expand into the ambient medium cannot be greater that the present internal energy of the radio lobes unless the conversion efficiency from gravitational infall energy to magnetic fields and CR's is nearly 100%.

The energy contents of the galaxy cluster-embedded sources give a quantitative, global estimate of the fraction of energy that is injected into the dense intracluster environment, into some other energy form that is less visible to synchrotron radiation detection. Shorter electron synchrotron lifetimes due to the higher field strengths in pressure-confined lobes are one effect that is operative. This means that cluster-embedded radio lobes will more quickly become (synchrotron) radio silent at all but the very lowest frequencies. In any case they are seen as sometimes radio quiet, sometimes radio loud, lobe-sized holes in detailed X-ray images of hot thermal gas (Fig. 12 below). Pressure balance between the magnetic field and gas (thermal and CR gas) within the holes, and the external thermal hot gas pressure implies typical $|\mathbf{B}|$ values of $\approx 20 - 30\,\mu$G within the 'holes'. Finally, the energy content 'gap' in Fig. 10 between the cluster core and giant sources is close to the independently estimated PdV work done on the intracluster thermal gas (Kronberg et al., 2001).

10 Magnetic Fields in Clusters of Galaxies

Galaxy clusters are both convenient, and unique laboratories for investigating the physics of the radio source–IGM interactions. Unique, in that they are the only extragalactic systems that have such a high ambient IGM density and pressure that the ambient medium can easily be detected by present X-ray telescopes, and its temperature and pressure can be quantified.

A detailed ICM magnetic field probe of a single cluster, Coma (Abell 1656) was undertaken by Kim et al. (1990), who used the VLA to compare Faraday rotation measures of radio sources within or directly behind the Coma cluster with those whose ray paths did not intersect the cluster. These were combined at the time with EINSTEIN IPC data from Abramopoulos and Ku (1983), and produced the first clear detection of μG level magnetic fields in a single cluster. Quantitatively, their result was

$$\langle |B_{\rm cl}| \rangle = (0.4 \pm 0.16)\, M^{0.5}\, h_{75}^{0.5}\ \mu{\rm G}\,, \tag{4}$$

where M is defined as the average number of field reversals over the cluster line of sight – a less easily measurable number, estimated of order 25 for which $\langle |B_{\rm cl}| \rangle \approx 1.8\,\mu$G. (I have renormalized the Kim et al. (1990) result to $H_0 = 75$ from their adopted $H_0 = 50$.) Kim et al. (1990) used the distributed depolarization ratio from multi-frequency polarization images over the cluster-embedded 'head-tail' source, 5C4.81 to create a model-based estimate of M, in which the model depends on the angular resolution of the radio source image. It is instructive for the purpose of understanding how the $\langle |B_{\rm cl}| \rangle$ estimates were arrived at to note that the lowest common resolution in their depolarization analysis was $21''\!.2$ ($11.2\,h_{75}^{-1}$ kpc), and the corresponding characteristic field reversal scale deduced was $7 \lesssim l \lesssim 27\,h_{75}^{-1}$ kpc. This corresponds to $M \approx 25$ reversals per X-ray core radius (assuming an isothermal cluster hot gas distribution). The result was $\langle |B_{\rm cl}| \rangle \approx 2\,\mu$G. A subsequent RM analysis of 5C4.81 by Feretti et al. (1995) but with nearly 10 times higher VLA resolution revealed B reversals on scales that were smaller by the ratio of the two resolutions i.e. $M \approx 250$, thereby raising $\langle |B_{\rm cl}| \rangle$ for Coma to the range 7–8 μG. These field strength estimates are subject to further scaling due to the unknown location of this cluster-embedded radio source along the line of sight through the cluster, and the thermal electron profile from the radio lobe to the nearside 'edge' of the cluster. An instructive discussion of the effects of field scaling models by Felten

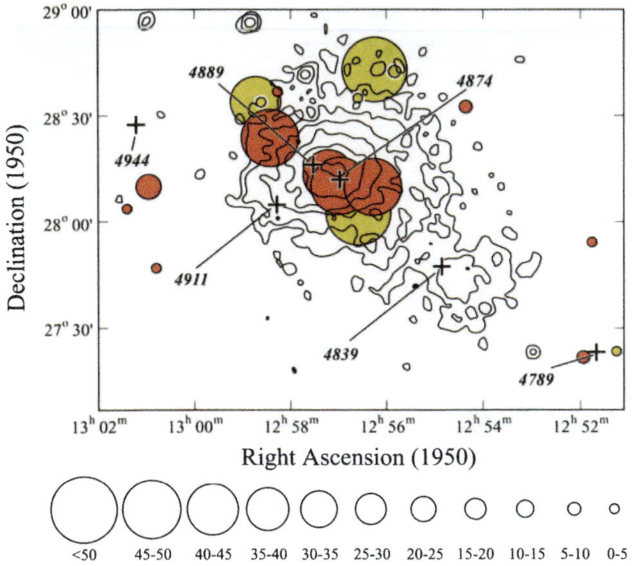

Fig. 11. The Faraday rotation measure probe of the Coma cluster of galaxies. Coloured circles distinguish the RM sign, and their sizes indicate their approximate RM. Overlaid are the Rosat X-ray contours measured by Briel et al. (1992), and the positions of some NGC galaxies in the field

(1996) derived the variance σ^2 (RM) for the case of a constant ICM plasma β at each cluster-centre radius. Figure 11 shows the original measured Faraday rotation measures of Kim et al. (1990) superimposed on a more recent ROSAT X-ray image of the cluster.

Current limitations of radio telescope sensitivity prevent a dense 2-D RM background source probing of other individual clusters in a way that we would wish, especially given the very detailed Chandra and XMM cluster X-ray images now available. The next best experimental approach is to undertake a global, 'statistical' probe of large numbers of clusters, each of which may typically have 1–3 background probe sources. This was first attempted by Lawler and Dennison (1982). A more extensive probe of this type was done for the (then) largest available RM sample of background sources at various impact parameter distances from centers of $\approx$40 clusters by Kim et al. (1991). They deduced a global $B_{\rm cl}$ in the range $(2-4)\,h_{75}^{0.5}$ μG, scaled to l–10 kpc and decreasing with cluster impact parameter out to $r \approx 700$ kpc. Similar results were obtained by Clarke et al. (2001) from a more strictly defined smaller sample of non-radio halo clusters without strong cooling flows, all of which had well-determined X-ray column densities for the radio RM lines of sight. The latter result, $\approx$5 μG for cluster radii $r_{\rm cl} \leq 500\,h_{75}^{-1}$ kpc appears consistent with the estimates described above, and not greatly different from the earlier predictions of Jaffe (1980).

All the above field estimates are based on the assumption a single reversal scale, l, in which $|\mathbf{B}|$ and $n_{\rm e}$ are assumed constant within each l. A more realistic description is to assume a power law distribution of magnetic cells based on

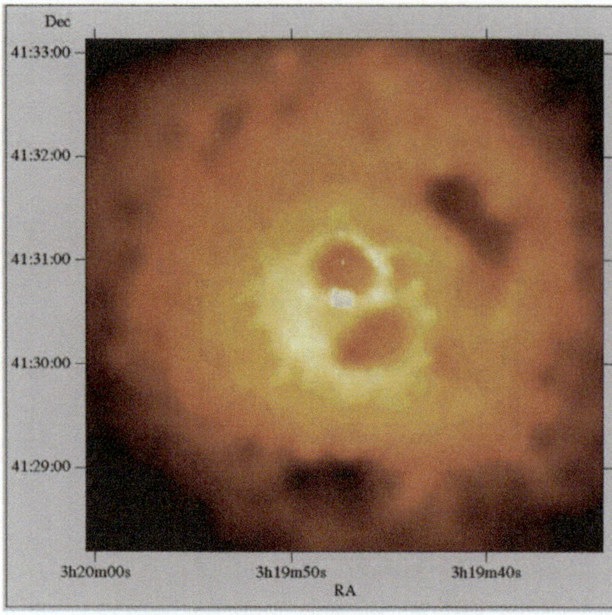

Fig. 12. Chandra image of the i.c. hot gas in the Perseus Cluster of galaxies, showing the X- ray 'cavities' corresponding to magnetized CR radio lobes of 3C 84 in the Perseus cluster of galaxies (NGC 1275) (Source: Fabian et al., 2000)

observations of a range of RM reversal scales over sample of extended background or cluster-embedded radio sources, i.e. a RM(x,y) over each cluster-embedded extended source: A refinement of this type has recently been undertaken by Govoni et al. (2003), and Vogt and Enßlin (2003) who use power spectrum models for the ICM field, e.g. $|B_{\mathrm{k}}|^2 \propto k^{-n}$. These two analyses of of RM images of cluster-embedded sources give n that is steeper than a $-5/3$ power Kolmogorov spectrum $\approx$3. An index, $n \approx 2.7$, was found by Dolag et al. (2002) in their simulations of cluster magnetic field structure from cosmological hydrodynamic simulations. From their analysis, which probes roughly a factor of 10 in magnetic l-scales in an extended cluster-embedded source RM(x,y) analysis in 3 clusters, Vogt and Enßlin (2003) deduce $\mathbf{B}_{\mathrm{ICM}}$ values in the range of 3–12 μG.

Minimum energy requirements suggest that B will not be locally uniform, as seen in the 'filamentation' of the synchrotron emissivity within radio lobes in well-imaged cluster-embedded systems such as Virgo A (see Owen et al., 2000). There is some observational indication from Faraday RM images of extended cluster-embedded radio lobes that the ICM is relatively free of 'low-B holes' on scales of $\approx$ 10 kpc and greater, i.e. the ICM volume in clusters appears reasonably 'filled' with magnetic field on these larger scales.

In the ideal 3-D description of the ICM, we would like to characterize $\mathbf{B}(x, y, z)$ down to the smallest possible l-scales (largest k-scales). Ideally, we would like a similar specification of the thermal gas density $2n_{\mathrm{e}}(x, y, z)$, and temperature $T(x, y, z)$ as might be model-constructed from high resolution X-ray emissivity and temperature distributions. Knowledge of all three would also permit an analysis of the physically important plasma β parameter throughout the ICM.

$$\beta(x,y,z) = \frac{n_e(x,y,z)kT(x,y,z)}{\frac{B^2(x,y,z)}{8\pi}} . \tag{5}$$

The inner, denser cluster 'cooling flow' zones ($r_{cl} \lesssim 100$ kpc) have been similarly probed with cluster-embedded radio sources such as Hydra A, and in X-ray bands. A review of the results for 14 cooling flow clusters by Taylor et al. (1999) has field strengths ranging from 10–100 μG in the relatively dense, X-ray intense cluster core regions. These suggest a plasma $\beta \lesssim 1$ in some of the cooling flow zones, i.e. the magnetic fields are a significant energy component, comparable to or exceeding $n_e kT$ in the cooling flow regions. This also may be true of less dense and lower B régimes at larger cluster radii.

Where clusters have a synchrotron-emitting halo, or cluster-scale relic emission (e.g. Abell 2256), the equipartition-level field to explain ICM emission implies field strengths near 1 μG or greater. Because such emission is usually faint, its degree of filamentation on small scales is so far poorly specified, so that the emission could represent a range of field strengths on different scales, as discussed above.

Recent Chandra and XMM Newton images of X-ray holes coinciding with radio lobes provide a new, if partly indirect, estimate of $\langle|B|\rangle_{cl}$. The holes appear to consist of predominantly relativistic plasma, whose pressure can be inferred from the parameters of the ambient hot gas, and which have expended a significant fraction of the BH energy release as PdV work in the ICM, as discussed above. Their internal pressure requires a well-distributed magnetic field within the X-ray hole to maintain its observed quasi-spherical shape, in quasi-static balance with the X-ray-observable hot gas. The former (CR gas + B pressure) is at least 3×10^{-11} dynes cm^{-2} (Zhao et al., 1993) in the case of 3C 317 in Abell 2052. For this source, ≈ 10 μG is required in the radio holes/lobes, whose typical size is $\approx$50 kpc – a non-trivial fraction of the inner ICM volume.

The confining thermal gas pressure, according to Blanton et al. (2001), appears $\approx$10 times greater than the above estimate (inferred from the radio-visible synchrotron electrons). This makes it possible that $\langle|B|\rangle$ within the X-ray holes is as much as ≈ 30 μG, or else an additional unseen pressure in the holes is due to a mix of magnetic field and CR proton pressure. Additionally there may be an extra hot thermal gas phase (e.g. Dogiel, 1999; Enßlin et al., 1999) to explain part of the unseen pressure component. It is reasonable to suppose that the accumulated magnetized CR gas from several generations (10–15) of such 'holes' generated by AGN's over a cluster's lifetime will spread magnetic fields throughout the cluster over the few dynamical crossing times of a cluster's past history. Similar numbers can be inferred for extended cluster embedded radio sources like 5C4.81 in Coma, whose equipartition field strengths are comparable to the above values and which, over time, get mixed to some degree into the ICM. These numbers converge on ICM field strengths of a few μG. They are also consistent with global predictions of intracluster magnetic field strengths due to the 'feeding' by AGN radio sources based on the global density of radio sources and their higher average volume density in clusters.

The flow 'path' of magnetic energy, CR energy and thermal heat seem to be different for different clusters: For example the morphology of the head-tail radio source 5C4.81 above suggests that it is being entrained into the internal flow of its host Coma cluster ICM (Fig. 13). A contrasting case is the radio source Hydra A (3C 218) in the Hydra cluster of galaxies which, as revealed by a recent low frequency radio image, continues well beyond the cluster core region all the way to Mpc scales. Integration of the energy content of the Hydra A radio lobes still within and outside, respectively, of the cluster core (see Fig. 14) suggests that the greater part of the $\approx 10^{59}$ erg produced by the galactic BH will escape from the cluster core.

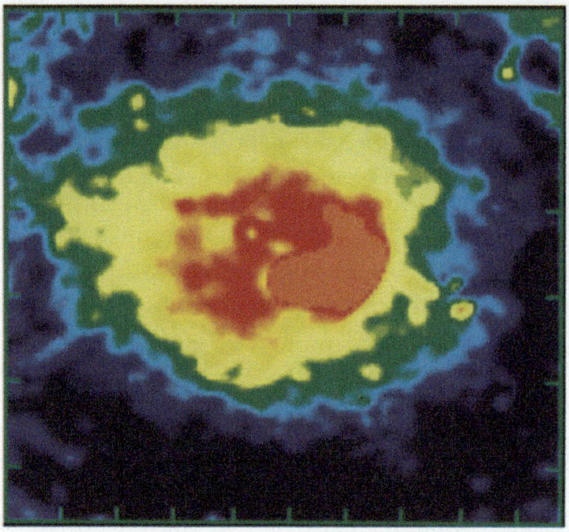

Fig. 13. Combined NRC-DRAO and NRAO-VLA 1.4 GHz image of Coma cluster with its prominently visible head-tail source, 5C4.81 (pink zone), showing how the energy of the CR + B gas in the synchrotron lobes is being subsumed into the ICM, in apparent contrast to the situation in the Hydra cluster (Adapted from Kim et al., 1990)

In cases where the radio-emitting CR clouds are confined within the observable hot gas reservoir of the cluster, the dimensions of the X-ray hole combined with the calculated gas pressure gives a 'laboratory-like' measurement of the central black hole's energy that is converted into PdV work against the ICM

$$E_{\mathrm{PdV}} \simeq nkTV \simeq 10^{60}\, n_{-2}\, T_8\, V_{70} \text{ erg} , \tag{6}$$

where n_{-2} denotes units of $10^{-2}\,\mathrm{cm}^{-3}$, T_8 in units of 10^8 K, and the lobe/hole volume V_{70} is normalized to $10^{70}\,\mathrm{cm}^3$, corresponding to a sphere of radius 45 kpc (Kronberg et al., 2001). This energy, and the lobe B+CR pressure can be quantified from X-ray cluster images such as in Fig. 12. Some of this energy will be deposited as additional heat in the hot intracluster plasma.

$\mathbf{B}_{\mathrm{ICM}}$, or a limiting value can also be estimated from the detection (or non-detection) of nonthermal, high energy X-ray (HEX) emission due to inverse Compton scattering of a population of relativistic electrons whose ICM radio emissivity

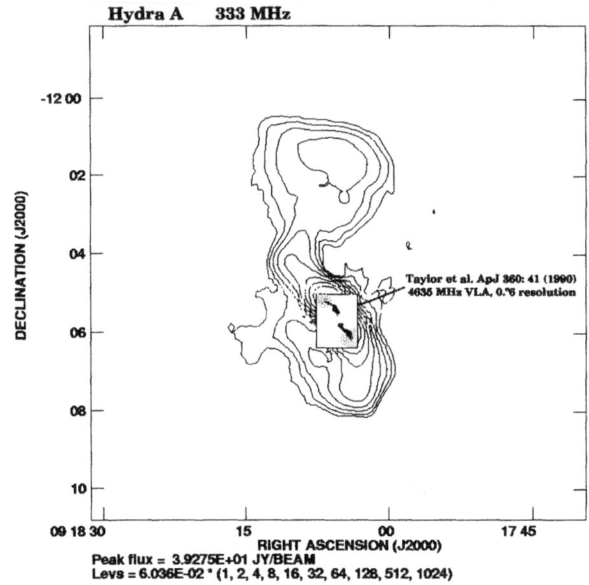

Fig. 14. Hydra A on both large and small scales (Taylor et al., 1990, and R. A. Perley, priv. comm.) relative to the X-ray image of the thermal gas distribution

and spectrum are measurable. Relativistic electrons in sufficiently large numbers will inverse-Compton (IC) scatter CMB photons into the X-ray bands if the magnetic fields are low, corresponding to a higher relativistic particle density n_e^r. n_e^r must be reconciled with both the observed synchrotron emissivity and HEX X-ray counts if they are inverse-Compton generated. The latter have been detected in the Coma cluster (Rephaeli et al., 1999), and also in other clusters e.g. A2256 (Fusco-Femiano et al., 2003). For Coma, the best studied cluster, Fusco-Femiano et al. deduce $\langle|B_{\rm cl}|\rangle \approx 0.2\,\mu$G from X-ray data, and suggest that the differences between X-ray and RM-determined intracluster fields is likely due to different ICM volumes being sampled. The above values represent an order of magnitude lower field strengths than we have discussed above, which represents a *two* orders of magnitude difference in the local ICM magnetic energy density.

The resolution of this apparent discrepancy in the $|\mathbf{B}_{\rm ICM}|$ deduced from Faraday rotation/X-ray data and IC X-ray/EUV emission will require observations in X-rays, the EUV and radio that better reveal the detailed spatial distribution of radiation in these bands. Either the volumes relevant to the IC HEX emission and synchrotron radio are different – e.g. on scales $\ll 3\,$kpc which are below the resolution limits of most current radio and X-ray telescopes, or there are other populations of electrons that mimick IC emission in the HEX bands by a thermal, or synchrotron component. The very short HEX loss times of the latter imply correspondingly short transport times from the sites of acceleration. A space-distributed acceleration mechanism such as magnetic reconnection is a possible candidate (Kronberg et al., 2004 and references therein). It will be important to test whether the HEX emission is associated more closely with the thermal intracluster gas, or with the CR clouds of cluster-embedded radio sources. The reader is referred at this point to Fusco-Femiano et al. (2003) for a recent assessment of detection claims from BeppoSAX of nonthermal high energy photons. It seems that a clear understanding of these apparent discrepancies may require observations beyond current instrumental capabilities of resolution and spectral coverage in the EUV, HEX and γ-ray bands.

11 Probes of Magnetic Fields at Larger Redshifts, Beyond Galaxies and Clusters

Faraday rotation measures can probe to significant cosmological look-back times, such as over the redshift range of quasars. The Galaxy-corrected Faraday rotation measure of a distant quasar can be broken into three unrelated components: (i) that due to a putative all-pervading, cosmologically scaled magneto-ionic medium and (ii) a component added by some **intervening** galaxy system(s) (e.g. galaxy disk or halo, galaxy group, or cluster of galaxies). Finally (iii) it is possible to probe the RM that is generated close to the synchrotron-radiating source at a large redshift. Using measured or estimated values of n_e, and the field reversal scale, an RM generated in any of these 3 situations could give an estimate, or limit to the associated magnetic field (e.g. Kronberg and Perry, 1982). We begin by discussing component (i).

11.1 Faraday Rotation Searches for All-pervading, Widespread Magnetic Fields

The availability of larger samples of extragalactic source rotation measures in the 1970's led to the first tests for a Faraday rotation from a widespread, and cosmologically scaled intergalactic magneto-ionic medium (Rees and Reinhardt, 1972; Nelson, 1973; Kronberg and Simard-Normandin, 1976). The density, $n_{\mathrm{i.g.}}(z)$, of a widespread intergalactic ionized gas can be parameterized as a fraction Φ of the average total matter density, which increases with cosmological epoch as $(1+z)^3$. A measured RM(z) out to some maximum (z_m) can be related to a widespread magnetic field as follows:

$$\mathrm{RM}(z_\mathrm{m}) = \frac{\Delta\chi_0}{\Delta\lambda_0^2} = 1.8\times10^5 \int_0^{z_\mathrm{m}} n_\mathrm{e}(z)B(z)(1+z)^{-2}dl(z)\ \mathrm{rad\,m^{-2}}\ , \qquad (7)$$

where χ_0 and λ_0 are the observed polarization orientation and wavelengths in the observer's frame (l is in pc). For a $\Lambda=0$ Friedmann Universe

$$dl(z) = 10^{-6}\frac{c}{H_0}(1+z)(1+\Omega z)^{1/2}\,dz\ \mathrm{Mpc}\,. \qquad (8)$$

n_e is in cm^{-3}, c in km s^{-1}, H_0 in km s^{-1} Mpc^{-1}, and B in Gauss.

Under these circumstances, χ_0 is invariant with redshift, whereas $\lambda(z)=\lambda_0(1+z)^{-1}$. This reduces the detectability of a rotation measure that is generated at redshift, z, by $(1+z)^{-2}$. On the other hand, due to cosmological expansion we expect

$n_e(z) \propto (1+z)^3$, and $|\mathbf{B}_{\rm IGM}| \propto (1+z)^{\varphi}$, where $\varphi \approx 2$ if we assume approximate flux conservation i.e. that $\mathbf{B}_{\rm i.g.m}$ is frozen into the intergalactic plasma. Inserting these into the equation above shows that $(n_e B)$ could, for large z, overwhelm the $(1+z)^{-2}$ watering-down effect on the observed RM's and thereby produce a detectable RM signal at $z \gg 1$ due to a co-expanding magneto-ionic medium.

Since these earlier papers were written we have learned that the baryonic matter is not uniformly distributed within large co-moving volumes but, along with the dark matter, has gravitationally condensed, concurrent with universal expansion into large-scale filaments or sheets. Thus (7) can be re-formulated and expanded when the evolution of the filamentary structure back to $\approx$6 is better known. Progress is rapidly being made with large-scale sky surveys such as the SDSS optical galaxy/redshift survey, and this will further improve with future large-scale diffuse X-ray surveys and absorption line surveys, both of which will directly detect the diffuse intergalactic gas, especially on gravitationally collapsed scales. The latter are relevant to collapsed intervening systems such as galaxies and clusters and Lyα clouds. This leads to a discussion below of how magnetic fields in discrete intervenor systems can be probed by combinations of Faraday rotation and spectroscopic measurements.

If such a field were ordered on the scale of the Universe, then an observed systematic increase of RM(z) would occur for a preferred direction in the sky which could, in principle also be determined (Woltjer, 1965; Zel'dovich, 1965; Brecher and Blumenthal, 1970). Early claims to the detection of such an aligned field were not substantiated in subsequent, better quasar RM data. These same data limit any systematic growth of $RM(z)$ to $\approx 5\,\mathrm{rad\,m^{-2}}$ or less at $z = 2.5$ (Kronberg and Simard-Normandin, 1976; Kronberg, 1977). This, for a Λ-CDM Universe with $\Omega m = 0.3$, $\Omega_\Lambda = 0.7$ and $H_0 = 75\,\mathrm{km\,s^{-1}\,Mpc^{-1}}$, places an upper limit on any **cosmologically aligned field** $|\mathbf{B}_{\rm IGM}| \lesssim 10^{-11}$ G at the present epoch (Vallée, 1975).

A field that is aligned on cosmological scales is unlikely. Given the large-scale homogeneity and isotropy of the Universe back to the last scattering surface at $z \approx 10^3$, one assumes that any widespread field in the Universe has a characteristic l_0 scaled to the present epoch, although this simple scaling may not be strictly true. Recent evidence from galaxy cluster RM's suggests that l_0 is crudely of order 1 Mpc. Assuming this scales by $(1+z)^{-1}$, and applying the observational limit to RM$(z_{\rm m})$ out to $z_{\rm m} = 2.5$ gives $|\mathbf{B}_{\rm IGM}| \lesssim 10^{-9}$ G at the current epoch for any widespread, all-pervading field in a homogeneous Universe. Future, more extensive RM data out to larger $z_{\rm m}$'s, and incorporating large-scale structure should make it possible to improve the sensitivity of this measurement, and eventually also measure the contrast in field strength between filaments and voids. 10^{-9} Gauss lies at the upper end of some recently calculated primordial field strengths generated in an inflation cosmology.

11.2 Magnetic Field Probes of Galactic Scale Intervening Systems Associated with Absorption Line Systems in Quasars

Having shown that the RM from a **widespread** IGM (i) is so far below current levels of detectability, we now focus on what has been learned about magneto-ionic gas in discrete intervening systems (ii) at intermediate redshifts ($z_{\rm a}$) between us and z_e, that of the emitters – the quasars. We ignore, for the moment, a *source-associated* component (iii).

Independent evidence from optical and 21 cm absorption lines at intermediate redshifts (z_a), reveals that quasar lines of sight pass through intervening galaxy systems whose column densities, and sometimes excitation conditions can be inferred from the equivalent widths at z_a. In a combined analysis of the quasar RM's and absorption line data, Kronberg and Perry (1982) found a correlation between high column depth absorbers in front of quasars, and the tendency for a quasar to have an excess RM (cf. also Watson and Perry, 1991). This confirmed the existence of magnetic fields in galaxy systems at large redshifts. For quasars with good estimates of *free electron* column density, N_e, Kronberg and Perry were able to make the first crude estimates of magnetic field *strengths* in these absorber systems (see below). These varied from a few μG to nearly a milligauss, sometimes with large uncertainties.

A small subset of quasar intervenors has high column density ($N_{\rm H\,I} \approx 10^{21}\,{\rm cm}^{-2}$) of neutral, or near-neutral hydrogen, which gives rise to damped Lyman-α absorption (Wolfe, 1988). This is usually an indicator of a galaxy disk. Combining these line equivalent width estimates with RM data, Wolfe (1988) obtained field estimates of a few μG for these systems – which are not much different from typical Galactic interstellar field values.

11.3 How Magnetic Field Strengths can be Estimated for Discrete Quasar Intervenor Systems

The observable RM and the observationally derivable N_e, the electron column density in an intervening cloud at z_a can be related to the cloud's magnetic field by

$$< B_{\|} >= 2.6 \times 10^{-13}\,({\rm RM}) N_{\rm e}(1+z_{\rm a})^2\,{\rm G}\,, \tag{9}$$

where $< B_{\|} >$ is defined by

$$< B_{\|} >= \frac{\int n_{\rm e} B_{\|} dl}{\int n_{\rm e} dl}\ . \tag{10}$$

For μG-level fields, $N_e \gtrsim 10^{20}\,{\rm cm}^{-2}$ at the current epoch is needed to produce a detectable rotation measure.

At most a small number $\mathcal{N}$ of large galaxies will intersect a typical quasar or radio galaxy line of sight out to $z_e \approx 3$.

$$\mathcal{N} = \int_0^{z_m} \rho_{\rm i}(z)\sigma_{\rm i}(z) dl(z)\ , \tag{11}$$

where ρ_i and σ_i are the co-moving density and cross-section, respectively, of the intervenors. The present-epoch density of galaxies having halos of $r \sim 45\,{\rm kpc}$ is $\rho \approx 0.017\,h_{75}\,{\rm Mpc}^{-3}$ (c.f Burbidge et al., 1977). We note that because $\mathcal{N}$ is a small number even for $z \approx 1$ sytems, lower redshift background radio sources have a very small probability of intersecting a large high column density galaxy. If we make the simplifying assumption that the few (m) magnetized intervening clouds have similar N_e, and $|\mathbf{B}|$ in a similar absorption redshift range, z_a, we can estimate the most likely actual magnetic field strength in the clouds that are intersected by a background quasar:

$$\langle |B| \rangle \simeq \frac{1.7 \times 10^{13} < \cos\theta > (1+ < z_{\mathrm{a}} >)^2 |\mathrm{RM}}{\sqrt{m} N_{\mathrm{e}}} \ \mathrm{G} \tag{12}$$

(Kronberg and Perry, 1982), where it is also assumed that θ in the clouds are randomly orientated so that $\cos\theta$ has an expectation value of 0.5.

The foregoing discussion implicitly assumes a unique line of sight, i.e. that the RM (and the column density) are not averaged over several adjacent sightlines to an extended illuminating quasar. In practice, the integrated RM might represent an average over several independent sightlines, if the radio source has extended structure. This fact can serve to *underestimate* $\langle |\mathbf{B}| \rangle$ as expressed in (12), in that the RM in one or more of the sight lines in an extended image might be significantly larger than the integrated RM (refer to Perry et al., 1993) for a discussion of this point). Another fact to take into account is that, because the radio structure sometimes extends over tens to hundreds of kpc, whereas the optical emission is typically $< 100\,\mathrm{pc}$ in quasars and AGN's, the line of sight which is relevant to N_{e} may not coincide with that along which the RM was generated (9 and 10). For statistical estimates of magnetic fields in quasar absorption line systems, only integrated RM's are currently available in large numbers. However RM images over extended radio sources is increasing in number for situations in which the intervening matter can be probed as we discuss below, and in the case of galaxy clusters discussed above where the local ambient medium is sufficiently dense that several different Faraday rotation sightlines can sometimes be probed in the cluster. We discuss some examples below where multiple, adjacent Faraday rotation paths at $z \gtrsim 1$, are probed. This type of observational method is suitable for probing magnetic field strengths in individual galaxy systems (iii), discussed next.

11.4 Transverse Magnetic Field Probes of Faraday Rotation Intervenors at High Redshift

Faraday rotation images made with sub-arcsecond resolution with the VLA have provided *transverse* probes in RM along various lines of sight through an intervenor. In such cases the discrete intervenors can be laterally probed if they lie in front of a background extended polarized source. An example is the high resolution RM image of the 4″ long polarized jet of 3C 191, a quasar at $z = 1$ by Kronberg et al. (1990). 3C 191 (Fig. 15) has a rich, 'associated' absorption line spectrum, i.e. absorption virtually *at* the redshift of the emission lines. It was proposed in this case that a QSO wind-driven shell of hot gas (seen in absorption and emission in this case) produces the relatively high gas column density. The magnetic field strength in this $z = 1.945$ absorber system was found to be in the range of 0.4–4 μG independent of the detailed system model, since the RM is sensitive to only the magnetic field-weighted column density (9), that is, independent of the line-of-sight distribution of n_{e}. The magnetic field's prevailing direction, another measurable, was found to be similar over $\sim 15\,\mathrm{kpc}$ – a substantial fraction of a galaxy dimension (Kronberg et al., 1990; Perry and Dyson, 1990).

Such associated-line systems are not the typical intervening galaxies discussed earlier (where $z_{\mathrm{a}} \ll z_{\mathrm{Q}}$), but the electron column density estimates from equivalent

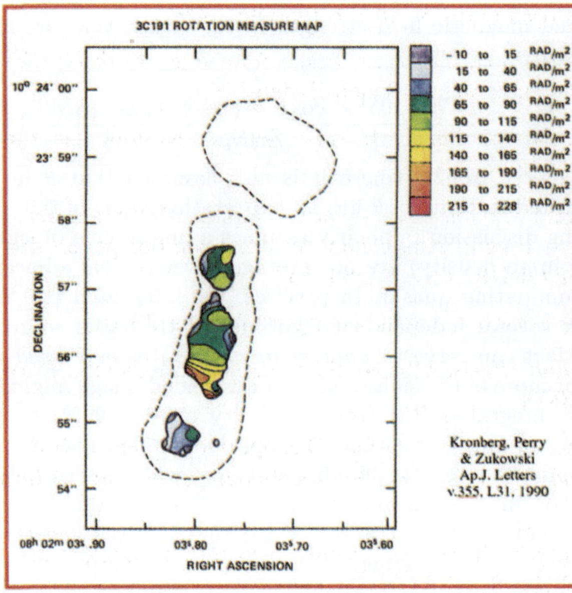

Fig. 15. Faraday rotation image of the 'associated absorption line' quasar 3C 191 at $z = 1.95$. The Faraday rotation image has a resolution of $0\farcs35$, and the jet is about $4''$ long in projection (Adapted from Kronberg et al., 1990)

width measurements of absorption lines in the spectrum of the parent QSO enable us to derive magnetic field estimates in a galaxy at an earlier cosmological epoch. The look-back time of 3C 9 is ≈9 Gyrs as we can estimate from the r.h. scale in Fig. 4.

A more 'conventional' intervening galaxy system that has been similarly imaged is an intervening galaxy at $z = 0.395$, in front of the extended, polarized jet of the quasar PKS 1229−021 at $z = 1.038$. The spectrum of the QSO exhibits damped Lyman-α absorption (indicative of a well-formed galaxy disk), and a combination of absorption transitions in which both a galaxy halo and disk are identified spectroscopically, and N_e in the disk is reasonably well measured. The combination of a detailed optical and 21 cm absorption spectrum analysis (Briggs et al., 1985) with the RM image (Kronberg et al., 1992) has yielded a fairly firm estimate of magnetic field strength in the PKS 1229−021 system of 1–4 μG. The latter found the field direction to reverse every ≈ $0\farcs7$, which is comparable to a spiral arm separation in a galaxy at that redshift. The combination of the radio RM imaging and the spectrally identifiable halo and disk components in the optical absorption spectrum confirm that this $z = 0.395$ quasar intervenor is very likely a spiral galaxy, as sketched in Fig. 16.

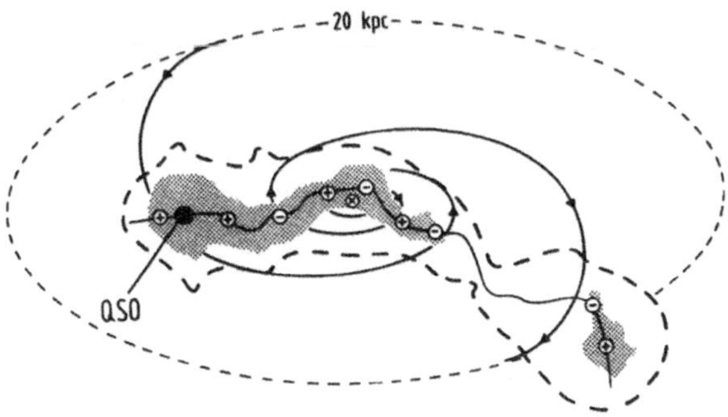

Fig. 16. Sketch of a spiral arm pattern of a galaxy at $z = 0.395$ derived from the Faraday rotation measure structure of the 4″ long inner radio jet of PKS 1229−021. The position of the QSO is indicated, and the zones of alternating positive and negative RM along the jet. The galaxy's inclination was estimated from the optical spectroscopic data

12 Summary

Evidence has grown over the past two decades that magnetic fields pervade all galaxy systems where hot gas and cosmic rays are detected. The intergalactic medium, appears also to be pervaded with magnetic fields that range from ≈10 μG in some galaxy halos and at the cores of some dense galaxy clusters to a few μG in galaxy clusters. There is theoretical expectation, and preliminary observational confirmation that beyond galaxy clusters, some fraction of the filaments and walls of galaxies are permeated by magnetic fields in the range 0.1–1 μG. Magnetic field strengths in the cosmic voids of the mature Universe are as yet unmeasured, but the magnetic field strengths in galaxy systems appear to have been at least as strong in the past, up to 10 Gyr or more, and were thus produced within a short period of proper time in the early Universe.

Acknowledgement

I wish to thank Peter Biermann, Stirling Colgate, Quentin Dufton, Torsten Enßlin, Nick Gnedin, Uli Klein, Giovanni Lapenta, Hui Li, Francesco Miniati and Richard Wielebinski for illuminating discussions, Gabi Breuer for her assistance and advice with the manuscript and Raul Cunha (Toronto) and Walter Fußhöller (Bonn) for assistance with the figure graphics. I am also grateful to the U.S. Department of Energy, Los Alamos National Laboratory (IGPP), and NSERC (Canada) for financial support.

References

Abramopoulos, F., Ku, W. H.-M.: 1983, ApJ 271, 446
Beckley, H. F., Colgate, S. A., Romero, Van D., Ferrel, R.: 2003, ApJ 599, 702
Birk, G. T., Wiechen, H., Lesch, H., Kronberg, P. P.: 2000, A&A 353, 108

Blanton, E. L., Sarazin, C. L., McNamara, B. R., Wise, M. W.: 2001, ApJ 558, L15
Brecher, K. I., Blumenthal, G. R.: 1970, Astrophys. Lett. 6, 169
Briel, U. G., Henry, J. P., Böhringer, H.: 1992, A&A 259, L31
Briggs, F. H., Turnshek, D. A., Schaeffer, J., Wolfe, A. M.: 1985, ApJ 293, 387
Burbidge, G. R.: 1856, ApJ 124, 416
Burbidge, G. R., O'Dell, S. L., Roberts, D. H., Smith, H. E.: 1977, ApJ 218, 33
Chokshi, A., Turner, E. L.: 1992, MNRAS 259, 421
Chyży, K. T., Beck, R., Kohle, S,. Klein, U., Urbanik, M.: 2000, A&A 355, 128
Chyży, K. T., Knapik, J., Bomans, D. J. et al.: 2003, A&A 405, 513
Clarke, T. E., Kronberg, P. P., Böhringer, H.: 2001, ApJ 547, L111
Colgate, S. A., Li, H., Pariev, V.: 2001, Physics of Plasmas 8, 2425
Colgate, S. A., Pariev, V., Beckley, H. F. et al.: 2002, Magnetohydrodynamics 38, 129
Dickey J. M., Salpeter, E. E.: 1984, AJ 284, 461
Dogiel, V.: 1999, in *Diffuse Thermal and Relativistic Plasma in Galaxy Clusters*, ed. H. Böhringer, L. Feretti, P. Schuecker (Garching: MPE) MPE Report No. 271, p. 259
Dolag, K., Bartelmann, M., Lesch, H.: 2002, A&A 387, 383
Duncan, D., Lambert, D., Lemke, D.: 1992, ApJ 401, 584
Enqvist, K.: 1998, Int. J. Mod. Physics D7, 331
Enßlin, T. A., Lieu, R., Biermann, P. L.: 1999, A&A 344, 409
Fabian, A. C., Sanders, A. S., Ettori, S. et al.: 2000, MNRAS 318, L65
Falcke H., Biermann, P. L.: 1995, A&A 293, 665
Feigelson E. D., Montmerle, T.: 1999, ARA&A 37, 363
Felten, J. E.: 1996, in *Clusters, Lensing, and the Future of the Universe*, ed. V. Trimble, A. Reisenegger, ASP Conf. Ser. 88, p. 271
Feretti, L., Dallacasa, D., Giovannini, G., Tagliani, A.: 1995, A&A 302, 680
Fermi, E.: 1949, Phys. Rev. 75, 1169
Furlanetto, S. R., Loeb, A.: 2001, ApJ 556, 619
Fusco-Femiano, R., Dal Fiume, D., Orlandini M. et al.: 2003, in *Matter and Energy in Clusters of Galaxies*, ed. S. Bowyer, C.-Y. Hwang, ASP Conf. Ser. 301, p. 109
Gnedin, N. Y., Ferrara, A., Zweibel, E. G.: 2000, ApJ 539, 505
Gopal-Krishna, Subramanian, P., Wiita, P. J., Becke, P. A.: 2001, A&A 377, 827
Govoni, F., Feretti, L., Murgia, M. et al.: 2003, in *Matter and Energy in Clusters of Galaxies*, ed. S. Bowyer, C.-Y. Hwang, ASP Conf. Ser. 301, p. 501
Harrison, E. R.: 1970, MNRAS 147, 279
Harwit, M., Pacini, F.: 1975, ApJ 200, L127
Helou, G., Bicay, M. D.: 1993, ApJ 415, 93
Howk, C., Savage, B.: 1997, AJ 114, 2463
Jaffe, W.: 1980, ApJ 241, 925
Kim, K.-T., Kronberg, P. P., Giovannini, G., Venturi, T.: 1989, Nat 341, 720
Kim, K.-T., Kronberg, P. P., Dewdney, P. E., Landecker, T. L.: 1990, ApJ 355, 29
Kim, K.-T., Tribble, P. C., Kronberg, P. P.: 1991, ApJ 379, 80
Koide, S., Shibata, K., Kudoh, T., Meier, D. L.: 2002, Science 295, 1688
Kronberg, P. P.: 1977, in *Radio Astronomy and Cosmology*, IAU Symp. 74, ed. D. L. Jauncey (Dordrecht: Reidel), p. 367
Kronberg, P. P.: 1986, Can. J. Phys. 64, 449
Kronberg, P. P.: 1994, Rep. Prog. Physics 57, 325

Kronberg, P. P., Perry, J. J.: 1982, ApJ 263, 518
Kronberg, P. P., Simard-Normandin, M.: 1976, Nat 263, 653
Kronberg, P. P., Biermann, P. L., Schwab, F. R.: 1985, ApJ 291, 693
Kronberg, P. P., Colgate, S. A., Li, H., Dufton, Q. W.: 2004, ApJL 604, 77
Kronberg, P. P., Dufton, Q. W., Li, H., Colgate, S. A.: 2001, ApJ 560, 178
Kronberg, P. P., Lesch, H., Hopp, U.: 1999, ApJ 511, 56
Kronberg, P. P., Perry, J. J., Zukowski, E. L. H.: 1990, ApJL, 355, L31
Kronberg, P. P., Perry, J. J., Zukowski, E. L. H.: 1992, ApJ, 387, 528
Lawler, J. M., Dennison, B.: 1982, ApJ 252 81
Lesch H., Hanasz M.: 2003, A&A 401, 809
Li, H., Colgate, S. A., Wendroff, R., Liska, R.: 2001, ApJ 551, 874
Lovelace, R. V. E.: 1976, Nat 262, 649
Nelson, A. H.: 1973, PASJ 25, 489
Owen, F. N., Eilek, J. A., Kassim, N.: 2000, ApJ 543, 611
Pacholczyk, A. G.: 1970, *Radio Astrophysics* (San Francisco: Freeman)
Parizot, E., Drury, L.: 2000, A&A 356, L166
Perry, J. J., Dyson, J. E.: 1990, ApJ 361, 362
Perry, J. J., Watson, A. M., Kronberg, P. P.: 1993, ApJ 406, 407
Prantzos, N., Aubert, O.: 1995, A&A 302, 69
Rawlings, S., Saunders, R.: 1991, Nat 349, 138
Rees, M. J.: 1987, QJRAS 28, 197
Rees, M. J., Reinhardt, M.: 1972, A&A 19, 189
Rephaeli, Y., Gruber, D. E., Blanco, P. R.: 1999, ApJ 511, L21
Reuter, H.-P., Klein, U., Lesch, H., Wielebinski, R., Kronberg, P. P.: 1994, A&A 282, 724
Richstone, D., Ajhar, E. A., Bender, R. et al.: 1998, Nat 395A, 14
Ryu, D. S., Kang, H. S., Biermann, P. L.: 1998, ApJ 335, 19
Ryu, D. S., Kang, H. S., Hallman, E., Jones, T. W.: 2003, ApJ 593, 599
Shklovskii, I. S.: 1953, Ast. Zhur. SSSR 30, 577
Small, T. A., Blandford, R. D.: 1992, MNRAS 259, 725
Soltan, A.: 1982, MNRAS 200, 115
Taylor, G. B., Perley, R. A., Inoue, M. et al.: 1990, ApJ 360, 41
Taylor, G. B., Allen, S. W., Fabian A. C.: 1999, in *Diffuse Thermal and Relativistic Plasma in Galaxy Clusters*, ed. H. Böhringer, L. Feretti, P. Schuecker (Garching: MPE) MPE Report No. 271, p. 77
Thorne, K. S.: 1974, ApJ 191, 507
Tremaine, S. D., Gebhardt, K., Bender, R. et al.: 2002, ApJ 574, 740
Vallée, J. P.: 1975, Nat 254, 23
Vogt, C., Enßlin, T. A.: 2003, A&A 412, 373
Völk, H. J.: 1989, A&A 218, 67
Völk, H. J., Atoyan, A. M.: 2000, ApJ 541, 88
Watson, A. M., Perry, J. J.: 1991, MNRAS 248, 58
Wolfe, A. M.: 1988, in *QSO Absorption Lines*, ed. C. J. Blades, D. Turnshek, C. A. Norman (Cambridge: Cambridge University Press), p. 297
Woltjer, L.: 1965, *The Structure and Evolution of Galaxies* (New York: Interscience)
Zel'dovich, Ya. B.: 1965, Sov. Phys. –JETP 21, 656
Zhao, J.-H., Sumi, D. M., Burns, J. O., Duric, N.: 1993, ApJ 416, 51

Magnetic Fields in Galaxies

Rainer Beck

Max-Planck-Institut für Radioastronomie, Auf dem Hügel 69, 53121 Bonn, Germany
rbeck@mpifr-bonn.mpg.de

1 Introduction

Magnetic fields are a major agent in the interstellar medium. They contribute significantly to the total pressure which balances the gas disk against gravitation. They affect the gas flows in spiral arms (Gómez and Cox, 2002). The effective sound speed of the gas is increased by the presence of strong fields which reduce the shock strength. The interstellar fields are closely connected to gas clouds. They affect the dynamics of the gas clouds (Elmegreen, 1981; de Avillez and Breitschwerdt, 2004). The stability and evolution of gas clouds are also influenced by magnetic fields, but it is not understood how (Crutcher, 1999; see Chap. 7). Magnetic fields are essential for the onset of star formation as they enable the removal of angular momentum from the protostellar cloud during its collapse (*magnetic braking*, Mouschovias, 1990). Strong fields may shift the stellar mass spectrum towards the more massive stars (Mestel, 1990). *MHD turbulence* distributes energy from supernova explosions within the ISM (Subramanian, 1998) and regenerates the field via *the dynamo process* (Wielebinski and Krause, 1993; Beck et al., 1996; Sect. 6). *Magnetic reconnection* is a possible heating source for the ISM and halo gas (Birk et al., 1998). Magnetic fields also control the density and distribution of cosmic rays in the ISM. A realistic model for any process in the ISM needs basic information about the magnetic field which has to be provided by observations.

Recent progress in observation of magnetic fields reveals the wide range of applications to large-scale MHD phenomena. Regular magnetic fields trace the gas flows in *barred galaxies* even if the gas can hardly be observed directly (Sect. 8). Regular fields can also help to feed their active galactic nuclei which may solve a long-standing problem. Vertical field lines above the disk of *edge-on galaxies* indicate strong galactic winds into a huge halo of hot gas, cosmic rays and magnetic fields (Sect. 9). *Interactions* between a galaxy and the intergalactic medium and between galaxies produce observable signatures in the magnetic field structure (Sect. 10). Finally, *jets* in spiral galaxies are rarely observed and spectacular (Sect. 11). Radio galaxies and their jets are discussed in Chap. 10.

2 Observing Extragalactic Magnetic Fields

Polarized emission at optical, infrared, submillimeter and radio wavelengths holds the clue to magnetic fields in galaxies. The only other method to detect magnetic fields is Zeeman splitting of spectral lines, which is restricted to observations of gas clouds in the Milky Way (Chap. 7).

Optical polarization is a result of extinction by elongated dust grains in the line of sight which are aligned in the interstellar magnetic field (the *Davis-Greenstein effect*). The **E**–vectors point parallel to the field. However, light can also be polarized by scattering, a process unrelated to magnetic fields and hence a contamination which is difficult to subtract (Fig. 1). Optical polarization surveys yielded the large-scale structure of the field in the local spiral arm of our Galaxy (Chaps. 5 and 7). The first extragalactic results by Hiltner (1958) were based on starlight polarization in M 31 and showed that the magnetic field is aligned along the galaxy's major axis. Appenzeller (1967) observed stars in several galaxies and found fields along the spiral arms. Significant progress was achieved by the group of Scarrott who detected polarization from diffuse light and could map the magnetic fields, e.g. in M 104 (Scarrott et al., 1997) and in M 51 (Scarrott et al., 1997). Figure 1 gives an example of more recent work.

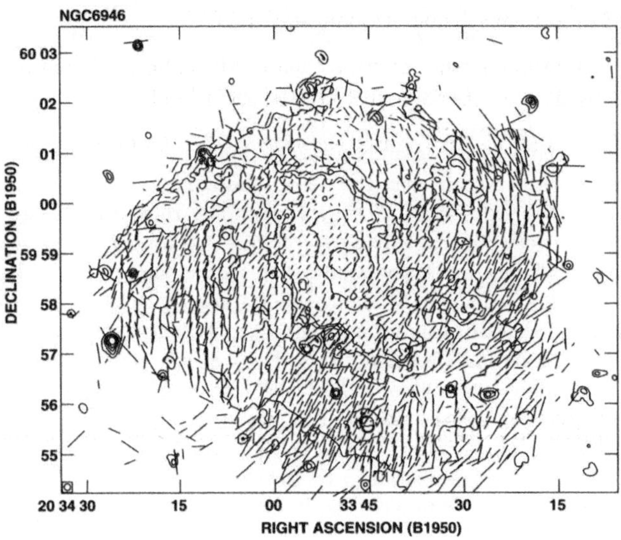

Fig. 1. Optical emission (contours) and polarization **E**–vectors of NGC 6946, observed with the polarimeter of the Landessternwarte Heidelberg at the MPIA Calar Alto 1-m telescope (Fendt et al., 1998). The polarization vectors are partly along the spiral arms (compare with radio **B**–vectors in Fig. 4). Polarization due to light scattering is obvious in the southern part

Polarization techniques in the infrared and submillimeter ranges, where emission from elongated dust grains is polarized and scattered light is no more a problem (Hildebrand, 1998; Chap. 7), are rapidly evolving (e.g. Greaves et al., 2000) and can be expected to contribute to our knowledge about interstellar magnetic fields in the near future.

In *radio continuum* the typical degrees of polarization are much higher than in the other spectral ranges, and we benefit from the development of large instruments and sensitive receivers. This is why most of our knowledge on interstellar magnetic fields in our Galaxy and in external galaxies is based on polarized radio emission and its Faraday rotation. A list of galaxies observed in polarization so far is given in Beck (2000).

The first detection of polarized radio emission in an external galaxy, M 51, came from Mathewson et al. (1972) using the then newly completed Westerbork Synthesis Radio Telescope, followed by Beck et al. (1978, 1980) who mapped the polarized emission of M 31 with the Effelsberg telescope. The huge Effelsberg dish has been equipped with excellent receivers and polarimeters which made it especially successful in mapping magnetic fields and to detect weak diffuse emission (Figs. 7, 14, 15, 20, 24, 26). To achieve higher resolution, Effelsberg data are combined with interferometric (synthesis) data from the Very Large Array or the Australia Compact Array (Figs. 3–6, 4, 13, 16, 22). The first detection of polarized emission from external spiral galaxies with the VLA was reported in 1982 for M 51 (van der Hulst, Kennicutt and Crane, unpublished) and for NGC 4258 (van Albada and van der Hulst, 1982), and with the ATCA for NGC 1566 by Ehle et al. (1996).

Interstellar magnetic fields are illuminated by cosmic-ray electrons emitting *synchrotron radiation*, the dominant contribution to the diffuse radio continuum emission at centimeter and decimeter wavelengths. Synchrotron emission is intrinsically highly linearly polarized, 70–75% in a completely regular magnetic field. The observable degree of polarization in galaxies is reduced by the contribution of unpolarized thermal emission which may dominate in star-forming regions, by *Faraday depolarization* (increasing with wavelength) and by wavelength-independent depolarization by unresolved field structures within the telescope beam (Sokoloff et al., 1998).

A map of the *total intensity*, mostly of synchrotron origin, reveals the strength of the total interstellar magnetic fields in the plane of the sky, while the *polarized intensity and polarization angle* reveal the strength and structure of the regular fields in the plane of the sky which are resolved by the telescope beam.

The orientation of polarization vectors is turned in a magnetic plasma by *Faraday rotation* which is proportional to the line-of-sight integral over the density of thermal electrons multiplied by the strength of the regular field component along the line of sight. The sense of Faraday rotation gives the *direction* of the average regular field. At centimeter wavelengths the Faraday rotation angle ($\Delta\chi$) of the polarization vectors varies with λ^2. ($\Delta\chi = RM\,\lambda^2$, where RM is called the *rotation measure.*) Typical average interstellar rotation measures in mildly inclined galaxies are $\pm$10–100 rad/m^2. This means that below about λ3 cm Faraday rotation is small and the **B**–vectors (i.e. the observed **E**–vectors rotated by 90°) directly trace the *orientation* of the regular fields in the sky plane. Larger Faraday rotation is observed in edge-on galaxies, in the plane and near the center of the Milky Way (Chap. 5), in nuclear jets (Fig. 25), and in radio galaxies (up to several $\pm$1000 rad/m^2).

Polarization angles are ambiguous by $\pm$180° and hence insensitive to field reversals. Compression or stretching of turbulent fields generates incoherent anisotropic fields which reverse direction frequently within the telescope beam, so that Faraday rotation is small while the degree of polarization can still be high. On the other hand, strong Faraday rotation is a signature of *coherent regular fields* (Sect. 6).

3 Measuring Magnetic Field Strengths

Estimates of the dynamical importance of magnetic fields are based on their energy density which increases with the square of the field strength. Hence, the determination of field strengths is a primary task for observations. As the dynamical effects of magnetic fields are anisotropic due to their vector nature, the field structure is of similar importance (see Sect. 5).

The average strength of the total $\langle B_{t,\perp}\rangle$ and the resolved regular field $\langle B_{reg,\perp}\rangle$ in the plane of the sky can be derived from the total and polarized radio synchrotron intensity, respectively, if energy-density equipartition between cosmic rays and magnetic fields is assumed. The revised formula by Beck and Krause (2005), based on integrating the energy spectrum of the cosmic-ray protons, may lead to significantly different field strengths than the classical textbook formula which is based on integration over the radio frequency spectrum. Other sources of systematic bias are spectral steepening due to energy losses of the cosmic rays (Beck and Krause, 2005) and small-scale field fluctuations (Beck et al., 2003).

In our Galaxy the accuracy of the equipartition assumption can be tested, because we have independent information about the local cosmic-ray energy density from in-situ measurements and from γ-ray data. Combination with the radio synchrotron data yields a local strength of the total field $\langle B_t\rangle$ of 6 μG and about 10 μG in the inner Galaxy (Strong et al., 2000), similar values as derived from energy equipartition (Berkhuijsen, in Beck, 2001).

The field strength is determined by the mean surface brightness (intensity) of radio synchrotron emission and thus does not depend on the size of a galaxy. The mean equipartition strength of the total field for a sample of 74 spiral galaxies (Niklas, 1995) is $\langle B_t\rangle \simeq 9$ μG. Dwarf galaxies may host fields of similar strength as spirals if their star-formation rate is high enough (Sect. 7). On the other hand, giant spirals with weak star-forming activity and low radio surface brightness like M 31 (Fig. 7) have $\langle B_t\rangle \simeq 6$ μG. In grand-design galaxies with massive star formation like M 51 (Fig. 3), M 83 (Fig. 6) and NGC 6946 (Fig. 4) $\simeq 15$ μG is a typical average strength of the total field. In the prominent spiral arms of M 51 the total field strength is 25–30 μG (Fletcher et al., 2004b). Field compression by external forces like interaction may also lead to very strong fields (Sect. 10). The strongest fields in spiral galaxies were found in starburst galaxies like M 82 with $\simeq 50$ μG strength (Klein et al., 1998), the 'Antennae' (Fig. 22), in nuclear rings like in NGC 1097 (Fig. 19) and NGC 7552 with $\simeq 100$ μG strength (Beck et al., 2004), and in nuclear jets (Fig. 25). If energy losses of electrons are significant in starburst regions or massive spiral arms, these values are lower limits (Beck and Krause, 2005).

The strength of resolved regular fields B_{reg} in spiral galaxies (observed with spatial resolutions of a few 100 pc) is typically 1–5 μG. Exceptionally strong regular fields are detected in the interarm regions of NGC 6946 of $\simeq 13$ μG (Beck and Hoernes, 1996; Fig. 4) and $\simeq 15$ μG at the inner edge of the inner spiral arms in M 51 (Fletcher et al., 2004b; Fig. 5). In spiral arms of external galaxies the resolved regular field is generally weaker and the tangled (unresolved) field is stronger due to turbulent gas motion in star-forming regions and the expansion of supernova remnants. In interarm regions the regular field is generally stronger than the tangled field.

The relative importance of various competing forces in the interstellar medium can be estimated by comparing the corresponding energy densities. In the local Milky Way, the energy densities of the stellar radiation field, turbulent gas motions, cosmic rays, and magnetic fields are similar (Boulares and Cox, 1990). A global study was performed in the spiral galaxy NGC 6946 (Fig. 2). The energy density of the warm ionized gas E_{th} in NGC 6946 is one order of magnitude smaller than that of the total magnetic field E_{magn}, i.e. the ISM is a low-β plasma ($\beta = E_{th}/E_{magn}$),

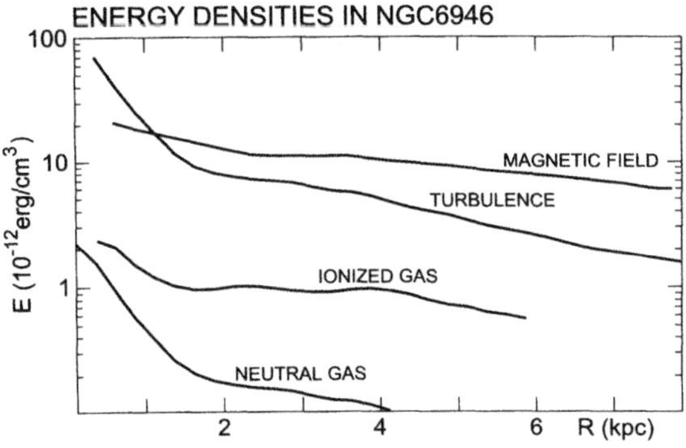

Fig. 2. Radial variation of the energy densities of the total magnetic field, turbulent motion of the neutral gas, thermal energy of the ionized gas, and thermal energy of the neutral gas in NGC 6946 – determined from observations of nonthermal and thermal radio continuum, and CO and HI line emission (Beck, 2004)

similar to the Milky Way (Boulares and Cox, 1990). The contribution of hot gas, which may increase β, is probably small (Ehle et al., 1998).

In the inner parts of NGC 6946 the energy densities of the total magnetic field and turbulent gas motion are similar, an argument for a close connection between field and gas clouds (see Sect. 4). The field dominates in the outer parts due to the large radial scale length of the total field B_t ($l_t \simeq 16$ kpc), compared to the scale length l_ρ of about 3 kpc for the neutral density ρ. This is in conflict with the models of turbulent generation of interstellar magnetic fields which predict similar scale lengths for B^2_{turb} and ρ. As the degree of polarization (and hence the field regularity) increases with increasing radius, the total field in the outer regions of galaxies becomes dominated by the regular field, so that the scale length of B_{turb} is smaller than that of B_t, but not small enough to match l_ρ.

The field in the outer region of galaxies may be amplified by a dynamo driven by the *magneto-rotational instability (MRI*; Sellwood and Balbus, 1999; Rüdiger and Hollerbach, 2004) which is believed to drive turbulent gas motion outside the star formation regime. MRI can generate magnetic energy densities beyond that of turbulent motions. Detailed models are being developed.

In the outermost parts of spiral galaxies the magnetic field energy density may even reach the level of global rotational gas motion and affect the rotation curve, as proposed by Battaner and Florido (2000). Field strengths in the outer parts of galaxies can be measured by Faraday rotation of polarized background sources. Han et al. (1997) found evidence for regular fields in M 31 out to 25 kpc radius (Sect. 6).

4 Magnetic Fields and Gas Clouds

Strongest total radio emission (i.e. total magnetic fields) generally coincide with highest emission from dust and cool gas in the spiral arms. Comparison of the maps of the total radio emission of M 51 and the mid-infrared dust emission at $\lambda 7$ µm and $\lambda 15$ µm (Sauvage et al., 1996) reveals a surprisingly close connection (Fig. 3). In NGC 6946, the highest correlation of all spectral ranges is between the total radio emission at $\lambda 6$ cm and the mid-infrared dust emission, while the correlation with the cold gas (as traced by the CO(1-0) transition) is less tight (Frick et al., 2001; Walsh et al., 2002).

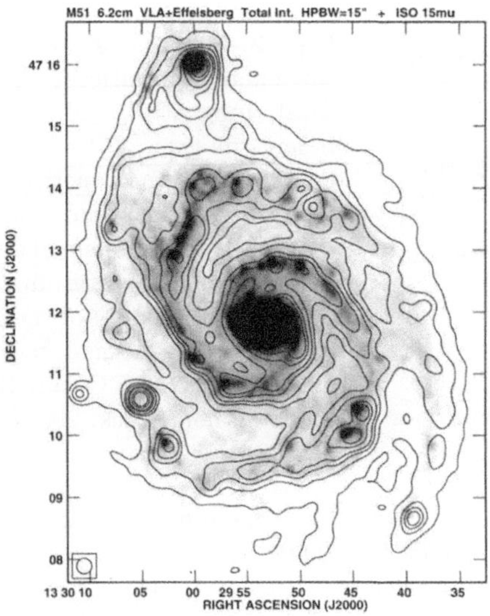

Fig. 3. Total radio continuum emission of M 51 at λ6 cm with 15″ beam size (contours), combined from Effelsberg and VLA observations (Fletcher et al., 2004b). The background image shows the λ15 μm infrared emission from observations of the ISO satellite (Sauvage et al., 1996)

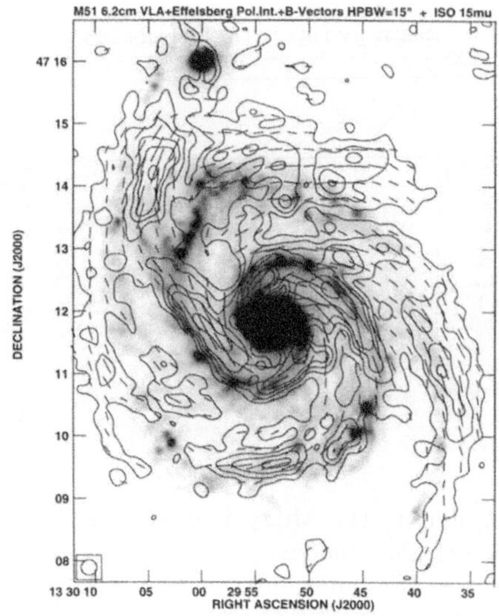

Fig. 4. Polarized radio emission (contours) and B–vectors of M 51 at λ6 cm, combined from Effelsberg and VLA observations (Fletcher et al., 2004b). The background image shows the λ15 μm infrared emission (Sauvage et al., 1996)

The physical background of the relation is hardly understood. Energetic photons from massive stars cause infrared and thermal radio emission. However, over most of the observable radio frequency range the emission is dominated by the non-thermal synchrotron process. Magnetic fields and star-formation processes must be connected. Probably the fields are coupled to the dense, mostly neutral *gas clouds*, which are mixed with the dust. The density of the hot gas is too low to contribute significantly to the anchoring of field lines in the galaxy disks. Niklas and Beck (1997) and Hoernes et al. (1998) proposed a scaling $B_t \propto \rho^{1/2}$ where ρ is the average density of the neutral gas (atomic + molecular) within the telescope beam, similar to the scaling in dense molecular clouds (Chap. 7). As the volume filling factor of the clouds is low, ρ mainly depends on the average *number density* of clouds within the volume observed by the beam, not on their internal density. In this case, the above scaling can be interpreted as magnetic flux freezing in the compressible 'fluid' of gas clouds. The total field strength B_t is highest in spiral arms where the number density of clouds is highest. This number may control the star-formation rate and the magnetic flux.

Photoionization may provide sufficient density of thermal electrons in the outer regions of gas clouds to hold the field lines, as indicated by C^+ emission from the warm surfaces of molecular clouds (Stacey et al., 1991). A Faraday screen of ionized gas has been detected in front of the Taurus molecular cloud complex in our

Galaxy (Wolleben and Reich, 2004). Polarization observations in the submillimeter range also indicate strong and distorted fields in the surroundings of molecular/dust clouds (Greaves et al., 1994).

5 Magnetic Field Structure

Radio *polarization* observations show that in most galaxies the regular field follows the spiral structure seen in the stars and the gas, e.g. in M 51 (Neininger, 1992; Neininger and Horellou, 1996; Fig. 4 and 5), M 81 (Krause et al., 1989b; Fig. 8), and M 83 (Neininger et al., 1991; Fig. 6), though the regions of strongest regular fields are generally *offset* from the spiral arms (see below). As the motion of gas and stars is not along the spiral density wave, but crosses the spiral arm, the magnetic field pattern does *not* follow the gas flow. If large-scale magnetic fields were frozen into the gas, differential rotation would have wound them up to very small pitch angles. The large observed pitch angles (10°–40°) indicate decoupling between gas and magnetic fields due to magnetic diffusivity which is essential for dynamo action (Sect. 6).

Fig. 5. Total radio emission (contours) and B–vectors from the inner disk of M 51 at $\lambda 6$ cm, combined from VLA and Effelsberg observations (Fletcher et al., 2004b). The field size is $4' \times 3'$, the beam size $6''$. The background image shows the optical emission observed with the Hubble Space Telescope (Hubble Heritage Project)

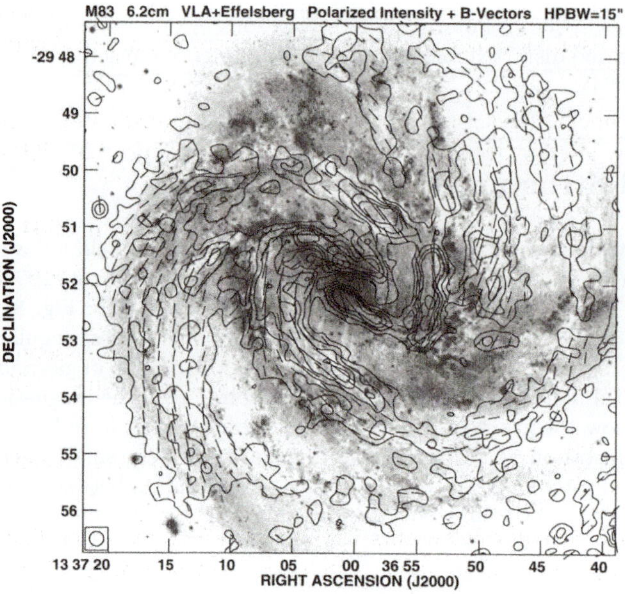

Fig. 6. Polarized radio emission (contours) and B–vectors of M 83 at $\lambda 6$ cm with $15''$ beam size, combined from VLA and Effelsberg observations (Beck, Ehle and Sukumar, unpublished); the background optical image is from the Anglo Australian Observatory (Malin, priv. comm.)

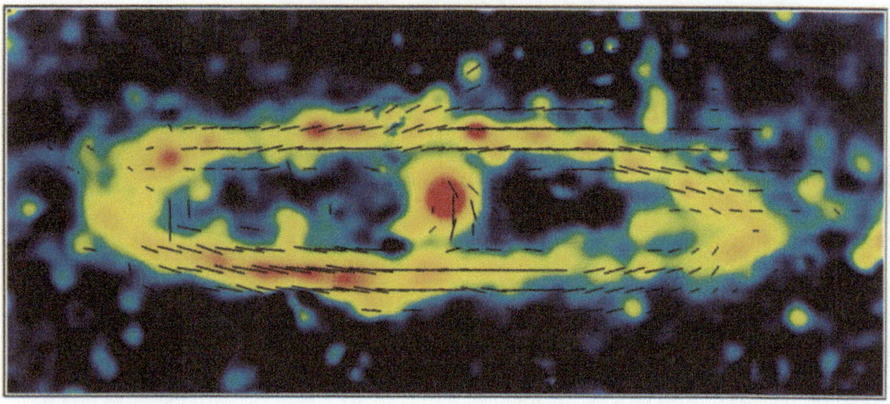

Fig. 7. Total radio emission (contours) and B–vectors (corrected for Faraday rotation) of M 31 at $\lambda 6$ cm, observed with the Effelsberg telescope (Berkhuijsen et al., 2003). The field size is $130'$ x $57'$, the beam size $3'$

In the spiral arms, polarized emission is generally weak, because the regular field is unresolved in the spiral arms due to field tangling by increased turbulent motions of gas clouds or by supernova shock fronts, and, at longer wavelengths, due to Faraday depolarization (Sokoloff et al., 1998). Turbulent fields, as indicated by unpolarized emission, are generally enhanced in spiral arms, following the general scaling with gas density (Sect. 4).

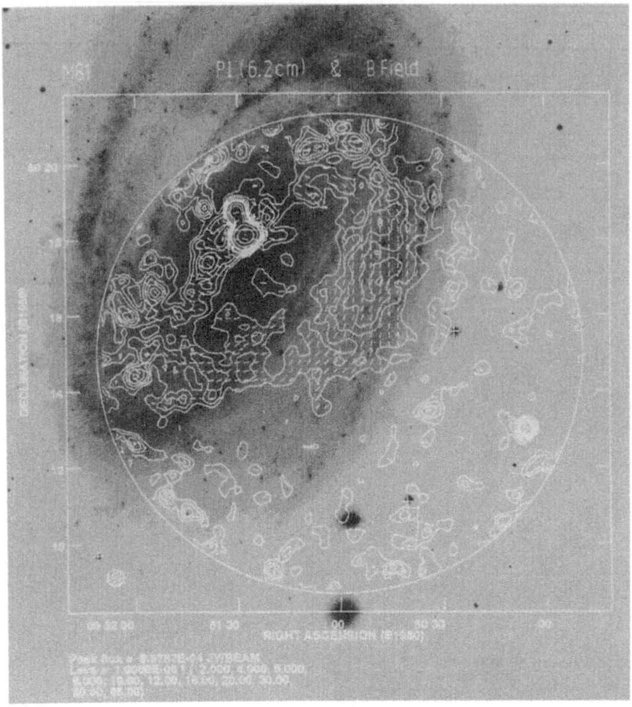

Fig. 8. Polarized radio emission (contours) and B–vectors of M 81 at $\lambda 6$ cm with $25''$ beam size, observed with the VLA (Schoofs, 1992). The circle shows the radius of the primary beam at the 25% intensity level. The intensities are corrected for primary-beam attenuation

In galaxies with strong density waves like M 51 three components of the regular field can be distinguished (Fig. 5). One component fills the interarm space, like in NGC 6946 (Fig. 4). The second component is strongest at the positions of the prominent dust lanes on the inner edge of the gaseous spiral arms, as expected from field alignment by compression. However, the arm-interarm contrast is low, which is in conflict with classical shock compression models and calls for new models of spiral arm formation (e.g Gómez and Cox, 2002). The third component of regular field coincides with the outer southern and southwestern spiral arms, without signs of compression (Fig. 4).

M 83 (Fig. 6) and NGC 2997 (Han et al., 1999) are cases similar to M 51, with enhanced regular fields at the inner edges of the inner optical arms, regular fields in some interarm regions, and also regular fields coinciding with the outer optical arms.

In the highly inclined galaxy M 31 (Fig. 7) the spiral arms are hard to distinguish. Star formation activity is restricted to a limited radial range around 10 kpc radius (the 'ring'). The strongest regular fields coincide with massive dust lanes, but regular fields were detected out to 25 kpc radius (Han et al., 1998). Other density-wave galaxies like M 81 (Krause et al., 1989b; Fig. 8) and NGC 1566 (Ehle et al., 1996) show little signs of field compression; regular fields occur mainly in *interarm regions*.

Observations of another gas-rich galaxy, NGC 6946, revealed a surprisingly regular distribution of polarized emission with two symmetric *magnetic arms* located in interarm regions, without any association with observable gas or stars, and running

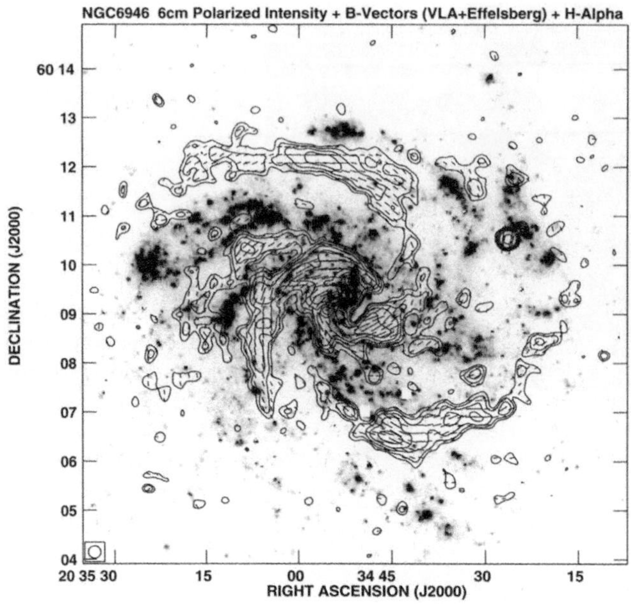

Fig. 9. Polarized radio emission (contours) and B–vectors of NGC 6946 at λ6 cm with 15″ beam size, combined from VLA and Effelsberg observations (Beck and Hoernes, 1996). The background image shows the Hα emission (Ferguson et al., 1998)

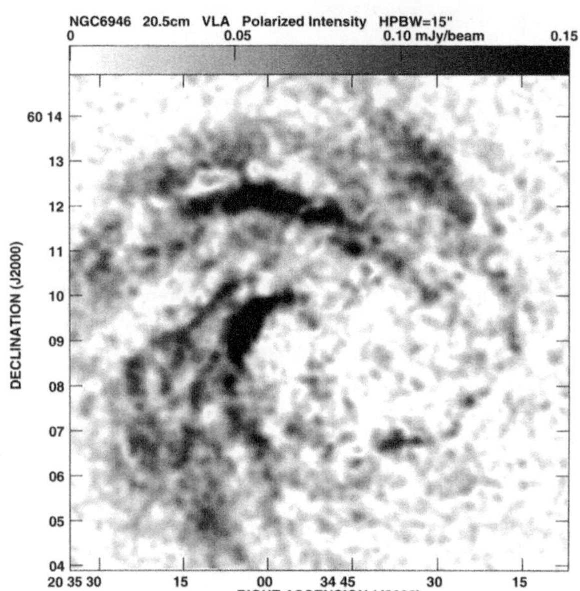

Fig. 10. Polarized radio emission of NGC 6946 at λ20 cm, combined from VLA C- and D-array observations. The field is the same as in Fig. 4, the beam size is 15″ (Beck, unpublished)

parallel to the adjacent optical spiral arms (Fig. 4). These magnetic arms do not fill the entire interarm spaces like the polarized emission in M81, but are less than 1 kpc wide. Their degree of polarization is exceptionally high (up to 50%); the field is *almost totally aligned*. With the higher sensitivity at λ20 cm (Fig. 10), more

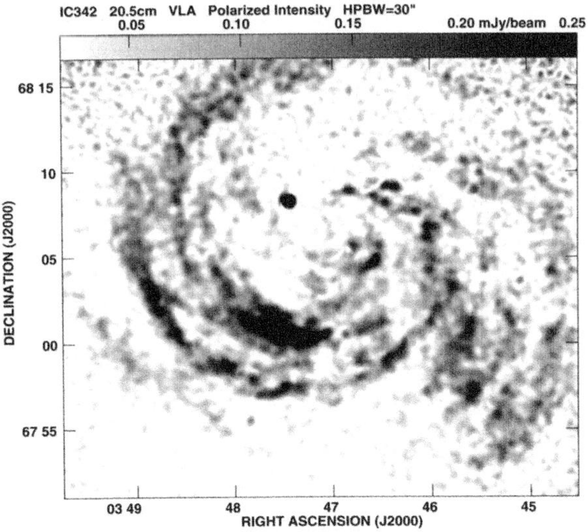

Fig. 11. Polarized radio emission of IC 342 at $\lambda 20\,\mathrm{cm}$, combined from VLA C- and D-array observations. The field size is $32' \times 32'$, the beam size $30''$ (Beck, unpublished)

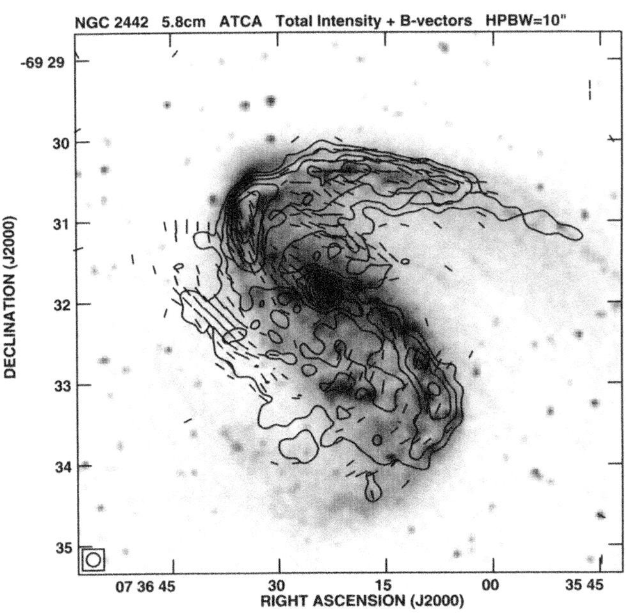

Fig. 12. Total radio emission (contours) and B–vectors of NGC 2442 at $\lambda 6\,\mathrm{cm}$ with $10''$ beam size, observed with the ATCA (Harnett et al., 2004). The background image shows the Hα emission, smoothed to the same resolution

magnetic arms appear in the northern half of NGC 6946, extending far beyond the optical arms, while the strong Faraday depolarization at this wavelength hides the southern magnetic arm. Magnetic arms outside of the optical arms have also been found in M 83 (Fig. 6), in NGC 2997 (Han et al., 1999), and in NGC 2442 (Fig. 12).

Several models have been proposed to explain the generation of magnetic arms. Fan and Lou (1997) suggested that they could be manifestations of slow MHD waves which may propagate in a rigidly rotating disk, with the maxima in field strength phase-shifted against those in gas density. However, all galaxies with magnetic arms rotate differentially beyond 1–2 kpc from the center. Differential rotation destabilizes slow waves so that they evolve into Parker instabilities, so that dynamo models are more promising to explain the magnetic arms (Chap. 6).

Regular magnetic fields form spiral patterns which can also be disconnected from the optical spiral arms. Long, highly polarized arms were first discovered in the outer regions of IC 342 where only faint arms of HI radio line emission exist (Krause et al., 1989a; Krause, 1993). More recent observations at $\lambda 20$ cm revealed a system of such features (Fig. 11; note that the inner part of IC 342 is Faraday-depolarized at this wavelength).

External forces may also compress the magnetic field. In NGC 2442 (Harnett et al., 2004) and in the Virgo cluster member NGC 4254 (Soida et al., 1996) the polarized emission on one side of the galaxy is shifted towards the *outer* edge of the spiral arm, an indication for ram pressure by the intracluster medium. In NGC 3627 (Fig. 13), member of the Leo triplett, the magnetic arm and the optical arm are totally misaligned on the eastern side, another sign of interaction (see Sect. 10).

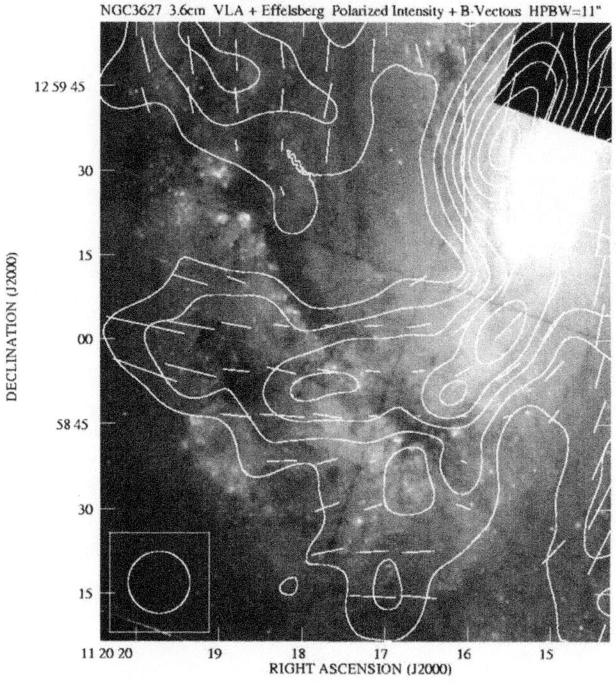

Fig. 13. Polarized radio emission (contours) and B–vectors of NGC 3627 at $\lambda 3.6$ cm with $11''$ beam size, combined from VLA and Effelsberg observations (Soida et al., 2001). The background optical image is from the Hubble Space Telescope

Large-scale *field reversals* were discovered in the Milky Way (Chaps. 5 and 6), but nothing similar has yet been detected in spiral galaxies. High-resolution maps of Faraday rotation, which measure the RMs of the diffuse polarized synchrotron emission and are sensitive to reversals, are available for a couple of spiral galaxies. In M 81 the dominating bisymmetric field structure implies two large-scale reversals (Krause et al., 1989b). The disk fields of M 51 and NGC 4414 can be described by a mixture of dynamo modes where reversals may emerge in a limited radial and azimuthal range of the disk (Berkhuijsen et al., 1997; Soida et al., 2002; Chap. 6). However, no multiple reversals along radius, like those in the Milky Way, were found so far in the disk of any external galaxy.

The discrepancy between Galactic and extragalactic data may be due to the different volumes traced by the observations. Results in the Galaxy are based on pulsar *RMs* which trace the warm ionized medium near the plane, while extragalactic *RMs* of the diffuse polarization emission are averages over the line of sight through the whole thick disk or the halo. Furthermore, some of the Galactic reversals may be due to local field distortions or loops (Mitra et al., 2003). *RM* data at high frequencies are needed to obtain a clear picture of the field structure in the Milky Way.

Present-day polarization observations cannot resolve the detailed field structure, especially in the spiral arms where the degrees of polarization are low due to beam smearing. Rotation measure data from Galactic pulsars (Chap. 5) and depolarization data in external galaxies (Beck et al., 1999b) indicate that the interstellar field is turbulent on scales of $\simeq$ 20 pc. The highest spatial resolutions obtained so far are $\simeq$ 50 pc in M 31 (Hoernes, 1997) and $\simeq$ 100 pc in IC 342 (Beck, in prep.). The new ATCA polarization survey of the Magellanic Clouds has about 10 pc resolution (Gaensler et al., 2005).

6 Dynamos

Observation of large-scale patterns in Faraday rotation measures, e.g. in M 31 (Berkhuijsen et al., 2003; Fletcher et al., 2004a; Fig. 14), NGC 6946 (Beck, 2001) and NGC 2997 (Han et al., 1999), shows that the regular field in these galaxies has a coherent direction and hence is not generated by compression or stretching of irregular fields in gas flows.

Coherent primordial fields, if they existed, are hard to be preserved over a galaxy's lifetime (see however Chap. 4). The dynamo mechanism (Chap. 6) is able to generate and preserve coherent magnetic fields, and these are of appropriate spiral shape (Beck, 1993; Beck et al., 1996). However, the pitch angle of the field spiral depends on the dynamo number, *not* on the pitch angle of the gaseous spiral (Rohde et al., 1999). The observed alignment of magnetic pitch angles with those of the gaseous arms can be achieved by inclusion of the shear flow around spiral arms (Linden et al., 1998; Elstner et al., 2000). The dynamo needs some seed field to start operation. The seed field can be turbulent, e.g. ejected from supernovae or stellar winds (Chap. 1) or from early starbursts (Chap. 2), or a primordial field (Chap. 4).

The field structure obtained in *mean-field* α–Ω *dynamo models* is described by modes of different azimuthal and vertical symmetry. The existing dynamo models (Beck et al., 1996) predict that several azimuthal modes can be excited, the strongest being $m = 0$ (an axisymmetric spiral field), followed by the weaker $m = 1$ (a bisymmetric spiral field), etc. These generate a Fourier spectrum of azimuthal RM patterns. The axisymmetric mode with even vertical symmetry (quadrupole) is excited most easily. For most of about 20 nearby galaxies observed so far, the RM data indicate a mixture of magnetic modes which cannot be reliably determined due to low angular resolution and/or low signal-to-noise ratios (Beck, 2000).

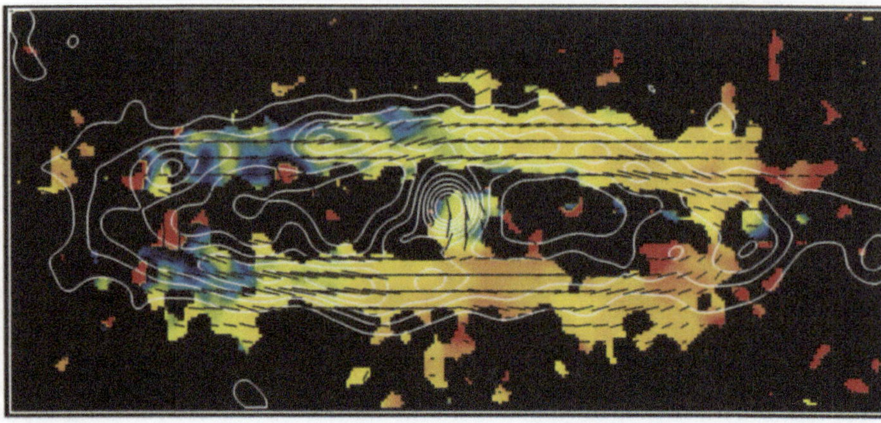

Fig. 14. Total radio emission of M 31 at $\lambda 6$ cm (contours) and Faraday rotation measures between $\lambda 6$ cm and $\lambda 11$ cm, observed with the Effelsberg telescope. The field size is $130'$ x $57'$, the beam size $5'$ (Berkhuijsen et al., 2003)

M 31 is an exception with a strongly dominating axisymmetric field (Beck, 1982; Berkhuijsen et al., 2003; Fig. 14). This field structure extends to at least 25 kpc radius (Han et al., 1998) and at least 1 kpc height above the galaxy's plane (Fletcher et al., 2004a). A bisymmetric mode dominates in M 81 (Krause et al., 1989b).

Magnetic arms (Sect. 5) can be understood to evolve between the optical arms if the dynamo number is smaller in the gaseous arms than between them, e.g. due to increased turbulent velocity of the gas in the arms (Moss, 1998; Shukurov, 1998; Chap. 6) or if turbulent diffusion is larger in the arms (Rohde et al., 1999); the magnetic arms in NGC 6946 are a superposition of an axisymmetric $m = 0$ and a quadrisymmetric $m = 2$ mode.

Spiral fields can be traced to within a few 100 pc from the centers of M 51 (Fig. 5), NGC 2997 (Han et al., 1999) and NGC 6946 (Fig. 4). Mean-field dynamo models can hardly reproduce this result because differential rotation is too weak near the centers of these galaxies. Enhancement of velocity shear by strong density waves (e.g. in M 51) or a non-axisymmetric gas flow around a nuclear bar (e.g. in NGC 6946) or inflow by magnetic stress (Sect. 8) are needed to increase dynamo action.

The strong fields in the outer regions of galaxies (Fig. 2) require non-standard dynamos in regions with low star formation. The *magneto-rotational instability* (MRI) transfers energy from the shear of differential rotation into turbulent and magnetic energy (Sellwood and Balbus, 1999; Rüdiger and Hollerbach, 2004). Other non-standard dynamos, which are faster than the mean-field α–Ω dynamo, have been proposed for young galaxies, e.g. driven by a *cross-helicity correlation* between the small-scale components of gas velocity and magnetic field (Brandenburg and Urpin, 1998), or by cosmic rays inflating buoyant *Parker loops* (Parker, 1992; Moss et al., 1999; Hanasz and Lesch, 2000; Hanasz et al., 2002).

Observation of a large sample of galaxies at medium and large distances with next-generation radio telescopes like the Square Kilometer Array will provide the data base to test dynamo against primordial theory (Beck and Gaensler, 2004).

7 Magnetic Fields in Flocculent and Irregular Galaxies

Regular magnetic fields with strengths similar to those in grand-design spiral galaxies have been detected in the flocculent galaxies M 33 (Buczilowski and Beck, 1991, Fig. 15), NGC 3521 and NGC 5055 (Knapik et al., 2000), and in NGC 4414 (Soida et al., 2002). The mean degree of polarization (corrected for different spatial resolutions) is similar between grand-design and flocculent galaxies (Knapik et al., 2000).

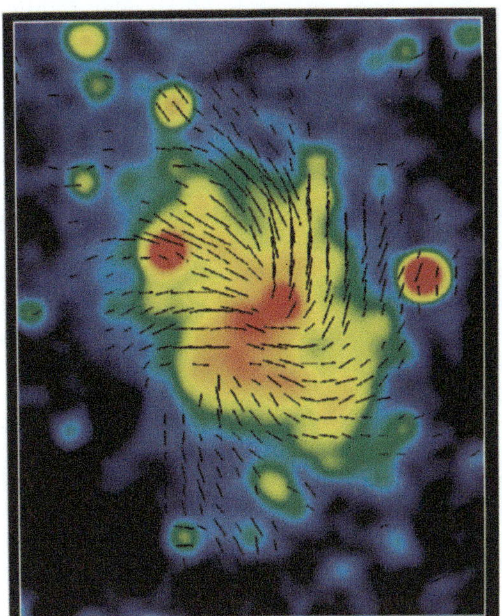

Fig. 15. Total radio emission (contours) and B–vectors of M 33 at λ6 cm, observed with the Effelsberg telescope. The field size is $51'$ x $66'$, the beam size $3'$ (Niklas and Beck, unpublished)

Spiral patterns are observed in all flocculent galaxies, indicative that the dynamo works without assistance from density waves, as expected from the classical α–Ω dynamo. However, the multi-wavelength data of M 33 and NGC 4414 call for a mixture of modes or an even more complicated field structure (Fletcher et al., 2000; Soida et al., 2002).

Radio continuum maps of irregular, slowly rotating galaxies of the Local Group reveal strong total magnetic fields of more than 10 μG in the galaxies NGC 4449 (Fig. 16) and IC 10 (Chyży et al., 2003). Even dwarf irregular galaxies with almost chaotic rotation host total fields with strengths comparable to spiral galaxies if their star formation activity is sufficiently high so that the fluctuation dynamo (see

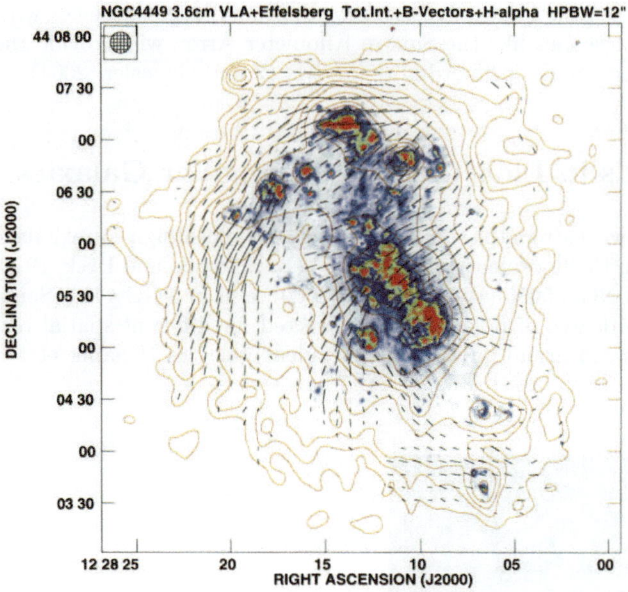

Fig. 16. Total radio emission (contours) and B–vectors of NGC 4449 at $\lambda 3.6$ cm with $12''$ beam size, combined from VLA and Effelsberg observations (Chyży et al., 2000). The background image shows the Hα emission

below) can operate. In these galaxies the energy density of the magnetic field is only slighly smaller than that of the (chaotic) rotation and thus may affect the evolution of the whole system. NGC 6822, on the other hand, has a low star-forming activity and only very weak total radio emission, i.e. a *total* field weaker than $5\,\mu$G, probably below the threshold for any type of dynamo action. As the production of cosmic rays is very low in NGC 6822, it is also possible that any existing magnetic field is not 'illuminated'.

In NGC 4449 (Fig. 16) the field is partly regular with some spiral pattern (Chyży et al., 2000), while in NGC 6822 the field is mostly tangled (Chyży et al., 2003). A few small spots of faint polarized emission indicate that the regular field in NGC 6822 is weaker than $5\,\mu$G, excluding the action of the classical α–Ω dynamo. Such galaxies require field amplification e.g. by the *fluctuation dynamo* (Kulsrud et al., 1997; Blackman, 1998; Subramanian, 1998; Schekochihin et al., 2004). Alternatively, some primordial field may have survived (Chap. 4), e.g. because velocity shear is smaller in irregular galaxies than in spiral galaxies.

The Magellanic Clouds are the closest irregular galaxies and deserve special attention. Polarization surveys with the Parkes telescope at several wavelengths revealed little polarized emission, only two magnetic filaments in the LMC south of the 30 Dor star-formation complex (Klein et al., 1998). The new ATCA polarization survey show that the LMC also hosts a large-scale magnetic field (Gaensler et al., 2005).

8 Magnetic Fields in Barred Galaxies

Gas and stars in strongly barred galaxies move in highly noncircular orbits. Numerical models show that gas streamlines are deflected in the bar region along shock fronts, behind which the gas is compressed in a fast shearing flow (Athanassoula, 1992; Piner et al., 1995). As the gas in the bar region rotates faster than the bar, compression regions traced by massive dust lanes develop along the edge of the bar that is leading with respect to the galaxy's rotation. Gas inflow along the compression region may fuel starburst activity near the galactic center. Magnetic fields have not been included in the models.

M 83 is the nearest barred galaxy and shows compressed magnetic fields at both leading edges of the bar (Fig. 6).

The total and polarized radio continuum emission of 20 galaxies with large bars was observed with the Very Large Array (VLA) at $\lambda 3$, 6, 18 and 22 cm and with the Australia Telescope Compact Array (ATCA) at $\lambda 6$ cm and 13 cm (Beck et al., 2002, 2005). The total radio luminosity (measuring the total magnetic field) is strongest in galaxies with high far-infrared luminosity (indicating high star-formation activity), a result similar to that in non-barred galaxies. The average radio intensity, radio luminosity and star-formation activity all correlate with *relative bar length.*

Polarized emission was detected in 17 of the 20 barred galaxies. The pattern of the regular field in the galaxies with long bars (NGC 1097, 1365, 1559, 1672, 2442 and 7552) is significantly different from that in non-barred galaxies: Field enhancements occur outside of the bar (*upstream*), and the field lines are oriented at large angles with respect to the bar. The symmetry of the velocity fields in these galaxies is distorted by the bar's gravitational potential, leading to enhanced velocity shear, which may enhance dynamo action (Moss et al., 2001).

NGC 1097 (Fig. 17) hosts a huge bar of 16 kpc length. The total radio intensity (not shown in the figure) is strongest in the region of the dust lanes, consistent

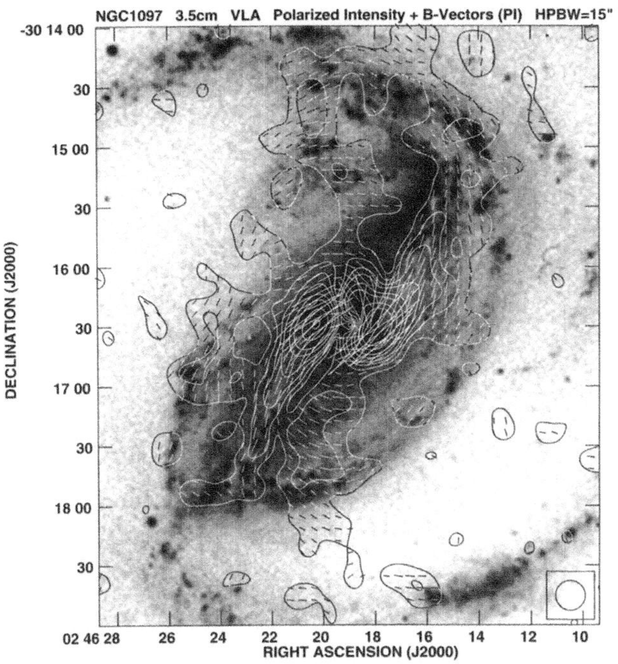

Fig. 17. Polarized radio emission (contours) and B–vectors of the barred galaxy NGC 1097 at $\lambda 3.5$ cm with $15''$ beam size, observed with the VLA (Beck et al., 1999a, 2005). The background optical image is from the Cerro Tololo Observatory (Arp, priv. comm.)

with compression by the bar's shock. The general similarity of the B–vectors in NGC 1097 and gas streamlines around the bar as obtained in numerical simulations (Athanassoula, 1992) is striking. This suggests that the regular magnetic field is aligned with the shearing flow (Beck et al., 1999a). In the southern bar, the upstream region (south of the center in in Fig. 17) and downstream region (southeast of the center, coinciding with the dust lanes) are separated in enhanced polarized emission by a strip of low polarized intensity where the regular field changes its orientation by almost 90°. This observation implies that the region of strongest shear in the magnetic field is located $\simeq 800\,\mathrm{pc}$ in front of the dust lanes, in contrast to the velocity field in the numerical simulations. Either the regular field is not coupled to the dense gas and avoids the shock, or the existing simulations are insufficient to model magnetized shocks. Remarkably, the optical image of NGC 1097 shows dust filaments in the upstream region which are almost perpendicular to the bar and thus aligned with the regular field.

NGC 1365 (Fig. 18) is similar to NGC 1097 in its overall properties, but the polarization data indicate that the magnetic shear due to the bar is weaker. The field bends even more smoothly from the upstream region into the bar, with no indication of a shock. NGC 1559, NGC 1672 and NGC 7552 show similar polarization features (Beck et al., 2002), but the spatial resolution is still insufficient to reveal the detailed structure of their regular fields.

The circumnuclear ring of NGC 1097 (Fig. 19) is a site of ongoing intense star formation, with an active nucleus in its centre. The orientation of the innermost spiral field agrees with that of the spiral dust filaments visible on optical images. Magnetic stress in the circumnuclear ring can drive mass inflow which is sufficient to fuel the activity of the nucleus (Beck et al., 1999a, 2005). Bright circumnuclear radio rings have also been found in the barred galaxies NGC 1672 and NGC 7552 (Beck et al., 2004).

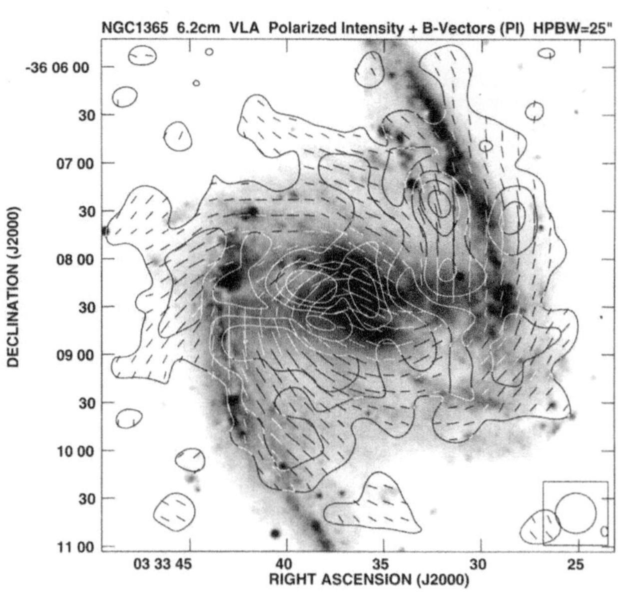

Fig. 18. Polarized radio emission (contours) and B–vectors of NGC 1365 at λ6 cm with 25″ beam size, observed with the VLA (Beck et al., 2002, 2005). The background optical image is from the ESO (Lindblad, priv. comm.)

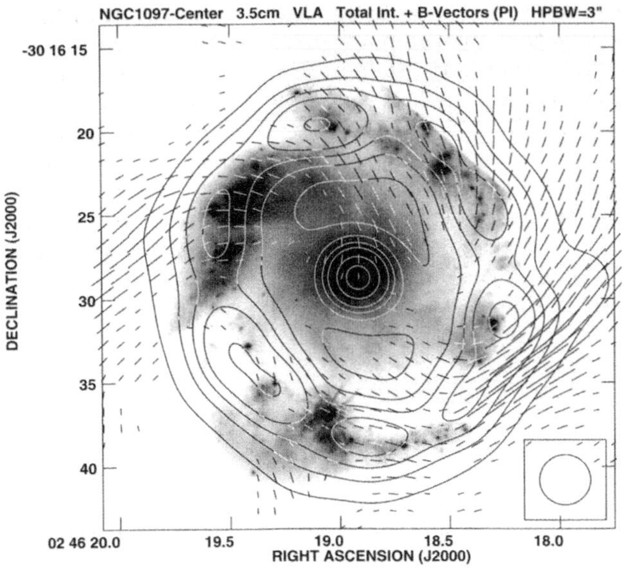

Fig. 19. Total radio emission (contours) and B–vectors from the central region of NGC 1097 at $\lambda 3.5$ cm with $3''$ beam size, observed with the VLA (Beck et al., 1999a, 2004). The background optical image is from the Hubble Space Telescope (Barth et al., 1995)

Radio polarization data have revealed a principal difference between the behaviours of magnetic fields in barred and non-barred galaxies. In non-barred galaxies the field lines are of overall spiral shape, they do not follow the gas flow and are probably controlled by dynamo action. In strongly barred galaxies the field mostly follows the gas flow, except in the upstream region where it (more or less) smoothly bends towards the bar. Polarized radio emission appears to be an excellent indicator of shearing motions.

9 Halos

Halo magnetic fields are important for the propagation of cosmic rays, the formation of a galactic wind and the stability of gas filaments. Their detection requires analysis of the Faraday effects on the polarized radio emission from the background disk of mildly inclined galaxies (e.g. in M 31, Fletcher et al., 2004a) or observation of edge-on galaxies.

The radio emission of most edge-on galaxies can be described by a thin disk plus a thick disk (halo), with similar scale heights of $\simeq$ 300 pc for the thin and $\simeq$ 1.5 kpc for the thick disk (Dumke and Krause, 1998; Dumke et al., 2000). The observed field orientations are mainly parallel to the disk (Dumke et al., 1995).

A prominent exception is NGC 4631 with the brightest and largest radio halo observed so far (Fig. 20), with a scale height of $\simeq$ 2.5 kpc. In case of energy equipartition between magnetic fields and cosmic rays, the scale height of the total field is $\simeq$ 10 kpc. The radio halo above the inner disk is composed of vertical magnetic spurs connected to star-forming regions in the disk (Golla and Hummel, 1994). The field is probably dragged out by a strong galactic wind. At larger radii where star formation is weaker, the field is parallel to the disk. Other galaxies with strong

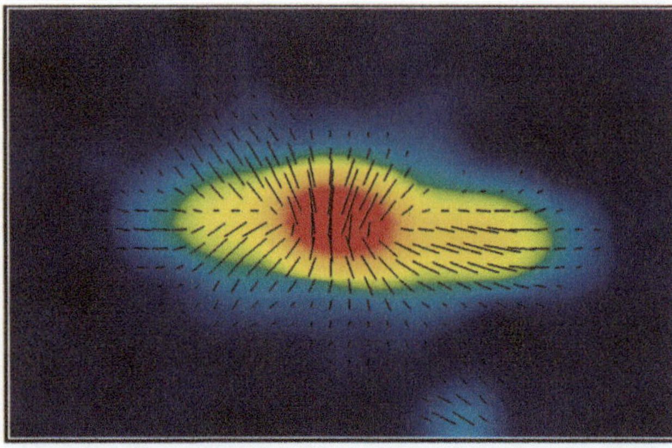

Fig. 20. Total radio emission (contours) and B–vectors of NGC 4631 at $\lambda 3.6$ cm, observed with the Effelsberg telescope. The field size is 17′ x 11′, the beam size 1.5′ (Krause et al., in prep.)

winds are M 82 (Klein et al., 1998; Reuter et al., 1994) and NGC 4666 (Dahlem et al., 1997). Starburst-driven outflows can be the origin of intergalactic magnetic fields (Chap. 2).

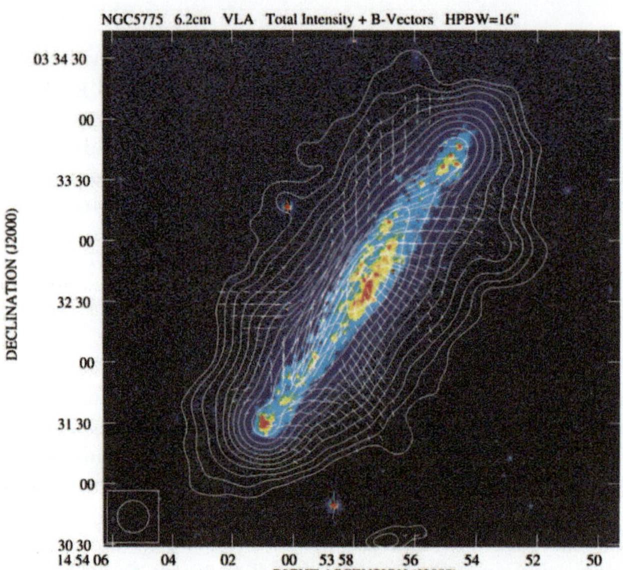

Fig. 21. Total radio emission (contours) and B–vectors of NGC 5775 at $\lambda 6$ cm with 16″ beam size, observed with the VLA (Tüllmann et al., 2000)

NGC 5775 is an intermediate case with parallel and vertical field components (Fig. 21). The magnetic energy density in the halo of, e.g. M 83, exceeds that of the hot gas (Ehle et al., 1998), so that halo magnetic fields are important for the formation of a galactic wind. Magnetic reconnection is a possible heating source of the halo gas (Birk et al., 1998).

Dynamo models predict the preferred generation of quadrupole fields where the toroidal component has the same sign above and below the plane, as claimed for the

Milky Way (Han et al., 1997). In external galaxies the vertical field symmetry could not yet been determined with sufficient accuracy. Indirect evidence for preferred quadrupole-type fields follows from the possible dominance of inward-directed radial field components (Krause and Beck, 1998). The dominance of one direction is in conflict with dipolar fields where the toroidal field reverses in the plane, so that the observed field direction depends on the aspect angle and no preference is expected for a galaxy sample. (Note that Faraday rotation is *not* zero along a line of sight passing through a disk containing a field reversal if cosmic-ray and thermal electrons are mixed in the disk. The sign of Faraday rotation traces the field direction in the layer which is *nearer* to the observer.)

10 Interacting Galaxies

Violently disrupted galaxies show strong departures from symmetric gas flows. Magnetic fields trace regions of gas compression, strong shear and enhanced turbulence.

The classical interacting galaxy pair is NGC 4038/39, the 'Antennae' (Fig. 22), with bright, extended radio emission filling the body of the whole system, with no dominant nuclear sources. Particularly strong emission comes from a star-forming region, hidden in dust, at the southern end of a massive cloud complex extending between the galaxies (the dark extended region in Fig. 22). In this region, highly tangled magnetic fields reach strengths of $\simeq 30\,\mu\mathrm{G}$, much larger than in both individual galaxies, probably the result of compression of original fields pulled out from the parent disks. Away from star-forming regions the magnetic field shows a coherent polarized structure with a strong regular component of $10\,\mu\mathrm{G}$, probably the result of gas shearing motions along the tidal tail. The mean total magnetic fields in both galaxies are about two times stronger than in normal spirals, but

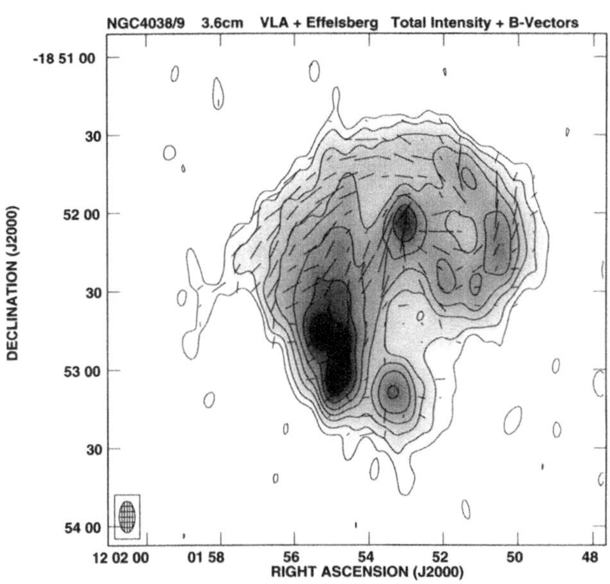

Fig. 22. Total radio emission (contours) and B–vectors of the 'Antennae' galaxy pair NGC 4038/39 at $\lambda 3.6\,\mathrm{cm}$ with $6''$ x $12''$ beam size, combined from VLA and Effelsberg observations (Chyży and Beck, 2004)

the degree of field regularity is unusually low, implying destruction of the regular component in regions with strong star formation due to the interaction.

Interaction with a dense intergalactic medium also imprints unique signatures onto magnetic fields and thus the radio emission. Tidal interactions within the Leo Triplet is the probable cause of the asymmetric appearance of NGC 3627. While the regular field in the western half is strong and precisely follows the dust lanes, a bright *magnetic arm* in the eastern half crosses the optical arm and its massive dust lane at a large angle (Fig. 13). No counterpart of this feature was detected in any other spectral range. Either the optical arms has been recently deformed due to interaction or ram pressure, or the magnetic arm is an out-of-plane feature generated by interaction.

Ram pressure of the intergalactic medium compresses the magnetic field, so that strong polarized emission is observed on one side of the galaxy, as e.g. in NGC 2276 (Hummel and Beck, 1995). The massive northern spiral arm of NGC 2442 with the polarized emission shifted towards the outer edge (Fig. 12) is also a result of ram pressure.

The Virgo cluster is a location of especially strong interaction effects. The magnetic field is of NGC 4522 is strongly compressed at the eastern side of the galaxy (Fig. 23). This kind of behaviour is also observed in another Virgo spiral galaxy, NGC 4254 (Soida et al., 1996), and might be sign of ram pressure by the intracluster medium (ICM).

Interaction may also induce violent star-formation activity in the nuclear region or in the disk which may produce huge radio lobes due to outflowing gas and magnetic field. The lobes of the Virgo spiral NGC 4569 reach out to 24 kpc from the disk and are highly polarized (Fig. 24). However, there is neither an active nucleus nor a recent starburst in the disk, so that the radio lobes probably are a signature of activity in the past.

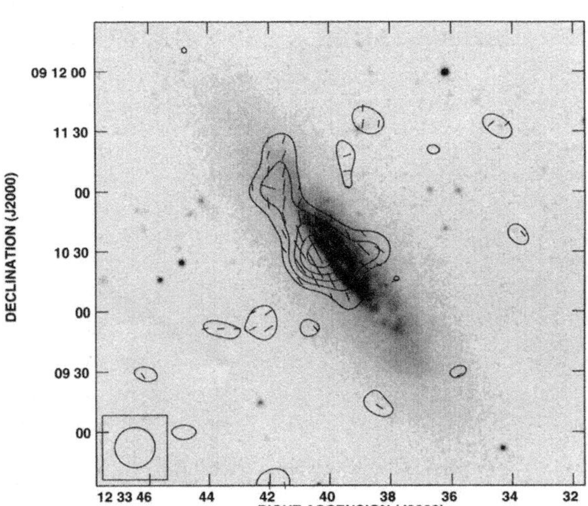

Fig. 23. Polarized radio emission (contours) and B–vectors of the Virgo galaxy NGC 4522 at $\lambda 6\,\mathrm{cm}$ with $20''$ beam size, observed with the VLA (Vollmer et al., 2004)

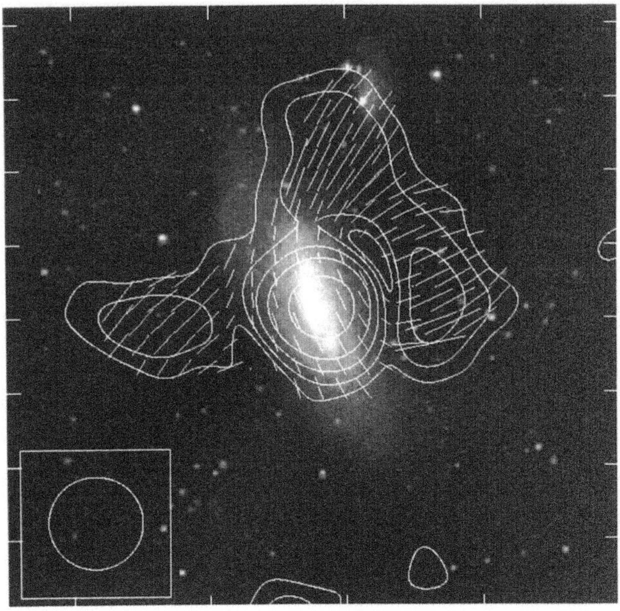

Fig. 24. Polarized radio emission (contours) and B–vectors of NGC 4569 at λ6.2 cm with 2.5′ beam size, observed with the Effelsberg telescope (Chyży et al., in prep)

11 Spiral Galaxies with Jets

Many (maybe most) spiral galaxies host nuclear jets. These are weak and small compared to those of radio galaxies and quasars. Detection is further hampered by the fact that they emerge at some angle with respect to the disk, so that little interaction with the ISM occurs. Only if the accretion disk is oriented almost perpendicular to the disk, the jet hits a significant amount of ISM matter. Such a geometry was first proven for NGC 4258 by observations of the water maser emission from the accretion disk (Greenhill et al., 1995). This is why NGC 4258 is one of the rare cases where a large-scale radio jet can be observed (van Albada and van der Hulst, 1982; Krause and Löhr, 2004).

The total intensity map of NGC 4258 (Fig. 25) reveals that the jets emerge from the galactic centre perpendicular to the accretion disk, which is oriented in east-west direction and is seen almost edge-on, and bend out to become the 'anomalous radio arms' visible out to the boundaries of the spiral galaxy. The magnetic field orientation is mainly along the jet direction. The observed tilt with respect to the jet axis may indicate an additional toroidal field component or a helical field around the jet. The equipartition strength is about 300 μG for a relativistic electron-proton jet and $\simeq$ 100 μG for a relativistic electron-positron jet. Two parallel CO ridges on both sides along the jets in NGC 4258 indicate a tunnel with walls made of molecular gas, filled with hot ionized gas that is entrained by the jet travelling along the axis of the tunnel (Krause et al., in prep.).

Highly polarized radio emission from kpc-sized jets has also been detected e.g. in NGC 3079 (Cecil et al., 2001), in the barred galaxy NGC 7479 (Fig. 26, with the field orientations perpendicular to the jet's axis), and in the outflow lobes of the Circinus Galaxy (Elmouttie et al., 1995).

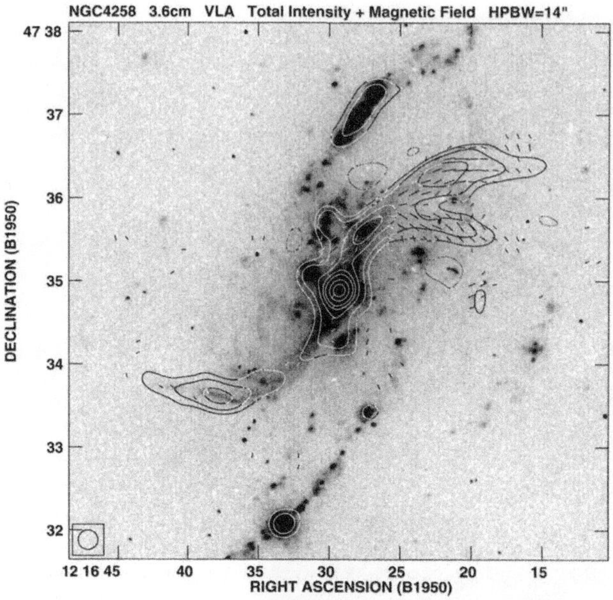

Fig. 25. Total radio emission (contours) and B–vectors (corrected for Faraday rotation) of NGC 4258 at $\lambda 3.6$ cm with $14''$ beam size, observed with the VLA (Krause and Löhr, 2004). The background image shows an $H\alpha$ image taken at the Hoher List Observatory of the University of Bonn

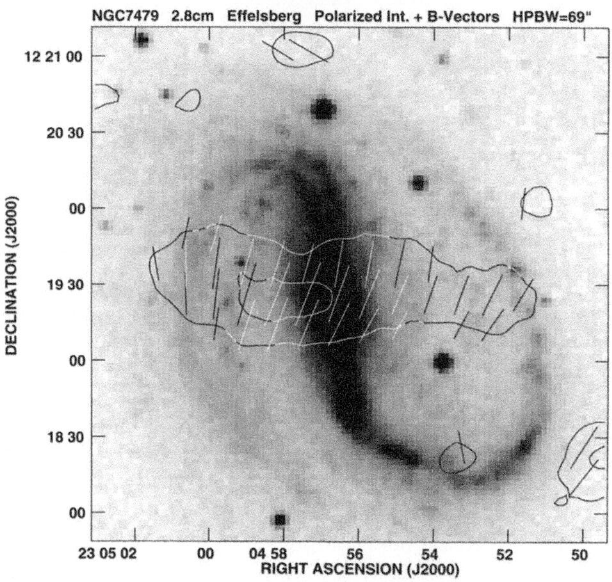

Fig. 26. Polarized radio emission (contours) and B–vectors of NGC 7479 at $\lambda 2.8$ cm with $69''$ beam size, observed with the Effelsberg telescope (Beck and Shoutenkov, unpublished)

12 Outlook

Thanks to radio polarization observations of external galaxies, much has been learnt about the global properties of interstellar magnetic fields, complemented by polarization observations in our Galaxy (Chap. 5) which trace structures of pc and sub-pc sizes. However, the physical connection between the features at large and small scales is not understood. Radio observations of external galaxies at better resolution are needed to see the full wealth of magnetic structures in galaxies.

In future, new polarimeters in front of bolometer arrays (Siringo et al., 2004) will allow observations of external galaxies in the submm range with unprecedented sensitivity.

In radio continuum, the EVLA will soon allow radio polarization observations with increased sensitivity and resolution. A much larger step is planned for the next decade. The Square Kilometer Array (SKA) will be able to map nearby galaxies with at least 10x better angular resolution compared to present-day radio telescopes, or 10x more distant galaxies with similar spatial resolution as today (Beck and Gaensler, 2004). Magnetic field structures will illuminate the dynamical interplay of cosmic forces, such as loops, twisted fibres and field reversals, which are of 0.1–10 pc width. Polarimetry of reconnection regions would help to understand the heating of the interstellar gas. Imaging of field loops with helical twist would clarify how the dynamo operates. With a field strength of 30 μG and 1 pc extent along the line of sight, a (distance-independent) polarization surface brightness of 0.2 μJy per arcsec beam at 5 GHz is expected. The SKA will detect such features in the Magellanic Clouds (0.24 pc/arcsec) and in M 31 and M 33 (3.5 pc/arcsec).

The SKA's sensitivity will even allow to detect synchrotron emission from the most distant galaxies in the earliest stage of evolution and to search for the earliest magnetic fields and their origin.

References

van Albada, G. D., van der Hulst, J. M.: 1982, A&A 115, 263

de Avillez, M., Breitschwerdt, D.: 2004, Ap&SS 292, 207

Appenzeller, I.: 1967, PASP 79, 600

Athanassoula, E.: 1992, MNRAS 259, 345

Barth, A. J., Ho, L. C., Filippenko, A. V., Sargent, W. L. W.: 1995, AJ 110, 1009

Battaner, E., Florido, E.: 2000, Fund. Cosmic Phys. 21, 1

Beck, R.: 1982, A&A 106, 121

Beck, R.: 1993, In: *The Cosmic Dynamo*, ed. F. Krause et al. (Dordrecht: Kluwer), p. 283

Beck, R.: 2000, Phil. Trans. R. Soc. Lond. A 358, 777

Beck, R.: 2001, Sp. Sci. Rev. 99, 243

Beck, R.: 2004, In: *How Does the Galaxy Work?*, ed. E. J. Alfaro et al. (Dordrecht: Kluwer), p. 277

Beck, R., Hoernes, P.: 1996, Nat 379, 47

Beck, R., Gaensler, B. M.: 2004, In: *Science with the Square Kilometer Array*, ed. C. Carilli & S. Rawlings (Amsterdam: Elsevier), New Astronomy Reviews 48, 1289, astro-ph/0409368

Beck, R., Krause, M.: 2005, AN 326, 414

Beck, R., Berkhuijsen, E. M., Wielebinski, R.: 1978, A&A 68, L27

Beck, R., Berkhuijsen, E. M., Wielebinski, R.: 1980, Nat 283, 272

Beck, R., Brandenburg, A., Moss, D., Shukurov, A., Sokoloff, D.: 1996, ARA&A 34, 155

Beck, R., Ehle, M., Shoutenkov, V., Shukurov, A., Sokoloff, D.: 1999a, Nat 397, 324

Beck, R., Berkhuijsen, E. M., Uyanıker, B.: 1999b, In: *Plasma Turbulence and Energetic Particles in Astrophysics*, ed. M. Ostrowski & R. Schlickeiser, Kraków, p. 5
Beck, R., Shoutenkov, V., Ehle, M., et al.: 2002, A&A 391, 83
Beck, R., Shukurov, A., Sokoloff, D., Wielebinski, R.: 2003, A&A 411, 99
Beck, R., Ehle, M., Fletcher, A., et al.: 2004, in: *The Evolution of Starbursts*, ed. S. Hüttemeister et al., AIP Conf. Proc., in press
Beck, R., Fletcher, A., Shukurov, A., et al.: 2005, A&A, submitted
Belyanin, M., Sokoloff, D., Shukurov, A.: 1993, Geophys. Astrophys. Fluid Dyn. 68, 227
Berkhuijsen, E. M., Horellou, C., Krause, M., et al.: 1997, A&A 318, 700
Berkhuijsen, E. M., Beck, R., Hoernes, P.: 2003, A&A 398, 937
Birk, G. T., Lesch, H., Neukirch, T.: 1998, MNRAS 296, 165
Blackman, E. G.: 1998, ApJ 496, L17
Boulares, A., Cox, D. P.: 1990, ApJ 365, 544
Brandenburg, A., Urpin, V.: 1998, A&A 332, L41
Buczilowski, U. R., Beck, R.: 1991, A&A 241, 47
Cecil, G., Bland-Hawthorn, J., Veilleux, S., Filippenko, A. V.: 2001, ApJ 555, 338
Chyży, K. T., Beck, R.: 2004, A&A 417, 541
Chyży, K. T., Beck, R., Kohle, S., et al.: 2000, A&A 356, 757
Chyży, K. T., Knapik, J., Bomans, D. J., et al.: 2003, A&A 405, 513
Crutcher, R.M.: 1999, ApJ 520, 706
Dahlem, M., Petr, M. G., Lehnert, M. D., et al.: 1997, A&A 320, 731
Dumke, M., Krause, M.: 1998, In: *The Local Bubble and Beyond*, ed. D. Breitschwerdt et al. (Berlin: Springer), p. 555
Dumke, M., Krause, M., Wielebinski, R., Klein, U.: 1995, A&A 302, 691
Dumke, M., Krause, M., Wielebinski, R.: 2000, A&A 355, 512
Ehle, M., Beck, R., Haynes, R. F., et al.: 1996, A&A 306, 73
Ehle, M., Pietsch, W., Beck, R., Klein, U.: 1998, A&A 329, 39
Elmegreen, B.G.: 1981, ApJ 243, 512
Elmouttie, M., Haynes, R. F., Jones, K. L., et al.: 1995, MNRAS 275, L53
Elstner, D., Otmianowska-Mazur, K., von Linden, S., Urbanik, M.: 2000, A&A 357, 129
Fan, Z., Lou, Y.-Q.: 1997, MNRAS 291, 91
Fendt, Ch., Beck, R., Neininger, N.: 1998, A&A 335, 123
Ferguson, A. M. N., Wyse, R. F. G., Gallagher, J. S., Hunter, D. A.: 1998, ApJ 506, L19
Fletcher, A., Beck, R., Berkhuijsen, E. M., Shukurov, A.: 2000, In: *The Interstellar Medium in M 31 and M 33*, ed. E. M. Berkhuijsen et al. (Aachen: Shaker), p. 201
Fletcher, A., Berkhuijsen, E. M., Beck, R., Shukurov, A.: 2004a, A&A 414, 53
Fletcher, A., Beck, R., Berkhuijsen, E. M., Horellou, C., Shukurov, A.: 2004b, In: *How Does the Galaxy Work?*, ed. E. J. Alfaro et al. (Dordrecht: Kluwer), p. 299
Frick, P., Beck, R., Berkhuijsen, E.M., Patrickeyev, I.: 2001, MNRAS 327, 1145
Gaensler, B. M., Haverkorn, M., Staveley-Smith, L., et al.: 2005, Science 307, 1610
Golla, G., Hummel, E.: 1994, A&A 284, 777
Gómez, G. C., Cox, D. P.: 2002, ApJ 580, 235
Greaves, J. S., Murray, A. G., Holland, W. S.: 1994, A&A 284, L19
Greaves, J. S., Holland, W. S., Jenness, T., Hawarden, T. G.: 2000, Nat 404, 732

Greenhill, L. J., Jiang, D. R., Moran, J. M., et al.: 1995, ApJ 440, 619
Han, J. L., Manchester, R. N., Berkhuijsen, E. M., Beck, R.: 1997, A&A 322, 98
Han, J.L., Beck, R., Berkhuijsen, E.M.: 1998, A&A 335, 1117
Han, J. L., Beck, R., Ehle, M., Haynes, R. F., Wielebinski, R.: 1999, A&A 348, 405
Hanasz, M., Lesch, H.: 2000, ApJ 543, 235
Hanasz, M., Otmianowska-Mazur, K., Lesch, H.: 2002, A&A 386, 347
Harnett, J., Ehle, M., Fletcher, A., et al.: 2004, A&A 421, 571
Hildebrand, R. H.: 1988, QJRAS 29, 327
Hiltner, W. A.: 1958, ApJ 128, 9
Hoernes, P.: 1997, PhD thesis, University of Bonn
Hoernes, P., Berkhuijsen, E. M., Xu, C.: 1998, A&A 334, 57
Hummel, E., Beck, R.: 1995, A&A 303, 691
Klein, U., Wielebinski, R., Morsi, H. W.: 1988, A&A 190, 41
Klein, U., Haynes, R. F., Wielebinski, R., Meinert, D.: 1993, A&A 271, 402
Knapik, J., Soida, M., Dettmar, R.-J., et al.: 2000, A&A 362, 910
Krause, F., Beck, R.: 1998, A&A 335, 789
Krause, M.: 1990, In: *Galactic and Intergalactic Magnetic Fields*, ed. R. Beck et al. (Dordrecht: Kluwer), p. 187
Krause, M.: 1993, In: *The Cosmic Dynamo*, ed. F. Krause et al. (Dordrecht: Kluwer), p. 305
Krause, M., Löhr, A.: 2004, A&A 420, 115
Krause, M., Hummel, E., Beck, R.: 1989a, A&A 217, 4
Krause, M., Beck, R., Hummel, E.: 1989b, A&A 217, 17
Kulsrud, R. M., Cen, R., Ostriker, J. P., Ryu, D.: 1997, ApJ 480, 481
von Linden, S., Otmianowska-Mazur, K., Lesch, H., Skupniewicz, G.: 1998, A&A 333, 79
Mathewson, D. S., van der Kruit, P. C., Brouw, W. N.: 1972, A&A 17, 468
Mestel, L.: 1990, In: *Galactic and Intergalactic Magnetic Fields*, ed. R. Beck et al. (Dordrecht: Kluwer), p. 259
Mitra, D., Wielebinski, R., Kramer, M., Jessner, A.: 2003, A&A 398, 993
Moss, D.: 1998, MNRAS 297, 860
Moss, D., Shukurov, A., Sokoloff, D.: 1999, A&A 343, 120
Moss, D., Shukurov, A., Sokoloff, D., Beck, R., Fletcher, A.: 2001, A&A 380, 55
Mouschovias, T. Ch.: 1990, In: *Galactic and Intergalactic Magnetic Fields*, ed. R. Beck et al. (Dordrecht: Kluwer), p. 269
Neininger, N.: 1992, A&A 263, 30
Neininger, N., Horellou, C.: 1996, In: *Polarimetry of the Interstellar Medium*, ed. W. G. Roberge & D. C. B. Whittet, ASP Conf. Ser. 97, p. 592
Neininger, N., Klein, U., Beck, R., Wielebinski, R.: 1991, Nat 352, 781
Niklas, S.: 1995, PhD thesis, University of Bonn
Niklas, S., Beck, R.: 1997, A&A 320, 54
Parker, E. N.: 1992, ApJ 401, 137
Piner, B. G., Stone, J. M., Teuben, P. J.: 1995, ApJ 449, 508
Priklonsky, V.I., Shukurov, A., Sokoloff, D., Soward, A.: 2000, Geophys. Astrophys. Fluid Dyn. 93, 97
Reuter, H.-P., Klein, U., Lesch, H., et al.: 1994, A&A 282, 724
Rohde, R., Beck, R., Elstner, D.: 1999, A&A 350, 423

Rüdiger, G., Hollerbach, R.: 2004, *The Magnetic Universe* (Weinheim: Wiley-VCH), p. 261
Sauvage, M., Blommaert, J., Boulanger, F., et al.: 1996, A&A 315, L89
Scarrott, S. M., White, C., Pallister, W. S., Solinger, A. B.: 1977, Nat 265, 32
Scarrott, S. M., Ward-Thompson, D., Warren-Smith, R. F.: 1987, MNRAS 224, 299
Schekochihin, A. A., Cowley, S. C., Taylor, S. F., et al.: 2004, ApJ 612, 276
Schoofs, S.: 1992, Diploma thesis, University of Bonn
Sellwood, J. A., Balbus, S. A.: 1999, ApJ 511, 660
Shukurov, A.: 1998, MNRAS 299, L21
Siringo, G., Kreysa, E., Reichertz, L. A., Menten, K. M.: 2004, In: *The Magnetized Interstellar Medium*, ed. B. Uyanıker et al. (Katlenburg-Lindau: Copernicus), p. 215
Soida, M., Urbanik, M., Beck, R.: 1996, A&A 312, 409
Soida, M., Urbanik, M., Beck, R., Wielebinski, R., Balkowski, C.: 2001, A&A 378, 40
Soida, M., Beck, R., Urbanik, M., Braine, J.: 2002, A&A 394, 47
Sokoloff, D. D., Bykov, A. A., Shukurov, A., et al.: 1998, MNRAS 299, 189, and MNRAS 303, 207 (Erratum)
Stacey, G. J., Geis, N., Genzel, R., et al.: 1991, ApJ 373, 423
Strong, A. W., Moskalenko, I. V., Reimer, O.: 2000, ApJ 537, 763
Subramanian, K.: 1998, MNRAS 294, 718
Tüllmann, R., Dettmar, R.J., Soida, M., Urbanik, M., Rossa, J.: 2000, A&A 364, L36
Vollmer, B., Beck, R., Kenney, J. D. P., van Gorkum, J. H.: 2004, AJ 127, 3375
Walsh, W., Beck, R., Thuma, G., et al.: 2002, A&A 388, 7
Wielebinski, R., Krause, F.: 1993, A&AR 4, 449
Wolleben, M., Reich, W.: 2004, A&A 427, 537

The Origin of Galactic Magnetic Fields

Russel M. Kulsrud

Princeton University, Plasma Physics Laboratory, Princeton NJ 08544, USA
rkulsrud@astro.princeton.edu

1 Introduction

In 1949 Hiltner (1949) and Hall (1949) independently discovered the large-scale magnetic field embedded in the interstellar medium. Actually, they inferred its existence from the polarization of star light and the fact that the amount of polarization was correlated with the amount of reddening of the star light. They had no way to gauge whether the field had a constant sign and net flux, or reversed its sign maintaining its direction with no net flux. It was only when the Faraday rotation of radio sources was measured in the late fifties and early sixties that it was shown observationally that our Galactic field indeed had a net flux on the Galactic scale. (The existence of net flux is important for the question of the origin of the magnetic field which is easier to answer if there is no net flux.)

The existence of such a field had previously been suspected during the late forties by those trying to determine the origin and propagation of cosmic rays. Without such a large-scale field, one would hardly expect cosmic rays to occupy and be confined in the entire galaxy.

A question that arose at once was, how such a field could be sustained against resistive decay. Fermi in his paper on the origin of cosmic rays (Fermi, 1949) stated that, because of the very large scale of the Galactic disc, its very large inductance would easily sustain it for a Hubble time. In fact, the current density needed to produce the extremely small field is $3 \times 10^{-6}/(4\pi)3 \times 10^{21} \approx 10^{-26}\,\mathrm{A/cm^2}$, while the resistivity of the Galactic plasma is the same as for normal metals on earth. Thus, the electric field needed to sustain this current against resistivity is easily supplied by a magnetic field decreasing over a period of 10^{26} y.

Assuming the field already exists, there is no problem at all in sustaining it. On the other hand, it is a very difficult problem to create it since there are no voltages present that could conceivably overcome the large inductive voltages that would result if the field increased over the much shorter Hubble time.

The above remarks apply to a galactic disc with no motions, and this is far from the case. There is large-scale turbulence in a galaxy stirred up by winds from hot stars and supernovae. In addition, there are the very large expansional motions of the supernova remnants and even the more powerful motions from multiple supernovae (superbubbles). Thus, to properly describe the evolution of magnetic field in matter with motions, one must employ the equation

$$\frac{\partial \mathbf{B}}{\partial t} = \nabla \times (\mathbf{v} \times \mathbf{B}) + \frac{\eta c}{4\pi} \nabla^2 \mathbf{B} \,. \tag{1}$$

The first term on the right-hand side, the 'dynamo' term, must be included in any discussion of magnetic field evolution. The last, resistive, term, discussed in Sect. 4, has a very long time constant and can safely be ignored. For example, if the electron temperature is 10^4 K, $\eta c/(4\pi = 10^7\,\mathrm{cm}^2/\mathrm{s}$ and for a scale of 100 pc the time scale is of order $L^2/\eta c/4\pi) \approx 10^{41-7} = 10^{34}\,\mathrm{s} \approx 3 \times 10^{25}\,\mathrm{y}$. Thus, the relevant equation for $\mathbf{B}$ is

$$\frac{\partial \mathbf{B}}{\partial t} = \nabla \times (\mathbf{v} \times \mathbf{B})\,. \tag{2}$$

Now, this equation has the well-known property that flux through any surface of moving plasma is constant. This condition is a very strong constraint on any theory of the origin of a magnetic field in our Galaxy or any other galaxy. It is now fairly certain that the magnetic field of our Galactic disc is in the toroidal direction and does not change sign as one passes vertically through the central plane of our disc. Within a radial extent $\Delta R \approx R/10$ there is at present a flux $\Phi = 2h\Delta RB$ where h is the half-thickness of the disc, the distance over which the field is greater than half its maximum value, and ΔR is a radial interval over which the Galactic field does not change sign. The field could be of opposite sign a distance ΔR towards the center of the galaxy, but let us for the moment assume that the plasma motions do not take the plasma a radial distance as large as ΔR during the lifetime of the Galaxy. Now, if there are also no strong vertical motions that take the plasma a distance large compared to h, then Φ must be a constant.

But, if the magnetic field were initially very weak, the initial flux through the same region would be very small and Φ would change appreciably, contradicting the conservation of flux. Thus, if the field has grown substantially since the time of formation of the disc, there must be substantial motion either in the radial or vertical directions.

How might general motions actually accomplish a growth in the flux from almost zero to its present value? If one just considers the question from a topological point of view, and assumes some initial small flux, say Φ_i, in the positive toroidal direction, then one could proceed as follows:

Take the toroidal lines and fold them back and forth in the toroidal direction so that there is $2\Phi_i$ flux in the positive direction and $-\Phi_i$ flux in the negative direction. Then extract the negative flux from the disc either in the vertical direction or the radial direction.

The question of course arises, how is one to distinguish whether the flux is positive or negative when extracting it. One would suppose that one would need a type of Maxwell demon to accomplish this. In fact, the separation of flux by sign is exactly accomplished by the so-called alpha–omega dynamo, as will be shown below. The remarkable fact that this dynamo accomplishes the separation was pointed out to me by George Field (1998).

2 The Alpha–Omega Disc Dynamo

The first theory of a cosmic dynamo was proposed by Parker (1955) to explain the long age of the Earth's magnetic field. Here the problem is actually opposite to the Galactic case. Without dynamo action the Earth's field would have decayed a long time ago. This is because the Earth is so small that the resistivity lifetime of its field is relatively short. Parker's theory is based on a regular velocity $\mathbf{v}$ due to the differential rotation of the Earth's core, and a turbulent velocity $\tilde{\mathbf{v}}$ due to convective motions inside the core. The important ingredient in the dynamo action is the twisting of the convection, due to Coriolis forces, which causes separation of the positive and negative flux with the negative flux at larger radius. Because the core is surrounded by a mantle which is insulating, the negative flux will easily float away from the core leaving only the positive flux. In fact, if the twisting is strong enough the positive flux could end up next to the mantle and negative flux would be left behind, resulting in the reversal of the field. This does occasionally happen. Parker's theory was qualitative and, although convincing, did not represent a full deductive theory for the Earth's dynamo.

An actual deductive theory was later constructed by Steenbeck et al. (1966). They applied what is usually termed a quasi-linear theory to the turbulent magnetic fields, to derive a simple differential equation for the evolution of the smoothed out, or mean, magnetic field. They chose the dynamo velocities

$$\mathbf{v} = \mathbf{V} + \tilde{\mathbf{v}} \; , \tag{3}$$

where $\mathbf{V}$ is the mean velocity and $\tilde{\mathbf{v}}$ is the turbulent velocity whose statistics are isotropic but has helical elements. For example, they took

$$\langle \tilde{\mathbf{v}}(\mathbf{r}', t)\tilde{\mathbf{v}}(\mathbf{r}, t)\rangle = [A(\rho)\mathbf{I} + B(\rho)\boldsymbol{\rho}\boldsymbol{\rho} + C(\rho)\boldsymbol{\rho} \times \mathbf{I}]\, f(t' - t) \; , \tag{4}$$

where $\boldsymbol{\rho} = (\mathbf{r}' - \mathbf{r})$, $\mathbf{I}$ is the unit dyadic, and the C term represents the helical part of the flow. They found that the mean field $\bar{\mathbf{B}}$ satisfies the mean field dynamo equation

$$\frac{\partial \bar{\mathbf{B}}}{\partial t} = \nabla \times \left(\mathbf{V} \times \bar{\mathbf{B}}\right) + \nabla(\alpha \bar{\mathbf{B}}) - \nabla \times \beta(\nabla \times \bar{\mathbf{B}}) + \frac{\eta c}{4\pi}\nabla^2 \bar{\mathbf{B}} \; , \tag{5}$$

where

$$\begin{aligned} \alpha = C(0)\tau &= -\frac{\tau}{3} < \tilde{\mathbf{v}} \cdot (\nabla \times \tilde{\mathbf{v}}) \\ \beta = B(0)\tau &= \frac{\tau}{3} < \tilde{\mathbf{v}}^2 > \end{aligned} \tag{6}$$

and where the correlation time τ is given by $\tau = \int f(t)dt, f(0) = 1$.

The theory turns out to mirror Parker's theory to a remarkable extent. α represents the twisting of the convection cells, while β is the diffusion which smoothes out the turbulent magnetic fields that Parker found. Thus, based on these equations, one could deduce a reasonable theory for the sustainment of the Earth's field, provided that the convection is sufficiently strong. For the Earth the boundary conditions applied at the mantle were natural vacuum conditions. The field outside the core is a vacuum solution.

In 1971 in two papers Parker (1971a,b) applied these ideas to a theory of the generation of the large-scale Galactic field embedded in the galactic disc. The galactic disc is a thin disc of half-thickness 200–300 pc, and radius 10–20 kpc. Parker assumed that earlier mechanisms such as a Biermann battery (Biermann, 1950) or interaction of electrons with the microwave background radiation, would generate a weak magnetic field of order 10^{-20} G. Compression into the Galactic disc would increase this to 10^{-16} G. This would form an initial condition for the dynamo equations. These, in turn, could be simplified and reduced to one-dimensional equations by making use of the thinness of the disc. In cylindrical coordinates only the Z derivative is important.

$$\frac{\partial B_{\mathrm{R}}}{\partial t} = \frac{\partial}{\partial Z}(\alpha(Z) B_\theta) + \beta \frac{\partial^2 B_{\mathrm{R}}}{\partial Z^2}$$
$$\frac{\partial B_\theta}{\partial t} = \Omega B_{\mathrm{r}} + \beta \frac{\partial^2 B_\theta}{\partial Z^2} \tag{7}$$

The boundary condition applied were again vacuum conditions as for the Earth's dynamo, $B_{\mathrm{R}} = B_\theta = 0$ at $Z = \pm h$. Parker recognized that α was an odd function of Z, reversing across the midplane, and he set $\alpha = \alpha_0 Z/h$.

Parker found unstable modes under these conditions, the most unstable mode being the even mode, i.e. B_{R} and B_θ even in Z. (This corresponds to the observed field.) However, because the plasma outside the disc is definitely not resistive, Parker (1971a) expressed considerable concern about these vacuum conditions applied to the Galactic field. The opposite conditions of a disc surrounded by a rigid infinitely conducting plane actually do not allow the symmetric instability solution since then the flux Φ of a mode must both be constant and nonzero. This condition however, need not inhibit an antisymmetric mode.

A similar theory of the origin of the galactic field was independently proposed by Vainshtein and Ruzmaikin (1972). Because of the success of the mean-field theory for the Earth's dynamo, this work and Parker's papers on the Galactic disc dynamo were generally regarded as the resolution of the problem of the origin of the cosmic magnetic fields. However, there remained difficulties that were bothersome and needed further resolution.

The instability only occurs if a certain dimensionless combination of the constants α_0, β, and Ω, the dynamo number

$$D = \frac{\alpha_0 \Omega h^3}{\beta^2} \tag{8}$$

exceeded 13. Now β was taken from the random motion of clouds ($\beta \approx (10\,\mathrm{km/s})^2/10^7\,\mathrm{y}$, Ω from the galactic rotation as $2\pi/250$ million y, h of order 200 pc. But α_0 was estimated from the effect of Coriolis forces on rising cells to be of order 0.5 km/s.

These numbers lead to a value of $D \approx 7$, below the critical value of 13. Even if D were larger than 13 by a substantial amount, one gets growth times of order 500 million years and during the age of the Galaxy, 10^{10} y this gives 20 e-foldings, an increase of 10^8, so that the initial field has to be 3×10^{-14} G not 10^{-16} G.

In order to raise the value of α, it is necessary to seek more intense turbulent motions. Ferrière (1992) was the first to quantitatively calculate the values of α characteristic of motions in expanding supernova remnants. These increased α somewhat but she further carried out her calculations for superbubbles, giant expanding remnants driven by multiple supernova explosions, whose observations were just being appreciated. These calculations led to values of α and D large enough to grow the Galactic field.

Curiously, her path to the generation of cosmic fields is rather more circuitous than it need be. She first examined the superbubble velocities under the influence of the Coriolis forces, and used them to derive effective values for the average α and β parameters. She did this by averaging the local values of α and β at a point over the many superbubbles expected at this point, during a dynamo growth time.

Then she plugged these averages into the results of the standard α–Ω mean-field dynamo analysis to derive the mean growth rate of the instabilities.

But one could take a shortcut by taking the change in the magnetic field itself during each supernova or superbubble and sum up the changes to find the growth of the Galactic field. This picture is more intuitively clear and actually avoids the assumptions of small-scale velocities employed in deriving the dynamo equations. It also makes clear how the dynamo selects out the regions of negative flux to be expelled. Finally, it makes clear the problems involved with the boundary conditions.

To see how this works refer to Figs. 1a–f.

Let the magnetic field be in the θ direction and consider, as in Fig. 1a, a single line of force just above a supernova that is about to explode. After the explosion, when the remnant has a radius ρ as in Fig. 1b, the line is pushed out into a loop over the remnant, with foot points A and B. Now, before the explosion the remnant material is initially rotating with the galactic rotation, and because the expansion

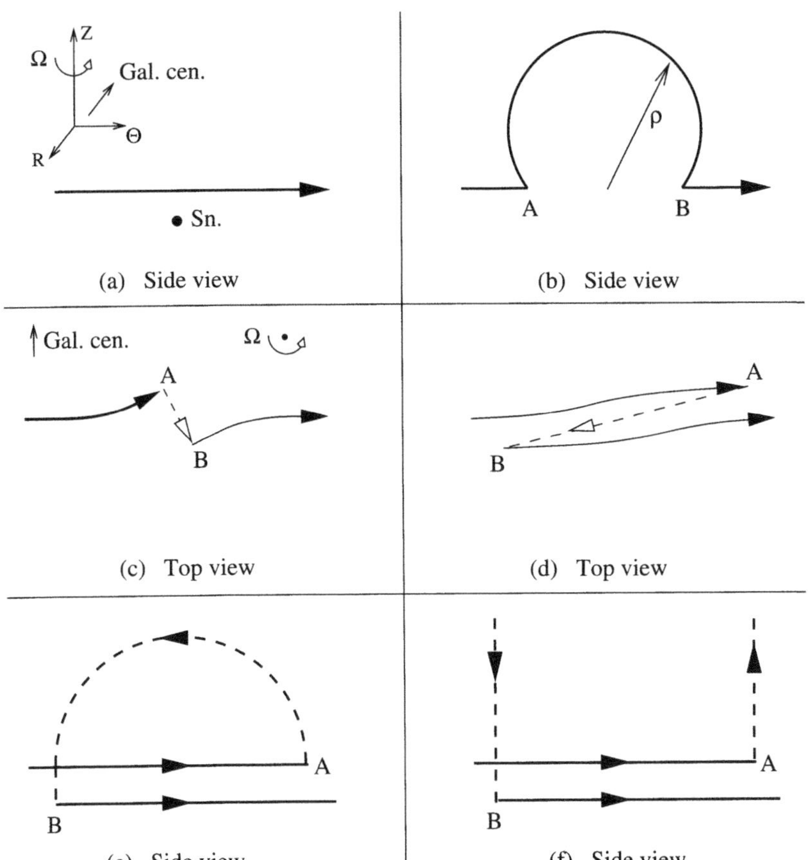

Fig. 1. The dynamo driven by a supernova or superbubble

leads to a greatly enhanced moment of inertia, the remnant rotation slows down in a fixed frame, which means that it rotates backward in the frame of the Galactic disc. This counter rotation leads to an inward motion of foot point A and and outward motion of foot point B. This is shown in Fig. 1c. After this time the differential Galactic rotation advances point A toroidally relative to foot point B since the angular velocity of the Galactic disc decreases outward. This is seen in Fig. 1d. Thus, between the foot points, A and B, in the original horizontal plane of the line, there are now two lines. This means that the toroidal field increases in this region.

But, the points A and B are still connected over the remnant by the line of force shown dotted in Fig. 1e, which has a negative component of toroidal field. Because of this the flux is still conserved. If this part of the line falls back down into the disc, the mean toroidal field will be reduced to its previous value, and no amplification will have resulted. However, if the supernova is powerful enough to blow the line of force (connecting A and B) entirely out of the Galaxy, as in Fig. 1f, then this part of the line can be ignored, and the field in the part of the field in the region surrounding the supernova will have doubled. The differential rotation will continue and the field will continue to amplify linearly in time. A second supernova can then act on the increased field and double it again.

This is the large-scale analogue of what is assumed, on a small scale, in the α–Ω dynamo theory. However, this picture is simple enough that there is really no need to refer to the mean-field dynamo equations to judge what is truly happening to amplify the field, although such a procedure is necessary to get a quantitative estimate of the rate of growth of the field.

Moreover, the simple picture does bring out the physics needed to completely justify the boundary conditions used in the calculation of the dynamo growth rate. In fact, it is essential for the flux overlapping the supernova remnant to be expelled from the disc. Also, one expects to end up with a field that is not as smooth and connected as the mean-field dynamo theory might lead us to expect. The field lines consist of finite segments of horizontal field interrupted by vertical field segments that leave the disc, go out of the Galaxy and then come back at a smaller value of the toroidal angle. A moment's thought shows that the vertical flux must equal the horizontal flux. However, since this upward vertical flux is spread out over an area πR^2 while the horizontal flux is spread out over the smaller area $2\pi hR$ the upward component of the vertical field is smaller than the horizontal field by the factor $2h/R \ll 1$. In addition, there is an equal downward vertical field of the same sign so that this essentially cancels the upward field leaving a very small net vertical field. But, in the absence of magnetic reconnection, the vertical field energy is definitely nonzero and would produce synchrotron radiation characteristic of the upward field strength. That is, the total vertical magnetic field energy rather than the purported energy of the mean vertical field determines the amount of synchrotron radiation. On the other hand it is possible that this vertical flux could entirely disappear by reconnection.

The essential question in the working of the dynamo is whether the top of the supernova remnant escapes from the disc or not. The general feeling is that supernova are not powerful enough to eject their remnants through the rest of the interstellar medium and out of the Galaxy.

One could turn to the more powerful superbubbles, driven by multiple supernovae from the nearly simultaneous explosions of all the massive stars of a galactic cluster. Even here it appears unlikely that the superbubble shells leave the disc (Rafikov and Kulsrud, 2002). If one concentrates on dynamo action when the field is extremely weak, then one can neglect any magnetic effects (such as buoyancy or Parker instabilities) during the dynamical evolution of these shells and restrict oneself to purely hydrodynamic simulations. The conclusion from these simulations is, that the expanding matter is decelerated by the gravitational field of the galaxy, and the matter falls back down carrying the magnetic flux with it (MacLow and McCray, 1989).

This is the case for the massive galaxies, such as our own. In smaller galaxies, with weaker gravitational fields, it is possible that purely hydrodynamic expulsion occurs, justifying the boundary conditions invoked in the mean-field α–Ω dynamo, and making a galactic origin possible there.

It seems unlikely in our Galaxy, that any appreciable fraction of the magnetic flux can be expelled when the field is weak. When the field becomes stronger, it will affect the dynamics of the supernova and superbubble. There could be downward slippage of the mass along a flux tube leaving the top unweighted and magnetic buoyancy or cosmic-ray pressure might then expel a part of the dashed line in Figs. 1e and 1f. This could allow the dynamo to work, once the field is strong enough, but it leaves the question of how the magnetic field becomes sufficiently strong. Since dynamo amplification does not work when the field is very weak, the field must have been placed in a galactic plasma before the disc formed. This origin is generally termed a primordial one.

In conclusion, the physical arguments point to a primordial origin of a substantial seed field, perhaps of order 10^{-8} G, before any subsequent evolution by the standardly accepted α–Ω theory occurs.

3 Evolution of a Primordial Magnetic Field

Let us consider that there is a weak primordial field present in the Galactic plasma prior to the formation of the Galactic disc. Parker (1969) has argued that such a field could not lead to the presently observed one because, first, it would diffuse away by turbulent diffusion, and second it would wind up tightly and reverse on such a small scale that it would be inconsistent with observation of systematic Faraday rotation measurements of pulsar emission. It is of interest to check the behavior of such a field when it is acted on by the various processes in the interstellar medium. We have argued in the previous section that horizontal flux cannot be expelled from the Galactic disc because it is embedded in the interstellar plasma by flux freezing, and if it were to be expelled it must take the interstellar medium with it. This is strongly resisted by the gravitational field of the Galaxy, so that as long as the field is horizontal this will not happen.

However, two things can mitigate the situation. First, when the field is strong it will control the plasma flow around supernova and superbubble so that when the field arches above the disc, plasma can flow downward along the field lines and the central parts of the field lines can be unloaded from the plasma and escape by buoyancy. Second, it must be remembered that the magnetic field is not actually frozen in the neutral component of the interstellar medium but only in the ionized part. The two components are held together by ion-neutral collisions so, in general, the separation is small. However, there are certain regions, such as molecular clouds and diffuse clouds, in which the degree of ionization is very low. Here any magnetic force will be exerted on the ionized component alone, and will be able to force the ionized component through the neutral component at a significant velocity.

This second process would seem to enforce Parker's argument that any primordial field, if it were initially purely horizontal, would diffuse out of the Galaxy. However, if the field were truly primordial there would generally be a vertical component to the field which would prevent the expulsion since this vertical component would not be changed by vertical motion. As seen in Fig. 2 many lines of force from the initial primordial field would, after the disc is formed, end up entering the disc from below and leaving above. Such a line could neither be expelled upward or downward. Now, if the scale length of the magnetic field is larger than the protogalaxy, then we see from the figure that the radial component of the field is positive on the right-hand side of the disc and is negative on the left-hand side. Then differential rotation will mix these two radial fields of different sign so the radial field will oscillate rapidly. The toroidal field resulting from the action of the differential rotation on the radial field will also oscillate rapidly every one or two hundred parsecs with Galactic radius and the resulting field will indeed average out in the mean, as Parker (1969) has suggested.

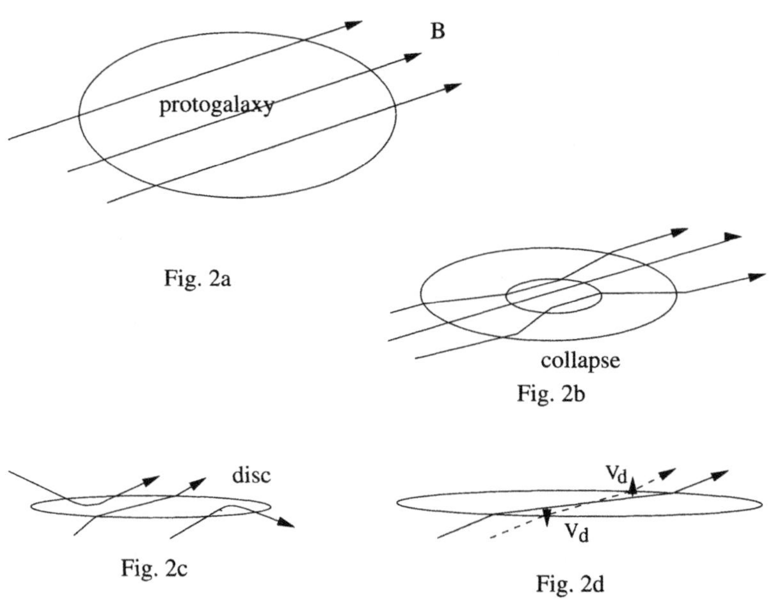

Fig. 2. The early evolution of a primordial field. Fig. 2a: The uniform field in the protogalaxy before collapse. Fig. 2b: The field after the protogalaxy has shrunk spherically and the field lines are drawn in from the extragalactic region. Fig. 2c: The field after the disc forms with some lines entering from above and leaving from above while some enter from below and leave from below. Fig. 2d: The field after only the lines threading the disc remain the others diffusing out by ambipolar diffusion

If the initial primordial field is constant over the scale of the Galaxy, then there will be no net flux and this cannot be consistent with the presently observed field. However, if the initial field is nonuniform over the scale of the galaxy, changing by a factor of two say, then it turns out as we shall see, that although the field will still oscillate rapidly in radius, it will not average out. If the toroidal field is averaged over a few hundred parsecs, as is done automatically by Faraday rotation measurements, the toroidal field will indeed appear as a constant toroidal field and will be consistent with present observational determinations of the Galactic field.

Thus, if one looks more realistically at the evolution of magnetic field in the disc, it is not so easy to dispense with the primordial origin hypothesis. The more difficult job is to explain the origin of such a primordial field.

Two possible origins which have been suggested are: One, the field may be injected into the protogalactic plasma by expanding radio lobes from radio galaxies (Daly and Loeb, 1990). These lobes seem to possess an adequate amount of flux to accomplish this. A second primordial origin lies in the powerful turbulence believed to be present in protogalactic plasmas, in the stage between the formation of the protogalaxy and its collapse to the disc. The latter hypothesis will be examined in the next section.

So far, in our discussion of the evolution of the primordial field, once it has been trapped in the disc, we have ignored the ambipolar diffusion mechanism. This mechanism was considered in detail by Howard and Kulsrud (1997). Here we only briefly discuss some of the more interesting aspects of the model they employed to investigate ambipolar diffusion.

Consider a line of force entering the Galaxy from below and leaving from above. As it passes through the disc it will intersect many interstellar clouds as in Fig. 3. In this discussion we ignore molecular clouds and consider only the diffuse interstellar clouds. In the upper half of the disc the field will have a negative Z gradient which exerts an upward force on the ionized component in each cloud. This will force the line to move through the cloud a given distance before the cloud is destroyed and a new cloud forms. As a result, we can assume that a certain fraction of the interstellar medium has been taken from above the line and replaced below it. This is a systematic motion of the line which moves it upward at a velocity which is proportional to the gradient of the square of the field strength and therefore to B^2

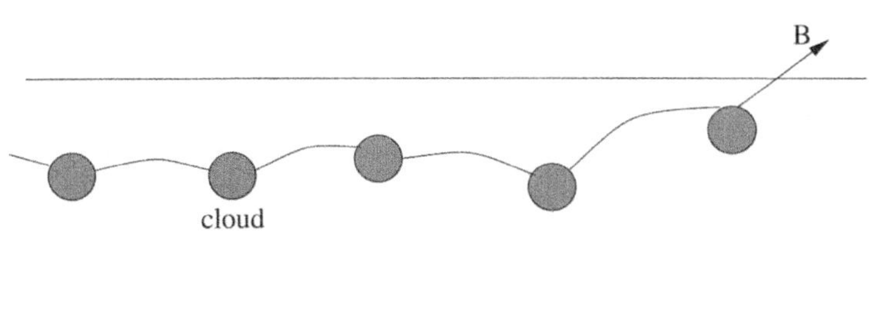

Fig. 3. A line of force penetrates many clouds and finally reaches a last cloud through which it diffuses and leaves the disc

$$V_Z = KB^2 . \tag{9}$$

One can estimate this velocity from the density and state of ionization of the clouds and their filling factor. (K depends mainly on cloud properties.) It turns out from estimates in the Howard–Kulsrud paper that V_Z for a 2 μG field will move a line by the disc thickness in about three billion years. In Fig. 3 we see there is generally a last cloud through which the line passes before it leaves the Galactic disc. When it slips through this last cloud, this part of the line then moves into the halo, the plasma slipping down it. This shortens the horizontal length of the line effectively weakening it.

With these considerations in mind we can write down simplified equations for the mean velocity of the ionized component and for the evolution of the field. We include only vertical motions and the differential rotation of the Galaxy ignoring dynamo terms. Then in cylindrical coordinates we have

$$\begin{aligned}
\mathbf{V} &= R\Omega(R)\hat{\theta} + V_Z\hat{\mathbf{z}} \\
V_Z &= -K\frac{\partial B^2/8\pi}{\partial Z} \\
K &= \frac{(1+\beta/\alpha)}{\rho_i \nu_{in} f} ,
\end{aligned} \tag{10}$$

where β/α is the ratio of cosmic-ray pressure to magnetic pressure. We assume that this remains constant as the field strength increases since the amount of cosmic-ray pressure that can be contained in the disc depends on the magnetic field. ρ_i and ν_{in} are ion density and the ion-neutral collision rates in the clouds, and f is the filling factor of the clouds. The factor f arises because the magnetic force acts on the larger volume between the clouds and this is focussed on the clouds by magnetic tension. Therefore, the force on the clouds is larger by $1/f$ (see Fig. 3).

Now, let us follow a given column of interstellar medium as it rotates around the galaxy and is sheared.

Assuming axisymmetry, the equations for B_R and B_θ are

$$\begin{aligned}
\frac{dB_R}{dt} &= \frac{\partial B_R}{\partial t} + V_\theta\frac{\partial B_R}{\partial \theta} = -\frac{\partial}{\partial Z}(V_Z B_R) \\
\frac{dB_\theta}{dt} &= \frac{\partial B_\theta}{\partial t} + V_\theta\frac{\partial B_\theta}{\partial \theta} = -\frac{\partial}{\partial Z}(V_Z B_\theta) + R\frac{d\Omega}{dR}B_R .
\end{aligned} \tag{11}$$

Now, to get an idea of the behavior of B_R and B_θ let us assume that they are parabolic in Z vanishing at the disc surface $Zs = \pm D$, i.e.

$$\begin{aligned}
B_R(Z) &= B_R(0)(1 - Z^2/D^2) \\
B_\theta(Z) &= B_\theta(0)(1 - Z^2/D^2) .
\end{aligned} \tag{12}$$

Now, substitute these expressions in (11) and evaluate the equations at $Z = 0$.

We thus get the ordinary differential equations

$$\begin{aligned}\frac{dB_{\mathrm{R}}(0)}{dt} &= -\frac{V_{\mathrm{D}}}{D}B_{\mathrm{R}}(0)\\ \frac{dB_{\theta}(0)}{dt} &= -\frac{V_{\mathrm{D}}}{D}B_{\theta}(0) - \Omega B_{\mathrm{R}}(0)\,,\end{aligned} \tag{13}$$

where from (10), $V_{\mathrm{Z}} = 0$ and $\partial V_{\mathrm{Z}}/\partial Z = 4V_{\mathrm{D}}/D$ at $Z = 0$, where

$$V_{\mathrm{D}} = \frac{K}{2\pi D}\left[B_{\mathrm{R}}^2(0) + B_{\theta}^2(0)\right] \approx \frac{K}{2\pi D}B_{\theta}^2(0)\,. \tag{14}$$

We can estimate the rough behavior from (11). Initially, take B_{R} small and $B_{\theta} = 0$ (where we drop the argument 0). Then V_{D} is negligible and B_{θ} increases linearly in time due to the second term in the B_{θ} equation. This is the differential stretching of the lines. As B_{θ} grows V_{D} becomes important and both B_{R} and B_{θ} weaken due to the ambipolar expansion in Z. But the weakening of B_{R} is more important. As it weakens, the amplification of B_{θ} by stretching (the second term on the right-hand side of the B_{θ} equation) becomes smaller and eventually B_{θ} decreases.

Equations (13) can be integrated exactly. The solution with B_{θ} initially zero is

$$\begin{aligned}B_{\mathrm{R}} &= \frac{B_1}{[1 + (2V_{\mathrm{D}1}/3D)\Omega^2 t^3]^{1/2}}\\ B_{\theta} &= \frac{-B_1\Omega t}{[1 + (2V_{\mathrm{D}1}/3D)\Omega^2 t^3]^{1/2}}\,,\end{aligned} \tag{15}$$

where B_1 is the initial value of B_{R} and $V_{\mathrm{D}1} = KB_1^2/2\pi D$ is the initial value of V_{D}. B_{θ} is plotted versus time in Fig. 4 for several values of B_1.

For small t we may replace the denominators by unity. Then B_{R} is constant and B_{θ} grows linearly in time. On the other hand, for large t the one in the denominator can be neglected, and we get

$$B_{\theta} \approx \pm\sqrt{\frac{3B_1^2 D}{2v_{\mathrm{D}1}}\frac{1}{t^{1/2}}}\,, \tag{16}$$

a field independent of B_1. Substituting this value of B_{θ} into (14) and using the above expression for $V_{\mathrm{D}1}$ we get that at large times t the vertical velocity V_{D} satisfies

$$V_{\mathrm{D}}t \approx \frac{3}{2}D\,. \tag{17}$$

The value of the field strength is just that needed to produce a drift distance of $3/2D$ in time t. Thus, the field becomes independent of the initial field strength B_1 and is a constant dependent only on the clouds. There is a critical field $B_{1\mathrm{c}}$ such that when $B_1 > B_{1\mathrm{c}}$ the field asymptotes to this value. In the paper of Howard and Kulsrud (1997), it is shown that for reasonable values for cloud properties, $B_{1\mathrm{c}} \approx 10^{-8}$ G.

We have derived the time evolution of the field following a single column of plasma of the disc, and have found its asymptotic value (16). For a time of order 10^{10} y, this value is of order 2 μG for the cloud properties employed by Howard and

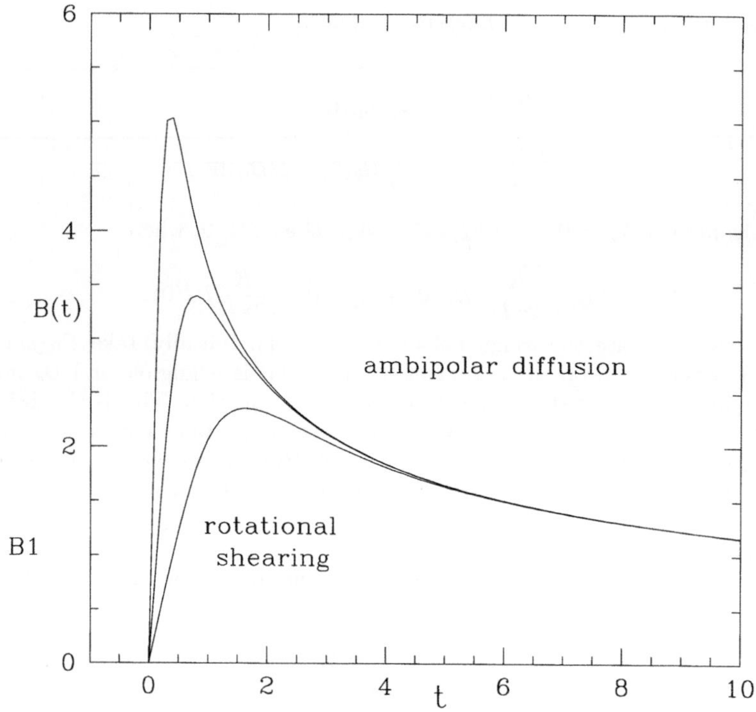

Fig. 4. The evolution of the field strength under differential rotation and ambipolar diffusion for an initial field strength of $B1 = 0.1, -0.3$ and $1\,\mu$G. The ordinate is B in μG and the abscissa is time in billions of years

Kulsrud, remarkably close to the present observed value. This value is independent of B_1, if B_1 is greater than 10^{-8} G, but its sign depends on the initial sign of B_{R}. If the initial sign of B_{R} is positive, the final value of B_θ is negative, and conversely if the initial value of B_{R} is negative, the final sign of B_θ is positive.

Now, in Fig. 2, B_{R} is negative on the left-hand side and positive on the right. But due to differential rotation any initial point on the disc at radius R is rotated by

$$\Delta\theta = \Omega(R)t \tag{18}$$

and $\Delta\theta$ will change by π when R changes by

$$\Delta R = -\frac{\pi}{t d\Omega/dR} \approx \frac{\pi R}{\Omega t}, \tag{19}$$

so for a galactic rotation period of 250 million years, and $t = 10^{10}$ y

$$\Delta R = \frac{1}{80} R \approx 100\,\mathrm{pc}\,. \tag{20}$$

Thus, the field reverses every 100 pc. Hence, if **B** is initially uniform, B_{R} consists of segments of numerically equal positive and negative signs in equal areas of the disc, and the final B_θ will oscillate like a square wave, and have zero mean value. Therefore, there should be zero net Faraday rotation arising from this primordial field.

On the other hand, if **B** is not initially uniform, so that the regions of positive and negative B_R are not of equal area, and the regions of negative B_R exceed those where B_R is positive, (independent of strength) then ambipolar diffusion and differential rotation will lead to a toroidal field that will oscillate but have a positive mean value. If the areas differ by a factor of 2, then B_θ plotted versus radius at constant θ will appear as in Fig. 5.

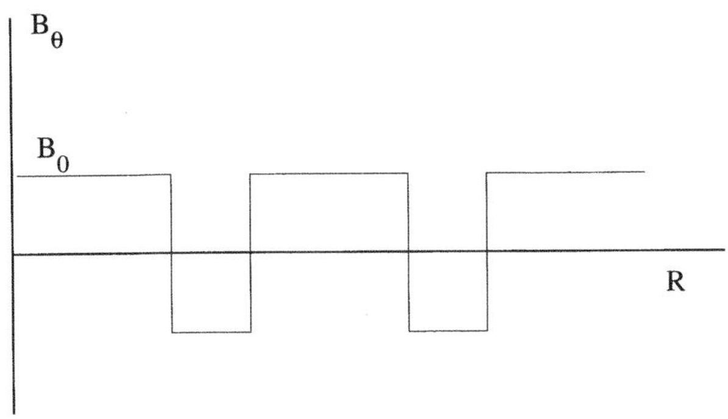

Fig. 5. The square wave behavior of B_θ of a saturated magnetic field coming from an nonuniform magnetic field

The field still oscillates, over a short distance, but will have a mean value of $B_0/3$ where B_0 is the saturated value of B_0. Its root mean-square component will be $2B_0/3$. These are typical of the quoted values for the ratio of the random component to the mean component (Rand and Kulkarni, 1989).

The estimate derived for the saturated field, $2\,\mu$G, is dependent on poorly established quantities, such as the state of ionization in different diffuse clouds. The contribution from molecular clouds, which we have not included in this chapter, will also change it. However, it is noteworthy that the predictive field based on the assumed cloud properties is in a range consistent with the observed values.

There have been attempts to observationally decide whether fields in other galaxies have a primordial origin or are generated by a disc dynamo. In order to do this, it has generally been assumed that a primordial field would end up as a bisymmetric spiral field, while a disc generated field would be an axisymmetric field. However, we see that the evolution of a field is more complicated than first supposed. We have shown that under the influence of ambipolar diffusion and differential rotation alone the primordial field would end up as an axisymmetric field when averaged over a few hundred parsecs. The field lines themselves would end up as short angular segments separated by vertical fields, but the segments would be so placed that the averaged field would appear smooth.

Inclusion of dynamo action would change the field still further from the simple intuitive picture making it difficult to be sure of the origin of any presently observed Galactic magnetic field.

4 The Protogalactic Dynamo

It appears that the galactic disc dynamo cannot develop a substantial magnetic field from a very weak seed field $\sim 10^{-16}$ G, although it certainly is able to from a stronger initial field $\sim 10^{-8}$ G. Thus, the question of the origin of this stronger initial field as a primordial field is still open. Two suggestions for its origin, which we have mentioned are: One, the fields are generated in the formation of extragalactic radio lobes which eventually expand and fill the plasma before it forms galaxies. Two, after the galaxy starts to form out of cosmic material, but is still in the protogalactic state, a powerful dynamo develops from its internal turbulence, which can produce a magnetic field of the required strength, and is coherent on scales comparable to those presently observed in the disc.

We discuss only this second possibility.

What are the requirements which the protogalactic dynamo must satisfy? The generated field after compression should be of order 10^{-8} G and coherent on scales of one tenth of a galactic radius. Therefore, before compression the field would be on a larger scale and weaker. For example, if we take 100 kpc as the radius of a spherical protogalaxy then the scale before compression must be 10 kpc and the field strength which will increase under compression by a factor of order 10^4 must be at least 10^{-12} G.

To see that the field strength must increase by 10^4 during the compression, take the disc radius as 10 kpc and its thickness as 100 pc. Then the field strength will first increase by 100 in shrinking to a sphere of radius 10 kpc and another factor of 100 when this smaller sphere shrinks to the disc of thickness 100 pc (with the same 10 kpc radius). Thus the minimum requirement on the protogalaxy galactic dynamo is that it produces a field of strength 10^{-12} G, and a coherent length of 10 kpc in the protogalaxy.

Now, it is observed from numerical simulations of cosmic structure formation that (1) turbulence is generated with a Kolmogoroff spectrum and (2) the energy of the turbulence is comparable to the binding energy of the baryonic material in the total gravitational field produced by the dark matter and the baryonic matter (Kulsrud et al., 1997). This turbulence will act on any initial field to amplify it strongly.

A question arises as to what initial field one should take, to analyze the field one expects to develop from this turbulence. Surprisingly, one gets an interesting result even if the initial magnetic field is zero, provided that one extends the magnetic differential equation to include the Biermann battery terms (Biermann, 1950). In fact, it is plausible to assume that both the magnetic field $\mathbf{B}$ and the vorticity $\omega = \nabla \times \mathbf{v}$ are initially zero. Then, up to the time when the turbulence is fully developed and has reached a steady state, one can show that the vorticity, $\boldsymbol{\omega}$ and the negative ion cyclotron frequency $-\boldsymbol{\Omega} = -e\mathbf{B}/Mc$ are essentially equal. M is the ion mass.

This is shown as follows:

First, we rederive the magnetic differential equation keeping the additional terms in Ohm's law. In fact, Ohm's law can be considered to be the equation of motion of the electron fluid.

$$n_e m \frac{d\mathbf{v}_e}{dt} = -n_e e \left(\mathbf{E} + \frac{\mathbf{v}_e \times \mathbf{B}}{c} \right) - \nabla p_e + n_e m \mathbf{g} + \mathbf{F}_{ei} \, , \tag{21}$$

where $\mathbf{F}_{\mathrm{ei}}$ is the electron ion frictional force which is related to the $\eta\mathbf{j}$ term. We drop this term. We also drop the inertial term $n_{\mathrm{e}}md\mathbf{v}_{\mathrm{e}}/dt$ and the gravitational term $n_{\mathrm{e}}m\mathbf{g}$ because of the smallness of the electron mass density. Dividing the resulting equation by $n_{\mathrm{e}}e$ we get

$$\mathbf{E} + \frac{\mathbf{v}_{\mathrm{e}} \times \mathbf{B}}{c} = -\frac{\nabla p_{\mathrm{e}}}{n_{\mathrm{e}}e} . \tag{22}$$

Now, the current is tiny so $\mathbf{v}_{\mathrm{e}} \approx \mathbf{v}_{\mathrm{i}}$ to a very high degree of approximation. Further, even if the plasma is very partially ionized one has the electron temperature very close to the neutral temperature so that $p_{\mathrm{e}}/n_{\mathrm{e}} = p/n(1+\chi) = Mp/\rho(1+\chi)$ where χ is the degree of ionization. Thus,

$$\mathbf{E} + \frac{\mathbf{v} \times \mathbf{B}}{c} = -\frac{M}{e(1+\chi)}\frac{\nabla p}{\rho} . \tag{23}$$

Finally, taking the curl of this equation and combining it with the induction equation, one gets the modified magnetic differential equation,

$$\frac{\partial \mathbf{B}}{\partial t} = \nabla \times (\mathbf{v} \times \mathbf{B}) + \frac{\nabla p \times \nabla \rho}{\rho^2}\frac{Mc}{e(1+\chi)} . \tag{24}$$

Now, it is well known that vorticity satisfies the very similar equation

$$\frac{\partial \boldsymbol{\omega}}{\partial t} = \nabla \times (\mathbf{v} \times \boldsymbol{\omega}) - \frac{\nabla p \times \nabla \rho}{\rho^2} + \nu\nabla^2\boldsymbol{\omega} . \tag{25}$$

As long as viscosity is small, these equations are identical up to the factor $-e/Mc(1+\chi)$ so $e\mathbf{B}/Mc = \boldsymbol{\Omega}$ and $-(1+\chi)\boldsymbol{\omega}$ satisfy the same equations. Since they have the same initial conditions ($\boldsymbol{\Omega} = \boldsymbol{\omega} = 0$) we find that

$$\boldsymbol{\Omega} = -\frac{\boldsymbol{\omega}}{(1+\chi)} . \tag{26}$$

Let the velocity spectrum of $\mathbf{v}$ be written

$$v^2 = \int I(k)dk \tag{27}$$

and let

$$v_{\mathrm{k}}^2 = kI(k) , \tag{28}$$

so that v_{k} is the rms value of the velocity in the logarithmic wave number interval k, $k + \Delta k$, $\Delta k \sim k$. Similarly let

$$\omega_{\mathrm{k}} = kv_{\mathrm{k}} \tag{29}$$

be the vorticity in the same wave number region.

The Kolmogoroff spectrum is the well-known inverse five thirds power law in the inertial region $k > k_0$ down to the inner or viscous scale k_ν, where $k_0 = 2\pi/R_0$ corresponding to the largest scale where R_0 is the protogalactic radius. Thus, in the inertial region

$$I(k) = \frac{2}{3}\frac{k_0^{2/3}v_0^2}{k^{5/3}} , \tag{30}$$

where v_0 is the rms of the total turbulent velocity. Thus, $v_{\mathrm{k}} \sim k^{-1/3}$ and $\omega_{\mathrm{k}} \sim k^{2/3}$.

The magnetic spectrum, at the end of the Biermann battery phase, before viscosity becomes important, is identical to the vorticity spectrum so

$$(1+\chi)\frac{e}{Mc}B(k) = \omega(k) = k_0 v_0 \left(\frac{k}{k_0}\right)^{2/3} . \tag{31}$$

We take the thermal energy comparable to the turbulent energy, which we assume in equipartition with the binding energy of the baryonic mass in the gravitational field of the dark matter. We take the baryonic mass equal to 10^{11} $M_\odot$ and the dark mass equal to 10^{12} $M_\odot$.

One finds $n \approx 10^{-3}/\text{cm}^3$ and the mean free path $\lambda \approx 10^{19}$ cm. The inner scale of the turbulence is then found to be 10^{-3} times the outer scale. With these numbers $B(k) \approx 10^{-20}$ G at the outer scale k_0^{-1} and $B(k) \approx 10^{-18}$ G at the inner scale falling off rapidly below this scale.

If one examines the modified magnetic differential equation (24), one sees that, when the magnetic field is very small, its strength is increased only by the last term and each scale is amplified linearly in time. As B grows the first term on the right-hand side $\nabla \times (\mathbf{v} \times \mathbf{B})$ then dominates and the field grows exponentially in time. In fact, one can show that the total magnetic energy grows as

$$\frac{dB^2}{dt} = 2\left[\int \omega_k^2 \tau_k d\ln k\right] B^2 , \tag{32}$$

where τ_k is the decorrelation time for eddies of scale k^{-1}. For Kolmogoroff turbulence this decorrelation time is the eddy turnover time ω_k^{-1} so

$$\frac{dB^2}{dt} = 2\left[\int \omega_k d\ln k\right] B^2 . \tag{33}$$

From this we see that each eddy adds independently to the growth rate of the magnetic energy. It is not implausible that each eddy k amplifies the Fourier component of the magnetic field with a similar wave number.

This is the case if we neglect mode coupling. Thus, we assume

$$\frac{dB(k)}{dt} = \omega_k B(k) . \tag{34}$$

As the field becomes stronger its tension tries to unwind the twisted magnetic field at a rate limited only by inertia. If we again ignore coupling between harmonics, this unwinding by itself would decrease the magnetic field strength $B(k)$ at the rate

$$\frac{dB(k)}{dt} = -kv_A(k)B(k) = k\sqrt{\frac{B^2(k)}{4\pi\rho}}B(k) = -k\frac{B^2(k)}{\sqrt{4\pi\rho}} . \tag{35}$$

When turbulence is present, the evolution of $B(k)$ will be dominated by these two effects and one has

$$\begin{aligned}\frac{dB(k)}{dt} &= \omega_k B(k) - \frac{kB^2(k)}{\sqrt{4\pi\rho}} \\ &= \omega_0\left(\frac{k}{k_0}\right)^{2/3} B(k) - \frac{kB^2(k)}{\sqrt{4\pi\rho}} .\end{aligned} \tag{36}$$

(After the turbulence dies away, one has (35) for the evolution of the magnetic field.)

For a given k, the exact solution of the full equation (with amplification), (36), is

$$B(k,t) = \frac{B_{\rm i}(k)e^{\omega_{\rm k}t}}{1 + \left[kB_{\rm i}(k)/\omega_{\rm k}\sqrt{4\pi\rho}\right] e^{\omega_{\rm k}t}} \tag{37}$$

$[B_{\rm i}(k)$ is the Biermann result given by (31)].

The behavior in time is easy to follow. For small t when $kB(k,t)/\sqrt{4\pi\rho} \ll \omega_{\rm k}$, one can drop the second term in the denominator and $B(k,t)$ increases exponentially at the rate $\omega_{\rm k}$. When $kB_{\rm i}(k,t)/\sqrt{4\pi\rho}e^{\omega_{\rm k}t} > 1$ then the field unwinds as fast as it winds up by the turbulence and $B(k,t)$ approaches a spectrum constant in time,

$$B(k,t) \to \sqrt{4\pi\rho}\frac{\omega_{\rm k}}{k} \sim k^{1/3} . \tag{38}$$

Further, from (28) and (30), $v_{\rm k} = v_0(k_0/k)^{1/3}$, $\omega_{\rm k} = k_0 v_0 (k/k_0)^{2/3} = \omega_0(k/k_0)^{2/3}$ where v_0 is the velocity and ω_0 the vorticity at the largest scale.

Initially, we have

$$B_{\rm i}(k) = \frac{Mc}{e}\Omega_{\rm k} = 10^{-4}\omega_{\rm k} = 10^{-4}k_0 v_0 \left(\frac{k}{k_0}\right)^{2/3} , \tag{39}$$

the magnetic field at the end of the Biermann battery phase. From equipartition of the kinetic energy of the turbulence with the gravitational energy we get

$$\frac{1}{2}v_0^2 = \frac{GM}{R} . \tag{40}$$

We take $M = 10^{12}\,{\rm M}_\odot$ for the dark mass and $R = 100\,{\rm kpc}$. Then we have $v_0 = 2 \times 10^7$ km/s. The vorticity at k_0

$$\omega_0 = \frac{v_0}{R} = 2 \times 10^{-16}{\rm s} . \tag{41}$$

Now, the turbulence decays in a time τ_1/ω_0 where τ_1 is a number of order unity. As the turbulence dies away, only the second term in (36) is important.

To form an idea of the resulting field let us assume for simplicity that the turbulence has a spectrum constant in time up to a time $t = \tau_1/\omega_0$, and then drops abruptly to zero. After this, let the protogalaxy remain constant in radius (with no turbulence) during an additional time τ_2/ω_0. Then let it collapse to a disc of radius 10 kpc. During this collapse the field will change its horizontal scale by 10 and increase its strength by 10^4 if the plasma is ideal.

Thus, the field goes through a number of stages as it evolves in the protogalaxy. This is illustrated in Fig. 5 by the curves for $B(k)$ at different stages. In the first stage it grows linearly in time from zero to $B_{\rm i}(k)$, due to the Biermann battery. In the second stage it grows in time to $B_2(k)$ due to dynamo action, driven by turbulence, until the turbulence stops. During this second stage the smaller scale fields grow faster (larger k), and reach saturation earlier due to the tension in the field that tends to unwind and reduce them. For larger scales (small k) the growth is slower and they do not reach saturation. The third stage commences after the turbulence stops and continues until the protogalaxy collapses. During this stage the field unwinds at all scales from the value $B_2(k)$ to the value $B_3(k)$, but the unwinding is faster at smaller scales due to the larger field strength and the larger value of k, in the unwinding rate $kv_{\mathrm{Ak}} = kB(k)/\sqrt{4\pi\rho}$. As a result, $B_3(k)$ peaks at an intermediate scale k_{p}. We again make a simplifying assumption that the protogalaxy is constant during this third stage. Finally, there is a fourth stage during which the protogalaxy collapses to the disc and the field is amplified and shortened. During this last stage we neglect any further unwinding and simply employ flux freezing to describe it.

We carry out the calculation of the field wind up to its value $B_2(k)$, at the end of stage 2, given by (37) with (31), at $t = \tau_1/\omega_0$.

$$B(k,t) = \frac{B_{\mathrm{i}}(k)e^{\tau_1 (k/k_0)^{2/3}}}{1 + \left[kB_{\mathrm{i}}(k)/\omega_{\mathrm{k}}\sqrt{4\pi\rho}\right] e^{\tau_1 (k/k_0)^{2/3}}} \,. \tag{42}$$

We assume that stage 3 lasts a time τ_2/ω_0. During this stage B is subject only to unwinding forces with no further turbulent amplification so that it satisfies (35). The solution of this equation, with initial condition $B_2(k)$ at $t = t_1 = \tau_1/\omega_0$, is

$$B(k,t) = \frac{B_2(k)}{1 + (kB_2(k)/\sqrt{4\pi\rho})(t - t_1)} \,. \tag{43}$$

Let stage 3 last for a time τ_2/ω_0. Then

$$B_3(k,t) = \frac{B_2(k)}{1 + (k/k_0)(B_2(k)/B_0)\tau_2} \,, \tag{44}$$

where B_0 is given by

$$\frac{kB_0}{\sqrt{4\pi\rho}} = \omega_0 \,. \tag{45}$$

This is the field such that the largest scale unwinds in a time $1/\omega_0$.

For illustration, take the parameters of the protogalaxy to be those after (40). Let us assume $k_0 = 2\pi/R$ and $v_0 = 2 \times 10^7$ cm/s as in equipartition. Let us further assume $\tau_1 = \tau_2 = 4$. For these parameters, the logarithm (to the base ten) of the relative field strengths of $B_{\mathrm{i}}(k)/B_0, B_2(k)/B_0$ and $B_3(k)/B_0$ are plotted in Fig. 5, with $B_0 = 2 \times 10^{-6}$ G. The ordinate has a large range. The resulting Biermann field is shown at the bottom of the plot and ranges from 10^{-20} to 10^{-18} G.

We see that the field $B_2(k)$ at the end of the second stage is saturated for scales up to 1/30 of the size of the protogalaxy and has a magnitude comparable with B_0, at its peak. The larger scales are unsaturated. At the end of stage 3 the smaller scales unwind to a value $B_3(k) \approx (k_0/k)B_0/\tau_2$ independent of its value at the beginning of stage 3, while for the larger scales, $B_3(k) \approx B_2(k)$ and there is scarcely any unwinding. The peak of the resulting field $B_3(k)$ drops from that for $B_2(k)$ to a value of $10^{-2}B_0$. After compression the field strength would be raised to $100B_0$ and the scale would be in the range of 1 kpc, if the compression were ideal. If τ_1 were smaller the final fields before and after compression would be considerably smaller and the scale would be somewhat smaller.

Of course, the numbers for the parameters of the protogalaxy and the lifetime parameters τ_1 and τ_2 are chosen purely for illustrative purposes, and are somewhat arbitrary. At this time, there are no observations to nail them down more precisely. However, if the field were as strong as that given in Fig. 6, we would expect ambipolar diffusion to reduce its strength.

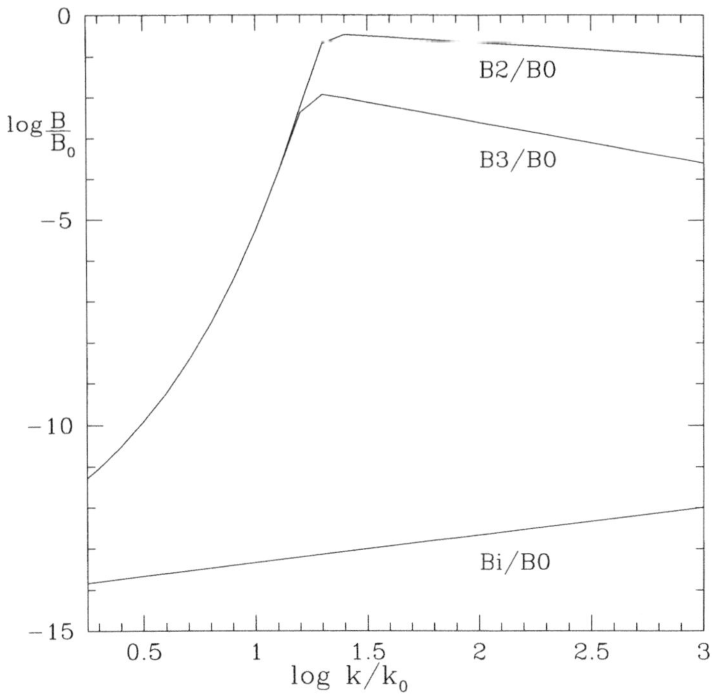

Fig. 6. The field strength at the end of the first three stages

To find out how the random field evolves beyond the end of stage four, one has to apply the various dynamo and ambipolar diffusion theories mentioned in the first three sections. Further, magnetic reconnection can lead to some smoothing of the field.

The main point is that by natural assumptions about the formation of our Galaxy (and other galaxies), one would expect very strong dynamo action producing very appreciable seed fields, even before the galactic disc dynamo processes set in.

5 Conclusion

One very striking observation of cosmic magnetic fields is their great strength and large coherence in our Galaxy and in many others. A dominant problem for plasma physics is the explanation of their origin and these outstanding properties.

The simplest and most elegant explanation is, that they arise from a weak seed field produced by a variety of possible mechanisms in the early Universe. The field is then amplified to its present state by the famous α–Ω mean-field dynamo. Unfortunately, there are two objections to this origin.

First the essential vacuum boundary conditions invoked in the solution of the dynamo equations, probably are not valid since flux lines cannot escape because they are embedded in heavy interstellar material, that is in turn bound to the disc by the Galaxy's very strong gravitational field as discussed in Sect. 2.

Second, the assumption of an initial weak field may not be valid because the Galactic plasma has already been through a protogalactic stage where it was subject to the dynamo action of the strong turbulence in the protogalaxy which together with compression is expected to produce a very substantial initial field. This is discussed in Sect. 4.

If the latter is the case, then the field must be evolved during the Galactic disc lifetime in a much different way than predicted by the kinematic $\alpha - \Omega$ dynamo treatment. During this evolution ambipolar diffusion processes probably play an important role which is discussed in Sect. 3.

In conclusion, the origin question for the Galactic magnetic field is still very open, but there are several promising ways to explain its existence.

References

Biermann, L.: 1950, Z. Naturforschung 5a, 65
Daly, R. A., Loeb, A.: 1990, ApJ 364, 451
Fermi, E.: 1949, Phys. Reviewer 75, 1169
Ferrière, K.: 1992, ApJ 389, 286
Field, G.: 1998, private communication
Hall, A. N.: 1949, Science 109, 166
Hiltner, W. A.: 1949, ApJ 109, 471
Howard, A., Kulsrud, R.: 1997, ApJ 483, 648
Kulsrud, R. M., Cen, R., Ostriker, J. P., Ryu, D.: 1997, ApJ 480, 481
MacLow, M. M., McCray, R.: 1989, ApJ 337, 141
Parker, E.: 1955, ApJ 122, 293
Parker, E.: 1969, ApJ 157, 1129
Parker, E.: 1971a, ApJ 163, 255
Parker, E.: 1971b, ApJ 163, 279
Rafikov, R. R., Kulsrud, R. M.: 2000, MNRAS 314, 839
Rand, R. J., Kulkarni, S. R.: 1989, ApJ 343, 760
Steenbeck, M., Krause, F., Rädler, K.-H.: 1966, Z. Naturforschung 21a, 369
Vainshtein, S., Ruzmaikin, A. A.: 1972, Sov. Astron. 16, 365

Magnetic Fields in the Milky Way, Derived from Radio Continuum Observations and Faraday Rotation Studies

Richard Wielebinski

Max-Planck-Institut für Radioastronomie, Auf dem Hügel 69, 53121 Bonn, Germany
rwielebinski@mpifr-bonn.mpg.de

Abstract. Magnetic fields are found everywhere in our Universe. We know that our Earth possesses a dipolar magnetic field. Magnetic fields have been observed on the Sun either as optical streamers during solar eclipses and by using remote sensing methods. Magnetic fields of the solar planets have been studied in situ by measurements made by magnetometers on various spacecraft. Stars, supernova remnants and pulsars have been shown to harbour magnetic fields. Also the diffuse Milky Way has been shown to possess magnetic fields by many methods. Possibly the whole Universe is intimately governed by the presence of magnetic fields. This review will summarize the observational results and discuss some of the implications of the presence of magnetic fields in the Milky Way.

1 Introduction

The first detection of a magnetic field beyond the Earth goes back on the optical Zeeman effect observations by G. E. Hale of the Sun in 1908. Detection of magnetic fields in peculiar 'magnetic stars' was made by H. W. Babcock in 1947, again using the Zeeman effect in the optical range. Early theoretical arguments for the presence of magnetic fields in the Milky Way were given by Alfvén (1937) and Fermi (1949). The realization that our Galaxy was permeated by magnetic fields came slowly as a result of the observations of the polarization of starlight, starting in 1949. The optical polarization data of Hiltner (1949); Hall (1949) were interpreted by Davis and Greenstein (1951) to be due to the alignment of dust grains in magnetic fields of the Milky Way. The interpretation of these observations led to considerable controversy in the optical community since many other polarizing effects were also known to be present. In retrospect, after radio data were obtained this original interpretation was shown to be correct. In particular large-scale surveys of the polarization of starlight, that are discussed in Chap. 7 by C. Heiles and R. Crutcher, agree with the radio results.

Radio astronomy has given us most of the information about the Galactic (and extragalactic) magnetic fields. Actually the low frequency cosmic radio waves, detected by Karl Jansky in 1932, were a result of magnetic fields in the Milky Way. The interpretation had to wait until Alfvén and Herlofson (1950); Kiepenheuer (1950) suggested that relativistic electrons in magnetic fields generate the observed radio waves by the synchrotron process. This emission theory was elaborated by many authors (e.g. Shklovsky, 1953; Ginzburg and Syrovatskii, 1965). The synchrotron emission theory has predicted that in a homogeneous magnetic field the linear polarization of the radio waves could be up to some 75% and be oriented with the 'E' vector perpendicular to the magnetic field direction. Thus the magnetic field orientation could be deduced from radio polarization observations. Through the equipartition arguments, where magnetic energy density is set equal to that of cosmic particles (e.g. Burbidge, 1956; Pacholczyk, 1970), the magnetic field strength could be derived from the radio intensity.

The first linearly polarized radio waves were detected by Mayer et al. (1957) from the Crab nebula. These observations were a follow-up of the detection of optical polarization in the Crab nebula (e.g. Dombrovsky, 1954; Oort and Walraven, 1957) and the suggestion of Shklovsky (1953) that this emission was due to the synchrotron process (and hence polarized). This was an important step in recognizing that magnetic fields must be present in Galactic radio sources. The linear polarization of the extended Galactic radio emission was detected by Westerhout et al. (1962) and Wielebinski et al. (1962) confirming that magnetic fields exist in the Milky Way. Early polarization observations have shown the Faraday effect at work due to Earth's ionosphere. The evidence for polarized radio emission from the radio galaxy Cygnus A was published by Mayer et al. (1962). The newly constructed Parkes radio telescope was also used to observe radio source polarization at several wavelengths (e.g. Bracewell et al., 1962). The Faraday rotation effect in the radio source Centaurus A was shown to be present by Cooper and Price (1962). Observing at two radio frequencies Muller et al. (1963) showed the action of the Faraday effect on the polarized radio waves passing through Galactic magnetic fields in the thermal Interstellar Medium (ISM). These early observations gave us new methods to observe magnetic fields in the Milky Way. With further radio polarization surveys and catalogues of linearly polarized radio sources the scene was set for detailed studies of Galactic magnetic fields. Finally pulsars joined extragalactic radio sources as excellent probes of the Galactic magnetic field.

The use of the radio Zeeman effect with the 21 cm H I line to measure Galactic magnetic fields has been proposed by Bolton and Wild (1957). However the detection of this effect, using the H I line, proved to be difficult, with the first positive result reported by Verschuur (1968). Subsequent observers reported detection of the Zeeman effect in the OH, H_2O, SO, CCS, CN molecular lines and the H30α recombination line. Details of the Zeeman effect measurements are described by C. Heiles and R. Crutcher in Chap. 7.

2 Observational Rationale

2.1 Synchrotron Emission

Although the details of the synchrotron theory have been presented numerous times, in view of their importance, they will be briefly summarized in the following text. A magnetic field is essential for the generation of radio waves by the synchrotron process. Charged particles moving at relativistic velocities in a magnetic fields (e.g. Ginzburg and Syrovatskii, 1965; Pacholczyk, 1970) generate polarized radio waves. The trajectory of a charged relativistic particle in a magnetic field is a spiral wound around the magnetic field. The radius of the spiral is proportional to the particle mass and hence electrons give the major contribution to radio emission. The relativistic electron distribution is usually given by the isotropic power law:

$$n(E)dE = N_0 \left(\frac{E}{E_0}\right)^{-\gamma} dE$$

$$\text{and } N_e = N_0 \int_{E_1}^{E_2} \left(\frac{E}{E_0}\right)^{-\gamma} \quad (1)$$

$$= N_0 \frac{1}{1-\gamma} \left[\left(\frac{E_2}{E_0}\right)^{1-\gamma} - \left(\frac{E_1}{E_0}\right)^{1-\gamma}\right] , \quad (2)$$

where N_e is the electron number density in the energy range $[E_1 - E_2]$, γ is the energy spectral index, and N_0 is the number of relativistic electrons per unit energy range E_0. Hence

$$N_0 = \frac{(\gamma - 1)N_e}{(E_1/E_0)^{(1-\gamma)}} \quad \text{for } E_2 >> E_1 . \quad (3)$$

The emissivity σ from unit volume in a magnetic field B is given by:

$$\sigma \propto N_0 \nu^{-(\gamma-1)/2} B_{\perp}^{(\gamma+1)/2} , \quad (4)$$

where ν is the frequency and $B_{\perp}$ is the magnetic field perpendicular to the line of sight.

By measuring the intensity of emission from a source of size L we see

$$B_{\perp}^{(\gamma+1)/2} N_0 L . \quad (5)$$

Usually $\gamma \sim 2.75$ for Galactic radio emission.

The equipartition energy between relativistic particles (cosmic rays) and magnetic fields allows an estimate of the magnetic field strength (e.g. Burbidge, 1956; Pacholczyk, 1970; Beck and Krause, 2005).

$$W_{cr} \sim \frac{B^2}{8\pi} \sim \frac{\rho u^2}{2} , \quad (6)$$

where W_{cr} is the energy density of cosmic rays, $B^2/8\pi$ is the magnetic field energy density, and $\rho u^2/2$ is the kinetic energy (turbulence) of gas with density ρ and velocity dispersion u.

2.2 Polarization of Synchrotron Emission

The linear polarization of the synchrotron emission has been studied by many authors. In general the emission from a single relativistic electron in a magnetic fields is elliptically polarized. When an ensemble of electrons is considered the averaged polarization is linear with the plane of the E vector normal to the magnetic field direction. The polarization degree p is given by (Ginzburg and Syrovatskii, 1965; Pacholczyk, 1970):

$$p = \frac{\gamma + 1}{\gamma + 7/3} . \quad (7)$$

Considering Galactic radio emission with $\gamma \sim 2.75$ we expect up to $p = 75\%$ linear polarization in a homogeneous field. Polarization percentage can be reduced by the orientation of the magnetic field distributions in the beam of the radio telescope and by the structure of the thermal electron clouds that cause the Faraday effect.

2.3 The Faraday Effect

The linearly polarized radio wave in its passage through an magnetoionic medium suffer the rotation of the intrinsic position angle ϕ_0 through the Faraday effect. After travelling a distance dl through a medium with n_e thermal electrons the position angle ϕ becomes:

$$\Phi = \frac{e^3}{2\pi m^2 c^4} \lambda^2 n_e B_{\parallel} dl + \phi_0 \; . \tag{8}$$

where $B_{\parallel}$ is the line-of-sight component of the coherent (unidirectional) magnetic field. In practice we use a parameter called Rotation Measure (RM) given by

$$\Phi = \mathrm{RM}\lambda^2 + \phi_0 \; . \tag{9}$$

By definition if RM > 0 the magnetic field is towards the observer. Also

$$\mathrm{RM} = 0.81 n_e B_{\parallel} dL \quad (\mathrm{rad\,m^{-2}}) \; , \tag{10}$$

where n_e is the number of thermal electrons in cm^3, λ is in metres, $B_{\parallel}$ in μG and L in pc.

The illustration of the emission and Faraday rotation effects is shown in Fig. 1.

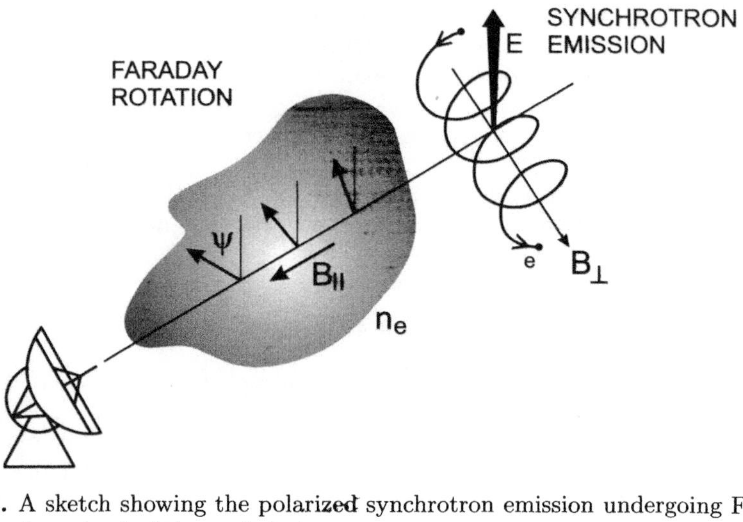

Fig. 1. A sketch showing the polarized synchrotron emission undergoing Faraday rotation in a cloud of thermal electrons

3 The Earliest Observations

3.1 Total Intensity Surveys

The low radio frequency total intensity emission is a signature of magnetic fields. The earliest data was gathered in the pioneering observations of Karl Jansky in 1932 at 20.5 MHz. The next significant data came from Reber (1944) who mapped the sky first at 160 MHz and later at 480 MHz. After the interpretation in terms of the synchrotron theory became accepted, it was clear that the total radio intensity traced the magnetic fields in our Galaxy. Surveys were made with low angular resolution in Australia (e.g. Bolton and Westfold (1950) at 100 MHz, Allen and Gum (1950) at 200 MHz) and in Cambridge (e.g. Baldwin (1955) at 81.5 MHz). The first all-sky map was constructed at 200 MHz by Dröge and Priester (1956) using their northern data and the Allen and Gum survey in the south. Mapping with larger antennas, that gave better angular resolution, continued in Cambridge (Turtle and Baldwin, 1962; Pauliny-Toth and Shakeshaft, 1962). Also the Dwingeloo telescope was used for surveying at 400 MHz by Seeger et al. (1965) and at 820 MHz by Berkhuijsen (1972). The southern skies were mapped by Hill et al. (1958), Mathewson et al. (1963), Yates et al. (1967); Landecker and Wielebinski (1970). The latter data was combined with the data from Turtle and Baldwin (1962) to make an all-sky map. The 30 MHz data of Mathewson et al. (1963) and the 38 MHz map of Milogradov-Turin and Smith (1973) were combined by Cane (1978) for a low-frequency all-sky map.

3.2 Polarization Observations

The consequence of the predictions of the synchrotron theory for polarization observations was quickly taken up by observers. Early attempts to detect polarization in the Crab nebula were made by e.g. Hanbury Brown et al. (1955). The early observations were made at rather low radio frequencies where polarization is easily depolarized by Faraday rotation in the beam. The first positive detection of linear polarization in Crab A was published by Mayer et al. (1957) who used the rather high frequency of 9.5 GHz (λ3.15 cm). The detection of linear polarization in the radio galaxy Cygnus A followed (Mayer et al., 1962). A search for linear polarization of the diffuse galactic emission was undertaken by Thomson (1957), Razin (1958) and by Pawsey and Harting (1961). All these observers used rather low radio frequencies and antennas of poor angular resolution and hence could not detect the linear polarization. Definite detection of the Galactic linear radio polarization was announced by Westerhout et al. (1962); Wielebinski et al. (1962) almost simultaneously who used the higher frequency of 408 MHz and the 25-m Dwingeloo telescope and the 7.5-m Würzburg dish in Cambridge, respectively. The early results of observations in the anticenter of our Galaxy are shown in Fig. 2. These observations gave an exact confirmation of the previous optical polarization studies of this region of the sky.

3.3 Faraday Rotation Studies

Ionospheric Faraday rotation was shown to be changing the position angle of the Galactic radio waves by Wielebinski and Shakeshaft (1962). Cooper and Price

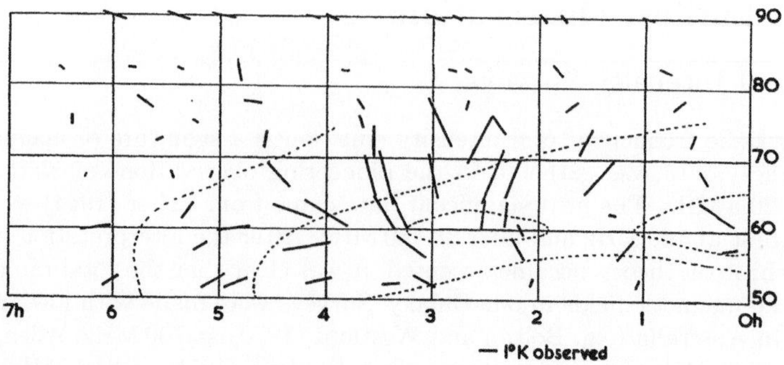

Fig. 2. Observations of the linear polarization at 408 MHz towards $l = 140°$, $b = 10°$ by Wielebinski et al. (1962)

(1962) demonstrated the $1/\lambda^2$ Faraday rotation effect in the radio galaxy Centaurus A by multi-frequency observations with the then newly commissioned 64-m Parkes radio telescope. Also unresolved radio sources showed Faraday rotation (e.g. Morris and Berge, 1964; Davies, 1967) and hence were potentially a probe of the Galactic magnetic fields. The Galactic polarization was shown to be subject to the Faraday effect in Galactic magnetic fields by Muller et al. (1963). Thus the basic discoveries made in these years set the scene for future investigations of Galactic magnetic fields. Finally it must be mentioned that the discovery of pulsars, that are highly polarized, by Hewish et al. (1968) gave us an additional excellent way to probe the magnetic fields of the Milky Way. Pulsars offer a unique way to determine the magnetic field intensity: from the combination of rotation measure and the dispersion measure average values of the magnetic field intensity in the line of sight can be deduced.

4 Radio Continuum Surveys of the Milky Way

4.1 Modern All-sky Surveys

A series of surveys at 408 MHz with the largest single dishes available (Jodrell Bank Mark I, Effelsberg and Parkes radio telescopes) resulted in an all-sky map (Haslam et al., 1982) that was the basis of many investigations up to the present. Higher frequency 1.4 GHz maps of the sky were made by Reich (1982), Reich and Reich (1986); Reich et al. (2001) to be combined to an all-sky map shown in Fig. 3. Using the southern polarization data of Testori et al. (2004) and combining it with a recent polarization survey of the northern sky by Wolleben et al. (2004), the all-sky polarization distribution has been mapped as shown in Fig. 4. In 1992 the COBE mission (Smoot et al., 1992) gave us all-sky data at 31.5, 53 and 90 GHz, albeit with low angular resolution. The 45 MHz surveys of Alvarez et al. (1997); Maeda et al. (1999) gave us a new view of the radio sky at this low frequency. A large part of the southern sky was mapped by Jonas et al. (1998) at 2.3 GHz. Most recently the WMAP mission (Bennett et al., 2003) gave us excellent all-sky data at 22.8, 33.0, 40.7, 60.8, and 93.5 GHz. The high-frequency maps delineate mainly the thermal

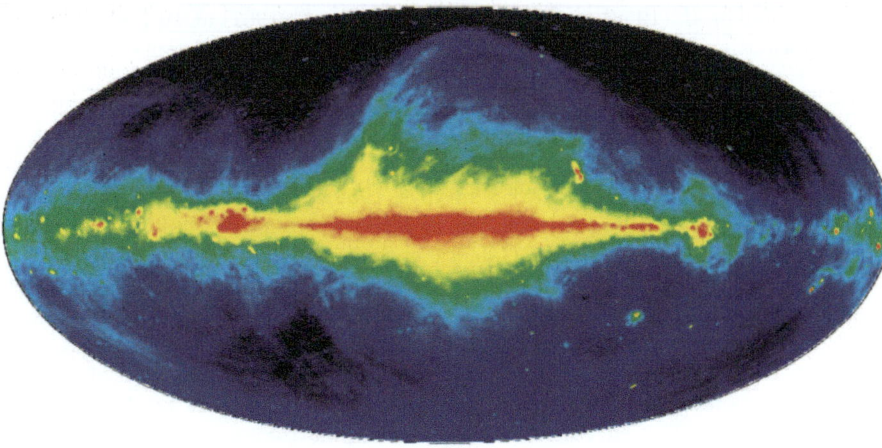

Fig. 3. All-sky 1400 MHz map (P. Reich et al., 2004)

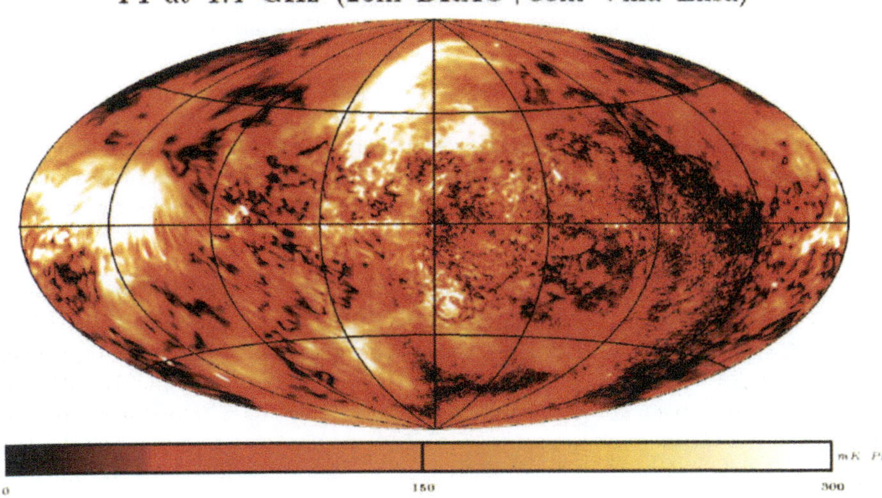

Fig. 4. All-sky polarization map (surveys from Wolleben et al. (2004) and Testori et al. (2004) combined by W. Reich)

(free-free) component of the Galactic emission, but are an important input for the separation of the two emission components.

4.2 Galactic Plane Surveys

Since the Galactic plane is much more intense than the diffuse Galactic emission it was possible to map the plane from the earliest days of radio astronomy. The early surveys of Westerhout (1958) at 1.4 GHz and Altenhoff et al. (1961) at 2.7 GHz gave us information mainly about the thermal sources, the H II regions. Low-frequency surveys of Hill et al. (1958) at 85.5 MHz or Shain et al. (1961) at 19.7 MHz gave us more information about the nonthermal sources (Supernova Remnants (SNR)) and showed H II regions in absorption. These surveys had an angular resolution of the order of 1°. The advent of arc minute angular resolution improved the ability to detect SNR (e.g. Green, 1974), a manifestation of magnetic fields in the Milky Way. With the advent of large radio telescopes, many higher angular resolution surveys were made (e.g. Hill, 1968; the Parkes surveys, 1969, 1970; Altenhoff et al. 1970) at cm wavelengths. A high-frequency survey at 10 GHz (λ3 cm) by Handa et al. (1987) probed preferentially the thermal Galactic emission regions. Modern surveys (e.g. Reich et al., 1990; Fürst et al., 1990; Duncan et al., 1995), that had an improved dynamic range, as well as mapping larger areas of the Galactic plane allowed us to study many details of Galactic emission. The most recent interferometer maps (e.g. Gaensler et al., 2001; Taylor et al., 2003) give us the angular resolution of $\sim 1'$.

The first polarization survey of the Galactic plane was made by Junkes et al. (1987) at 2.7 GHz. This survey showed numerous polarized sources in the Galactic plane and suggested that polarized emission was observable from distances that are on Galactic scales, possibly from near the Galactic center itself. The southern polarization survey of Duncan et al. (1997) at 2.4 GHz covered a large section of the Galactic plane. The reducttion of an Effelsberg 2.7 GHz survey by Duncan et al. (1999) showed many unusual polarized regions in the northern Galactic plane. The map of total intensity and polarization vectors shown in Fig. 5 suggests that Faraday depolarization is present in the inner Galactic plane.

The all-sky surveys and the Galactic plane surveys, at low radio frequencies, contain considerable information about the morphology of Galactic magnetic fields. The problem of interpretation comes from the fact that the nonthermal emission that is observed is the superposition of emission regions in the line of sight. Earliest

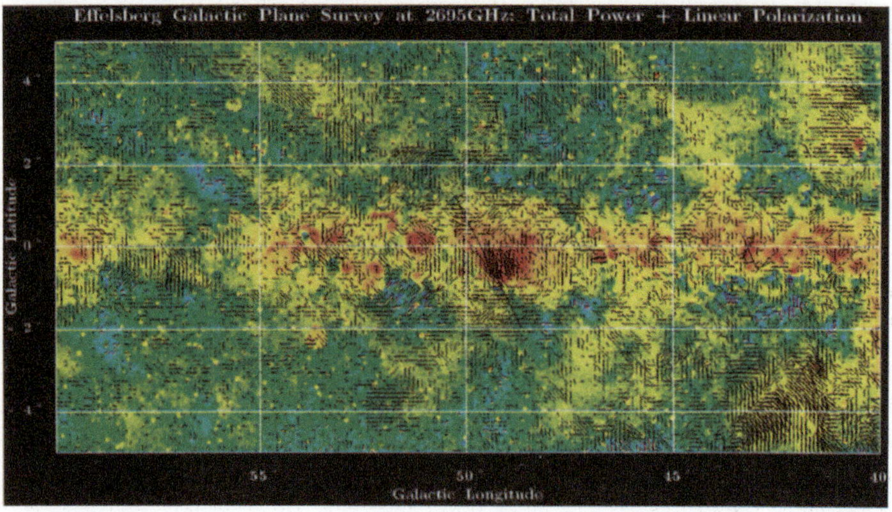

Fig. 5. A section of the 2.7 GHz Galactic plane survey of Reich et al. (1990) with superposed polarization vectors from Duncan et al. (1997) (courtesy of E. Fürst)

surveys, especially in the southern hemisphere (e.g. Mills, 1964; Wielebinski et al., 1968), noted that tangential directions of the known spiral arms show also 'steps' in radio emission. These steps were attributed to the increased emission seen in the line of sight. Similar analysis that used more data was performed by Wilkinson and Smith (1974); Heiles (1976). The 408 MHz all-sky survey was analyzed by Haslam et al. (1981); Phillips et al. (1981); Beuermann et al. (1985) who could reproduce the spiral structure after using some assumptions. In particular the analysis of Georgelin and Georgelin (1976) that defined the distribution of H II regions was needed for the fitting procedures. This is shown in Fig. 6.

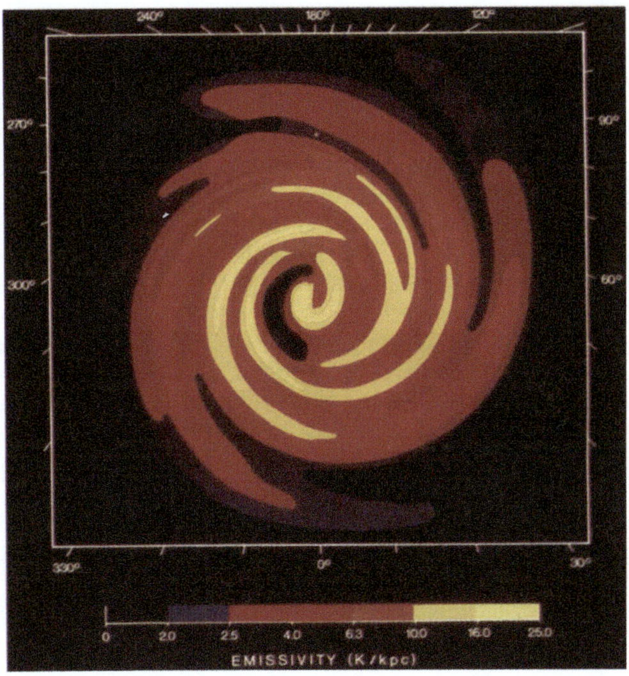

Fig. 6. A model of the Galactic radio emission based on 408 MHz observations (from Beuermann et al., 1985)

4.3 Polarization Surveys

Surveys of the linear polarization of the northern sky were made at 408 MHz by Berkhuijsen and Brouw (1963) and Wielebinski and Shakeshaft (1964). The initial result was that the local magnetic field of the Galaxy had negligible Faraday rotation towards $l = 140°$ and hence was oriented normal to this direction (i.e. towards $l = 50°$). Also two polarization maxima were found: towards the region ($l = 140°$, $b = 10°$) and towards the North Polar Spur ($l = 30°$, $b = 45°$). Also observers in the southern hemisphere took up polarization mapping with a series of papers by Mathewson and Milne (1964, 1965); Mathewson et al. (1966). In the southern hemisphere no distinct polarization maxima were seen, except possibly towards ($l = 340°$, $b = -30°$). The first combined map of linear polarization of the whole sky was presented by Mathewson and Milne (1965). The logical move to higher radio frequencies led to surveys by Bingham (1966); Baker and Smith (1971); Spoelstra (1971); Brouw and Spoelstra (1976). Based on the five frequency maps between 408 MHz and 1411 MHz in Brouw and Spoelstra (1976) the rotation measure was derived by Spoelstra (1984). The RM was found to be low, namely $-18 < \mathrm{RM} < +12\,\mathrm{rad\,m^{-2}}$. This result was accepted for many years, however in view of most recent observations, it was only valid for low angular resolution RM determinations.

The Galactic Plane and the Galactic Center

The advent of new radio telescopes in the 1970's (Westerbork Synthesis Array, Effelsberg 100-m radio telescope, the Very Large Array in Soccoro and the Australia Telescope Compact Array) gave us a quantum jump in the angular resolution. All these new telescopes were equipped to measure polarization. At first only smaller areas were mapped. Large SNR could by mapped (e.g. Wilson and Weiler, 1976; Sieber et al., 1979) with unprecedented angular resolution showing detailed structure. Also selected areas of the Galactic plane (e.g. Downes et al., 1981) were studied showing highly polarized emission where no distinct radio continuum source seemed to be present. Spectacular maps of the Galactic Center were made (Yusef-Zadeh et al., 1984; Seiradakis et al., 1985, 1989; Reich, 2003) showing that thin filaments of polarized emission, with unusual spectral index, were distributed normal to the Galactic plane. A recent map of the Galactic center region made with the Effelsberg radio telescope by W. Reich is shown in Fig. 7. The Faraday rotation found in some of the Galactic center filaments (e.g. Inoue et al., 1984; Tsuboi et al., 1985) was huge, more like $\mathrm{RM} \sim 1600\,\mathrm{rad\,m^{-2}}$, not the $\mathrm{RM} < 18\,\mathrm{rad\,m^{-2}}$ as derived by Spoelstra (1984). These results made observers aware of the need to have good angular resolution to determine the RM in the Galactic plane.

Studies of Galactic Emission with High Angular Resolution

The observations by Wieringa et al. (1993) showed that there was considerable small-scale polarization structure in the diffuse Galactic emission at 325 MHz at an arc minute resolution. Observing with the Westerbork synthesis radio telescope thin polarized filaments were observed by Wieringa et al. (1993) of unknown origin. It was clear, in view of the low observing frequency, that Faraday effects were playing a dominant role. The authors also have pointed out that there are great problems in handling the large-scale polarization structures that are rejected in interferometric observations. Follow-up observations (W. Reich, unpublished) with $9'$ resolution in Effelsberg of the same area at the frequency of 1.4 GHz confirmed the presence of extended emission regions but not of the filamentary structure seen at 325 MHz. Obviously there is a rapid depolarization effect by the Faraday rotation between these two frequencies. Surveys of the Galactic plane, in polarization, were published by Duncan et al. (1997) at 2.4 GHz made with the Parkes telescope and at 2.7 GHz Duncan et al. (1999) observed with the Effelsberg telescope. These two surveys give

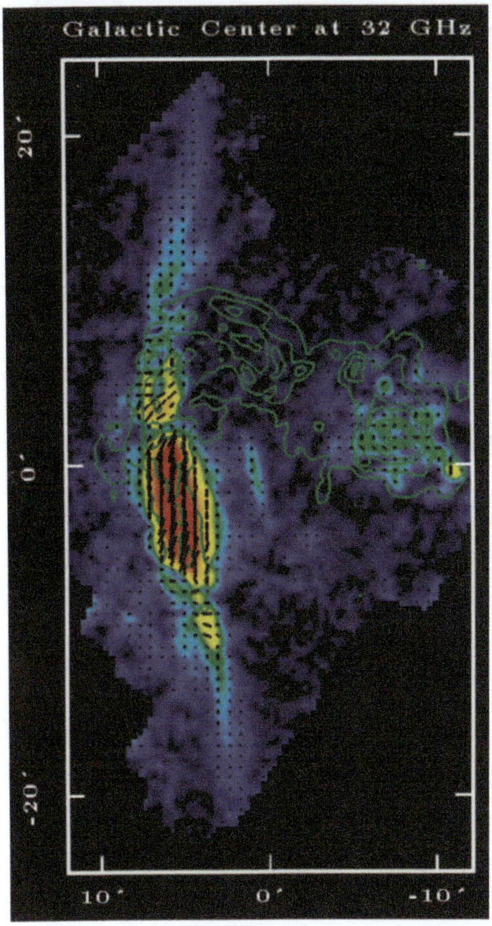

Fig. 7. The Galactic center at 32 GHz. The highly polarized Galactic center Arc runs almost perpendicular to the Galactic plane tracing a vertical magnetic field (courtesy of W. Reich)

us information about the polarization of the Galactic plane ($b \pm 5°$) from $b = 74°$ through the Galactic center to $b = 238°$. The one clear feature of these surveys are the depolarization effects seen towards $b = 0°$. However, in some directions the depolarization along the Galactic plane is low, possibly allowing us to se distant features.

The observations at 1.4 GHz with arcminute angular resolution showed large-scale Faraday structures (e.g Gray et al., 1998; Uyanıker et al., 1998, 1999) towards the prominent H II region W5. Observing with the Dominion Radio Astrophysical Observatory's 1.4 GHz interferometer Gray et al. (1998) found a structure in polarized intensity towards ($l = 137.^\circ5$, $b = 0.^\circ5$) that was interpreted as a lens, due to a magnetoionic Faraday screen. The Effelsberg Medium Latitude Survey (EMLS) of a broad section ($b = \pm 20°$) of the Galactic plane was started by Uyanıker et al. (1998, 1999) showing a multitude of Faraday structures (see Fig. 8). The EMLS is now nearing completion (Reich et al., 2004, Fig. 9) and will give us a new insight into the magnetic field of the Galaxy. In particular, Uyanıker (1998) has investigated

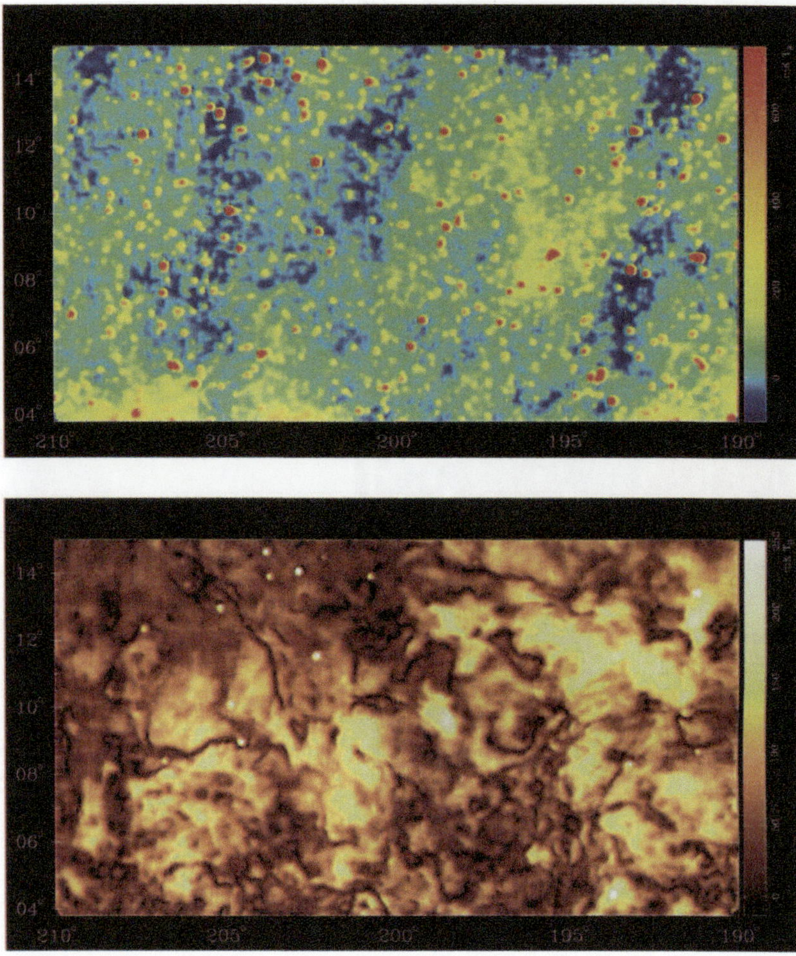

Fig. 8. A section of the Galactic anticenter in total power (above) and polarized intensity (below) showing a multitude of Faraday structures (Uyanıker et al., 1999)

the problems of calibration of the polarization data and of the addition of polarization baselines to limited areas observed at high resolution. Failure to make absolute calibration of polarization maps may lead to spurious features. A big step in the mapping of the detailed polarization and rotation measure distribution was taken by Gaensler et al. (2001) who studied a field in the direction of the Norma spiral arm with the ATNF compact array. This is the first map in a series that will map the southern Galactic plane at 1.4 GHz. It shows a section of the Galactic plane near the Norma spiral arm with high RM values (RM $= \pm 150\,\mathrm{rad\,m^{-2}}$) seen towards H$\alpha$ regions. The Canadian Galactic Plane Survey (CGPS) at 408 MHz and 1.4 GHz has been published (Taylor et al., 2003). This large data set, when combined with the EMLS observations at 1.4 GHz, will be a basis of future investigations of Galactic

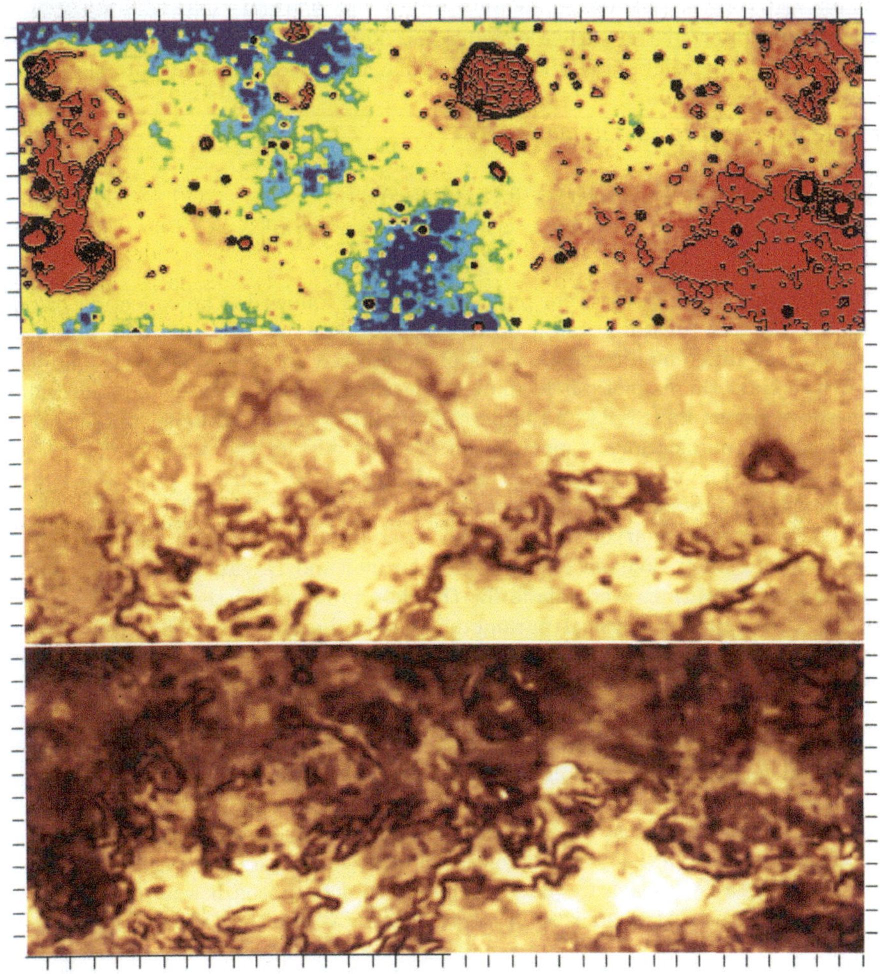

Fig. 9. A $24^\circ \times 9^\circ$ section of the 1.4 GHz Effelsberg Medium Latitude Survey centered at $l, b = 162^\circ, 0^\circ$ (from Reich et al., 2004). Absolutely calibrated total intensities are shown on top with color-coded intensities running from 4.5 K to 5 K. Contours are shown for intensities above 5 K in steps of 0.25 K T_b. Polarized intensities with preliminary absolute calibration are displayed from 0 mK to 850 mK (middle). The Effelsberg data with missing large-scale structures are shown for comparison (bottom). Polarized intensities run from 0 mK to 500 mK

magnetic fields. A new method of determining the magnetic fields in the direction of molecular clouds in the Perseus spiral arm has been proposed by Wolleben and Reich (2004).

The low-frequency studies, started by Wieringa et al. (1993), were continued by Haverkorn et al. (2000) who found very low rotation measures, suggesting that the observed structure was due to a very local (< 500 pc) ISM. Follow-up observations of selected regions (Haverkorn et al., 2003a,b) at 350 MHz with the Westerbork telescope supported the earlier conclusions. These observations show the importance of low-frequency data with good angular resolution for studies of the nearby magnetoionic medium.

5 Surveys of Rotation Measure of Extragalactic Radio Sources (EGRS)

Once the basic discoveries were published surveys followed. The collection of data was a great task, it still goes on up to the present time. The source surveys soon divided into different directions. Lists of unresolved sources were made and used to derive Galactic magnetic field models. In parallel maps of extended sources like supernova remnants (SNR) and large radio galaxies were also being published. To quote all the publications, especially on SNR, is beyond the scope of the present contribution. The development of the studies of EGRS will be sketched with important milestones discussed in some detail.

The point-source surveys started with observations of the strongest sources (e.g. Mayer et al., 1964; Bologna et al., 1965; Gardner and Davies, 1966). Soon multi-frequency observations were made for the determination of the rotation measure (RM), an important physical parameter. Also it became clear that these multi-frequency observations had to be closely spaced to avoid the 2π ambiguity in RM (e.g. see Ruzmaikin and Sokoloff, 1977). In addition to many individual papers, several major catalogues were published (e.g. Tabara and Inoue, 1980; Simard-Normandin et al., 1980, 1981; Broten et al., 1988). It is important to note that the published sources with unambiguous RM steadily became less due to the fact that RM values were often found to be affected by various instrumental or observational effects. Observations must be made with single-dish telescopes that contain all the polarized intensity in the beam or, in case of array telescopes, the angular resolution must be the same at each frequency. One additional approach was to concentrate on observing the RM of sources in interesting areas like the Orion–Eridanus area MacLeod et al. (1988), the Gum nebula Vallée and Bignell (1983) or the Bootes void (Vallée, 1991). In the early surveys RM of extragalactic sources distributed across the whole sky were collected, but not at low Galactic latitudes. This was in part rectified by Clegg et al. (1992) who observed sources within $b < 5°$ of the Galactic plane. This work was more recently dramatically extended by Brown et al. (2003b) who have provided RM of a large sample of sources in a section of the Canadian Galactic Plane Survey (CGPS).

The interpretation of the data sets in terms of large-scale magnetic field structure of the Milky Way was taken up by many authors. Gardner and Whiteoak (1963) have shown that the magnitude of the RM was correlated with Galactic latitude, hence due to the magnetic fields in the plane of the Galaxy. Based on this result other authors (e.g. Morris and Berge, 1964; Davies, 1967) started to model the Galactic magnetic field. Vallée and Kronberg (1975) investigated 251 RM using a 'slab-model' technique. The major observational work, that remained a standard for many years, was published by Simard-Normandin et al. (1980). The interpretation of this data set was that a very large magnetic anomaly exists south of the Galactic plane at $60° < l < 140°$. Furthermore effects of Galactic loops were shown to enhance the RM of EGRS in these directions. This data set of 552 RM, distributed unequally over the whole sky, was used by several authors to model the 'Galactic magnetic field' (e.g. Inoue and Tabara, 1981; Vallée, 1983, 1988, 1996). The model proposed was that of a clockwise direction of the magnetic field in the local and Perseus arms and anticlockwise direction in the inner Sagittarius spiral arm. The question of the relation between a regular magnetic field and a random component has been investigated by Ohno and Shibata (1993) based on RM of EGRS. Recently a new method of analysis, the wavelet technique, was applied to this somewhat limited data sample (e.g. Frick et al., 2001). Sofue and Fujimoto (1983) proposed a bisymmetric spiral model, driven mainly by theoretical considerations published by Piddington (1964). Most recently Johnston-Hollit et al. (2004) reanalysed the data by an interpolation of the all-sky RM distributions finding no support for the bisymmetric spiral model. The interpretation of the data from the CGPS (Brown and Taylor, 2001; Brown et al., 2003a,b) concentrated on the study of the magnetic field reversal in the direction of the Perseus spiral arm. The work of Brown et al. (2003a) showed the RM of EGRS is reasonably constant along the Perseus spiral arm (see Fig. 11) with some reversals occurring near H II regions. Recently Klein et al. (2003) measured the RM of sources in the B3 VLA sample. All these results are combined in Fig. 10 showing that a lot of work is still necessary to obtain a well-sampled distribution across the sky. A recent extension of the southern survey of Gaensler et al. (2001) shows that RM of up to 1000 are present in the Galactic plane (J.C. Brown, private communication).

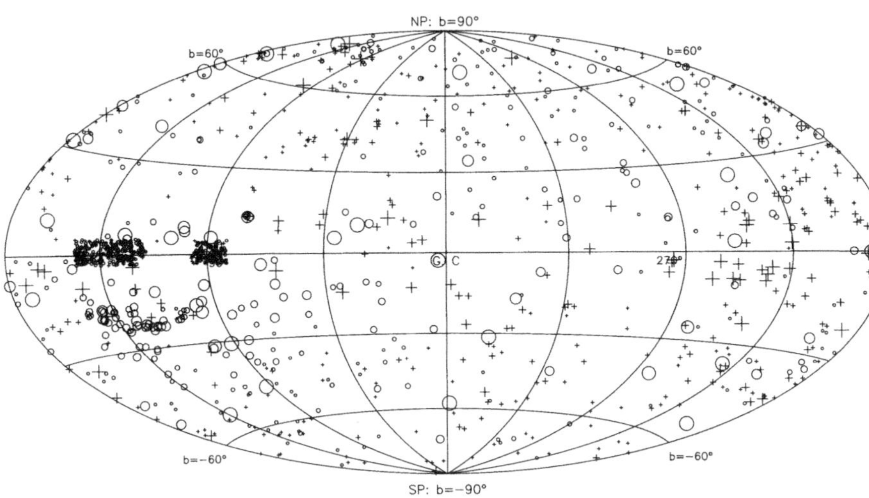

Fig. 10. Rotation measures of extragalactic sources. Crosses are positive RM, open circles negative RM. The scale goes from 12 < RM < 300. 674 sources from catalogues, 380 sources from Brown (2003) and Brown et al. (2003b), 143 sources from Klein et al. (2003) (courtesy of J.L. Han)

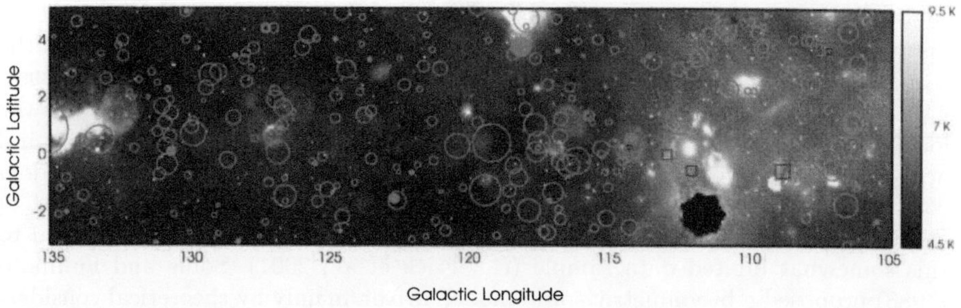

Fig. 11. RM sources superposed on a 21 cm total intensity image from the CGPS for $105° < l < 135°$. Circles: CGPS sources; squares: pulsars. Filled symbols represent positive RM, and open symbols represent negative RM. Symbol sizes are proportional to the magnitude of the RM and are scaled linearly from 100 to 750 rad m^{-2}. The missing field at $l = 111.^{\circ}5$, $b = -2°$ contains the supernova remnant Cas A. (Brown et al., 2003a)

6 Pulsars as Probes of the Magnetic Fields of the Galaxy

Pulsars are excellent probes of Galactic magnetic fields because we can observe both the rotation measure and the dispersion measure (e.g. Smith, 1968; Manchester, 1972, 1974; Manchester and Taylor, 1977). By combining the two parameters and assuming the small-scale fluctuations in magnetic field and electron density to be uncorrelated (Beck et al., 2003), we can obtain a value of the average magnetic field $< B_{\|} >$ in the line of sight:

$$\langle B_{\|} \rangle = 1.232 \frac{\mathrm{RM}}{\mathrm{DM}} \ \mu G \ . \tag{11}$$

Once a larger number of pulsars have been observed, attempts to model the Galactic magnetic field could be pursued. An analysis of the RM of 38 pulsars by Thomson and Nelson (1980) suggested that the Galactic magnetic field is confined to a narrow region of the Galactic plane ($|Z| > 120$ pc) and exhibits a reversal towards the inner spiral arm. The direction of the local magnetic fields was given as towards $l = 94° \pm 11°$ and with an average field strength of $< B >= 2.2 \pm 0.4\,\mu$G. A later paper by Lyne and Smith (1989) considered 185 pulsars and derived a value for the local magnetic field of $< B > \sim 2 - 3\,\mu$G directed towards $l = 90°$. Towards the Galactic center a field reversal was postulated. These authors also argued that since some extragalactic sources had comparatively lower values of RM than nearby pulsars, field reversals must be present in the outer Galaxy. Rand and Kulkarni (1989) extended their analysis to consider random fields superposed on the local uniform magnetic field component. Further observers collected more RM data for selected pulsars: e.g. Rand and Lyne (1994) for pulsars with high dispersion measure, Han and Qiao (1994) and Qiao et al. (1995) for additional southern hemisphere pulsars. In particular in view of the favored BSS field model many papers tried to fit this model to the observed data. Independent investigations by Vallée (1996) and Han et al. (1997) tried to fit the data to a BSS model.

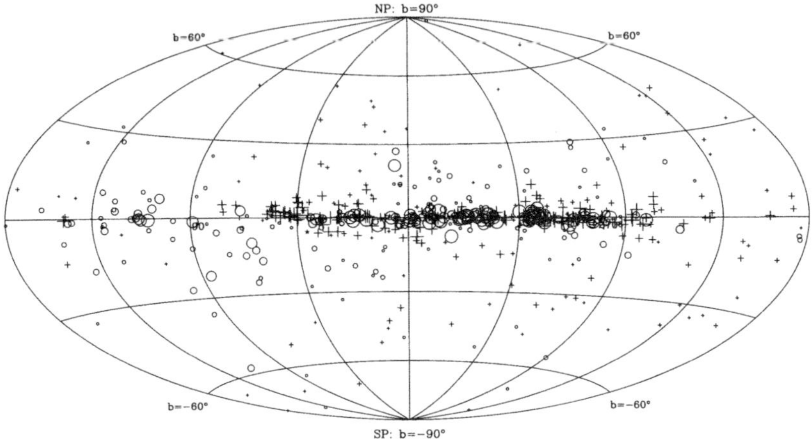

Fig. 12. Rotation measure of 537 pulsars. Crosses are positive RM, open circles negative RM. The scale goes 12 < |RM| < 300 (courtesy of J.L. Han)

Most recent developments in this field concerned the addition of more and carefully selected data point and the combination with the observation of EGRS. For one Han et al. (1999, 2002) have proceeded to suggest that we have several large-scale reversals of the magnetic field in successive spiral arms. In fact, recently a new field reversal towards the Norma spiral arm was suggested by Han et al. (2002) based on recent southern data. On the other hand Mitra et al. (2003) have investigated pulsars in the direction of the Perseus (anticenter) spiral arm and found that foreground H II regions effect the values of RM and DM and in some instances lead to a local magnetic field reversal. This area was investigated previously by Vallée (1983) by studying EGRS and in the studies of RM of Galactic emission towards the Norma arm by Gaensler et al. (2001) and Wielebinski and Mitra (2004). Magnetic field reversals are often associated with Hα regions that also increase the dispersion measure in pulsars. The advent of the recent Hα surveys (e.g. WHAM, SHASSA, VTSS, SHS) help us in these investigations. The important fact is that so far less than half of the 1300 pulsars known have measured RM due to sensitivity problems. The efforts of Han JinLin (private communication) have raised the sample to 537 sources with measured RM as shown in Fig. 12. An increased effort in this direction should help us to understand the Galactic magnetic field.

7 The Magnetic Fields of the Milky Way

The combination of all the data presented above should allow us to construct a model of the Galactic magnetic field. Some years ago this seemed feasible even with the limited data then available. However with the complexity of the new data it has become obvious that a complete model is not possible. Only some statements about the observational 'facts' can be made which should us guide to theoretical modeling.

1. Magnetic fields are present everywhere in the Milky Way, in Galactic sources such as supernova remnants, stars and pulsars. Also the inter-arm space is permeated by magnetic fields. High-frequency Galactic plane polarization surveys can contribute to our knowledge in this area.
2. Faraday rotation effects, seen from low to high radio frequencies in the Galactic plane, suggest a very turbulent magneto-ionic medium (MIM) to be present. Observations at medium galactic latitudes show that this MIM is very turbulent even away from the Galactic plane in the anti-center direction.

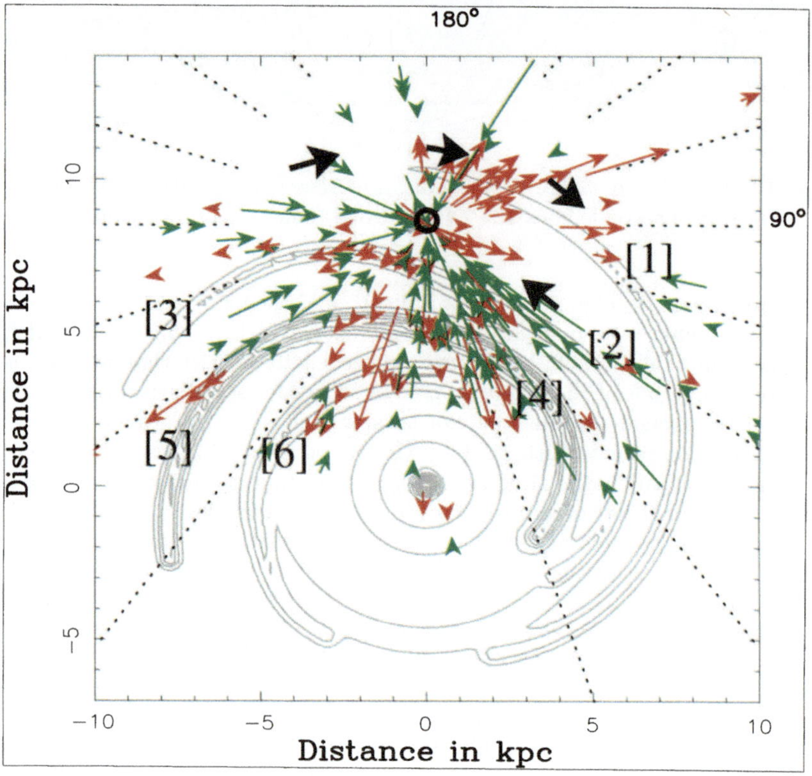

Fig. 13. The grey contours represents the electron density distribution of the top view of our Galaxy as given by the (Taylor and Cordes, 1993) electron density model. Numbers correspond to spiral arms named as [1] Perseus, [2] Sagittarius, [3] Carina, [4] Scutum, [5] Crux and [6] Norma. The dark open circle is the location of the Sun. Red and green arrows are negative and positive RM of pulsars with pulsars located at the center of these arrows and the size of the arrows corresponding to the magnitude of RM. The dark arrows correspond to the direction of the average magnetic field towards the Perseus and the Sagittarius arms. Note that these are only two directions where such coherent direction of the field can be inferred from the pulsar data. Other directions are limited by statistics and insignificant pattern in pulsar RM distribution.

3. The all-sky rotation measure surveys of extragalactic sources are important information but in view of the observed Faraday structure we need extensive samples to probe the magnetic fields. New surveys of 1000s of sources are needed. This data should allow us to model the magnetic fields in the Galactic halo.
4. The RM of extragalactic radio sources along the Galactic plane is important information about the magnetic fields in the spiral arms. Detailed studies of each source must be made to understand the RM effects contributed by the Hα emission in the line of sight.
5. The studies of pulsars are crucial in the understanding of the Galactic magnetic fields. We need again a much larger number of objects with measured RM, especially away from the inner Galactic regions. A combination of the pulsar data and the surveys of EGRS mentioned above is essential.
6. A large-scale magnetic field, directed clockwise (see Fig. 13), exists in the Perseus spiral arm. The Perseus spiral arm seems to be predestined to give us important information about the morphology of the Galactic magnetic fields, since in this direction we have only one Faraday screen. The effects of individual clouds can be investigated in this direction. Additional information can come from Zeeman studies of the molecular clouds.
7. There is a large-scale magnetic field reversal in the Sagittarius spiral arm where the magnetic field direction is anti-clockwise. This field orientation does not seem to continue in the Carina segment of the spiral arm (Fig. 13).
8. The observations of magnetic fields towards the inner Galactic region are strongly affected by the intense Hα emission. In some directions there are 'windows' where distant magnetic fields can be observed but most of the inner Galaxy is not transparent to polarized emission, even at the higher radio frequencies.
9. Magnetic field reversals are often seen on edges of Hα regions.

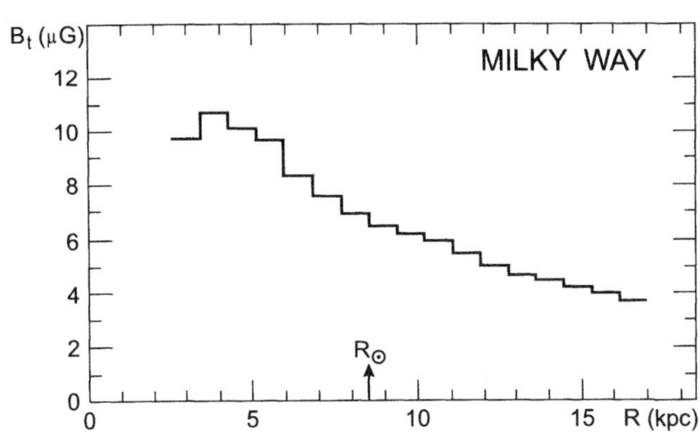

Fig. 14. Strength of the total magnetic field in the Galaxy, averaged from the deconvolved surface brightness of the synchrotron emission at 408 MHz (Beuermann et al., 1985), assuming energy equipartition between magnetic field and cosmic ray energy densities (Berkhuijsen, private communication). The accuracy is about 30%. The Sun is assumed to be located at $R = 8.5\,\mathrm{kpc}$.

10. The magnetic field strength has in general a regular component with $B_{\rm reg} \sim 5\,\mu$G. The random field component has a comparable, or possibly greater magnitude, i.e. $B_{\rm ran} > 5\,\mu$G (see Beck, 2001).
11. The strength of the total magnetic field in the Galaxy falls from the inner Galaxy to the outer reaches as shown in Fig. 14.
12. Comparison with observations of extragalactic magnetic fields (Chap. 3) should help us in modeling, but these observations must be made with the highest angular resolution, to give comparable information. The advent of new telescopes, like the VLA E-array and the Square Kilometer Array one day should allow us an enormous step in this direction.

Acknowledgement

The radio data on Galactic magnetic fields have been collected over the past 40 years. I am indebted to many colleagues and students who made observations, discussed the data and developed this field of research. In writing this article I have had important discussions with Wolfgang Reich, Rainer Beck and Han Jin Lin. I also thank Gabriele Breuer who made this article possible.

References

Alfvén, H.: 1937, Ark. Mat. Astr. Fys. 25B, No. 29
Alfvén, H., Herlofson, N.: 1950, Phys. Rev. 78, 616
Allen, C. W., Gum, C. S.: 1950, Aust. J. Phys. 3, 224
Altenhoff, W. J., Mezger, P. G., Strassl, H. et al.: 1961, Veröff. Universitäts-Sternwarte Bonn Nr. 59
Altenhoff, W. J, Downes, D., Goad, L. et al.: 1970, A&AS 1, 319
Alvarez, H., Aparici, J., May, J., Olmos, F.: 1997, A&AS 124, 315
Baker, J. R., Smith, F. G.: 1971, MNRAS 152, 361
Baldwin, J. E.: 1955, MNRAS 115, 684
Beck, R.: 2001, Sp. Sci. Rev. 99, 243
Beck, R., Krause, M.: 2005, AN 326, 414
Beck, R., Shukurov, A., Sokoloff, D., Wielebinski, R.: 2003, A&A 411, 99
Bennett, C. L., Hill, R. S., Hinshaw, G. et al.: 2003, ApJS 148, 97
Berkhuijsen, E. M.: 1972, A&AS 5, 263
Berkhuijsen, E. M., Brouw, W. N.: 1963, BAN 17, 185
Beuermann, K., Kanbach, G., Berkhuijsen, E. M.: 1985, A&A 153, 17
Bingham, R. G.: 1966, MNRAS 134, 327
Bologna, J. M., McClain, E. F., Rose, W. K., Sloananker, R. M.: 1965, ApJ 142, 106
Bolton J. G., Westfold, K. C.: 1950, Aust, J. Sci. Res. A3, 19
Bolton, J. G., Wild, J. P.: 1957, ApJ 125, 296
Bracewell, R. N., Cooper, B. F. C., Cousins, T. E.: 1962, Nat 195, 1289
Broten, N. W., MacLeod, J. M., Vallée, J. P.: 1988, Ap&SS 141, 303
Brouw, W. N., Spoelstra, T. A. Th.: 1976, A&AS 26, 129
Brown, J. C.: 2003, PhD thesis, University of Calgary

Brown, J. C., Taylor, A. R.: 2001, ApJ 563, L31
Brown, J. C., Taylor, A. R., Wielebinski, R., Müller, P.: 2003a, ApJ 592, L29
Brown, J. C., Taylor, A. R., Jackel, B. J.: 2003b, ApJS 145, 213
Burbidge, G. R.: 1956, ApJ 124, 416
Cane, H. V.: 1978, Aust. J. Phys. 31, 561
Clegg, A. W., Cordes, J. M., Simonetti, J. H., Kulkarni, S. R.: 1992, ApJ 386, 143
Cooper, B. F. C., Price, R. M.: 1962, Nat 195, 1084
Davies, R. D.: 1967, in IAU Symp 31 *Radio Astronomy and the Galactic System*, ed. H. van Woerden (London: Academic Press), p. 391
Davis, L. J., Greenstein, J. L.: 1951, ApJ 114, 206
Dombrovsky, V. A.: 1954, Dokl. Akad. Nauk 94, 1021
Downes, A. J. B., Pauls, T., Salter, C. J.: 1981, A&A 97, 296
Dröge, F, Priester, W.: 1956, Zeitschr. f. Astrophys. 40, 236
Duncan, A. R., Stewart, R. T., Haynes, R. F., Jones, K. L.: 1995, MNRAS 277, 36
Duncan, A. R., Haynes, R. F., Jones, K. L., Stewart, R. T.: 1997, MNRAS 291, 279
Duncan, A. R., Reich, P., Reich, W., Fürst, E.: 1999, A&A 350, 447
Fermi, E.: 1949, Phys. Rev. 75, 1169
Frick, P., Stepanov, R., Shukurov, A., Sokoloff, D.: 2001, MNRAS 325, 649
Fürst, E., Reich, W., Reich, P., Reif, K.: 1990, A&AS 85, 805
Gardner, F. F., Whiteoak, J. B.: 1963, Nat 197, 1162
Gardner, F. F., Davies, R. D.: 1966, Aust. J. Phys. 19, 129
Gaensler, B. M., Dickey, J. M., McClure-Griffiths, N. M. et al.: 2001, ApJ 549, 959
Georgelin, Y. M., Georgelin, Y. P.: 1976, A&A 49, 57
Ginzburg, V. L., Syrovatskii, S. I.: 1965, ARA&A 3, 297
Gray, A. D., Landecker, T. L., Dewdney, P. E., Taylor, A. R.: 1998, Nat 393, 660
Green, A. J.: 1974, A&AS 18, 267
Hall, J. S.: 1949, Science 109, 166
Han, J. L., Qiao, G. J.: 1994, A&A 288, 759
Han, J. L., Manchester, R. N., Berkhuijsen, E. M., Beck, R.: 1997, A&A 322, 98
Han, J. L., Manchester, R. N., Qiao, G. J.: 1999, MNRAS 306, 371
Han, J. L., Manchester, R. N., Lyne, A. G., Qiao, G. J.: 2002, ApJ 570, L17
Hanbury Brown, R., Palmer, H. P., Thompson, A. R.: 1955, MNRAS 115, 487
Handa, T., Sofue, Y., Nakai, N. et al.: 1987, PASJ 39, 709
Haslam, C. G. T., Kearsey, S., Osborne, J. L. et al.: 1981, Nat 289, 470
Haslam, C. G. T., Stoffel, H., Salter, C. J., Wilson, W. E.: 1982, A&AS 47, 1
Haverkorn, M., Katgert, P., de Bruyn, A. G.: 2000, A&A 356, L13
Haverkorn, M., Katgert, P., de Bruyn, A. G.: 2003a, A&A 403, 1045
Haverkorn, M., Katgert, P., de Bruyn, A. G.: 2003b, A&A 404, 233
Heiles, C.: 1976, ARA&A 197, 14
Hewish, A., Bell, S. J., Pilkington, J. D. H. et al.: 1968, Nat 217, 709
Hill, E. R.: 1968, Aust. J. Phys. 21, 735
Hill, E. R., Slee, O. B., Mills, B. Y.: 1958, Aust. J. Phys. 11, 530
Hiltner, W. A.: 1949, Science 109, 165
Inoue, M., Tabara, H.: 1981, PASJ 33, 603
Inoue, M., Takahashi, T., Tabara, H. et al.: 1984, PASJ 36, 633
Johnston-Hollit, M., Hollitt, C. P., Ekers, R. D.: 2004, in *The Magnetized Interstellar Medium*, ed. B. Uyanıker, W. Reich, R. Wielebinski (Katlenburg-Lindau: Copernicus, p. 13

Jonas, J. L., Baart, E. E., Nicolson, G. D.: 1998, MNRAS 297, 1998
Junkes, N., Fürst, E., Reich, W.: 1987, A&AS 69, 451
Kiepenheuer, K. O.: 1950, Phys. Rev. 79, 738
Klein U., Mack, K.-H., Gregorini, L., Vigotti, M.: 2003, A&A 406, 579
Landecker, T. L., Wielebinski, R.: 1970, Aust. J. Phys. Suppl. 16, 1
Lyne, A. G., Smith, F. G.: 1989, MNRAS 237, 533
MacLeod, J. M., Vallée, J. P., Broten, N. W.: 1988, A&AS 74, 63
Maeda, K., Alvarez, H., Aparici, J. et al.: 1999, A&AS 140, 145
Manchester, R. N.: 1972, ApJ 172, 43
Manchester, R. N.: 1974, ApJ 188, 636
Manchester, R. N., Taylor, J. H.: 1977, *Pulsars* (San Francisco: W.H. Freeman)
Mathewson, D. S., Milne, D. K.: 1964, Nat 203, 1273
Mathewson, D. S., Milne, D. K.: 1965, Aust. J. Phys. 18, 635
Mathewson, D. S., Broten, N. W., Cole, D. J.: 1963, Aust. J. Phys. 18, 665
Mathewson, D. S., Broten, N. W., Cole, D. J.: 1966, Aust. J. Phys 19, 93
Mayer, C. H., McCullough, T. P., Sloanaker, R. M.: 1957, ApJ 126, 468
Mayer, C. H., McCullough, T. P., Sloanaker, R. M.: 1962, ApJ 135, 656
Mayer, C. H., McCullough, T. P., Sloanaker, R. M.: 1964, ApJ 139, 248
Mills, B. Y.: 1964, ARA&A 2, 185
Milogradov-Turin, J., Smith, F. G.: 1973, MNRAS 161, 269
Mitra, D., Wielebinski, R., Kramer, M., Jessner, A.: 2003, A&A 398, 993
Morris, D., Berge, G. L.: 1964, ApJ 139, 1388
Muller, C. A., Berkhuijsen, E. M., Brouw, W. N., Tinbergen, J.: 1963, Nat 200, 155
Ohno, H., Shibata, S.: 1993, MNRAS 262, 953
Oort, J. H., Walraven, T.: 1957, BAN 12, 285
Pacholczyk, A. G.: 1970, *Radio Astrophysics* (San Francisco: W. H. Freeman)
Parkes surveys at 2.7 GHz: 1969, Aust. J. Phys. Suppl. No. 11
Parkes surveys at 2.7 GHz: 1970, Aust. J. Phys. Suppl. No. 13
Pauliny-Toth, I. I. K., Shakeshaft, J. R.: 1962, MNRAS 124, 61
Pauliny-Toth, I. I. K., Baldwin, J. E., Shakeshaft, J. R.: 1961, MNRAS 122, 279
Pawsey, J. L. Harting, E.: 1961, Aust. J. Phys. 13, 740
Phillips, S., Kearsey, S., Osborne, J. L. et al.: 1981, A&A 98, 286
Piddington, J. H.: 1964, MNRAS 128, 345
Qiao, G. J., Manchester, R. N., Lyne, A. G., Gould, D. M.: 1995, MNRAS 274, 572
Rand, R. J., Kulkarni, S. R.: 1989, ApJ 343, 760
Rand, R. J., Lyne, A. G.: 1994, MNRAS 268, 497
Razin, V. A.: 1958, Soviet Astronomy 2, 216
Reber, G.: 1944, ApJ 100, 279
Reich, P., Reich, W.: 1986, A&AS 63, 205
Reich, P., Reich, W., Fürst, E.: 1997, A&AS 126, 413
Reich, P., Testori, J. C., Reich, W.: 2001, A&A 376, 861
Reich, P., Reich, W., Testori, J. C.: 2004, in *The Magnetized Interstellar Medium*, ed. B. Uyanıker, W. Reich, R. Wielebinski (Katlenburg-Lindau: Copernicus), p. 63
Reich, W.: 1982, A&AS 48, 219
Reich, W.: 2003, A&A 401, 1023
Reich, W., Fürst, E., Steffen, P. et al.: 1984, A&AS 58, 197
Reich, W., Sofue, Y., Wielebinski, R., Seiradakis, J. H.: 1988, A&A 191, 303

Reich, W., Fürst, E., Reich, P., Reif, K.: 1990, A&AS 85, 633
Reich, W., Fürst, E., Reich, P. et al.: 2004, in *The Magnetized Interstellar Medium*, ed. B. Uyanıker, W. Reich, R. Wielebinski (Katlenburg-Lindau: Copernicus), p. 45
Ruzmaikin, A. A., Sokoloff, D. D.: 1977, A&A 58, 247
Seeger, Ch. L., Westerhout, G., Conway, R. G., Hoekema, T.: 1965, BAN 18, 11
Seiradakis, J. H., Lasenby, A. N., Yusef-Zadeh, F. et al.: 1985, Nat 317, 697
Seiradakis, J. H., Reich, W., Wielebinski, R. et al.: 1989, A&AS 81, 291
Shain, C. A., Komessaroff, M. M., Higgins, C. S.: 1961, Aust. J. Phys. 14, 508
Shklovsky, I. S.: 1953, Dokl. Akad. Nauk 90, 983
Sieber, W., Haslam, C. G. T., Salter, C. J.: 1979, A&A 74, 361
Simard-Normandin, M., Kronberg, P. P.: 1980, ApJ 242, 74
Simard-Normandin, M., Kronberg, P. P., Button, S.: 1981, ApJS 45, 97
Smith, F.G.: 1968, Nat 218, 325
Smoot, G. F., Bennett, C. L., Kogut, A. et al.: 1992, ApJ 396, L1
Sofue, Y., Fujimoto, M.: 1983, ApJ 265, 722
Spoelstra, T. A. Th.: 1971, A&A 13, 237
Spoelstra, T. A. Th.: 1984, A&A 135, 238
Tabara, H., Inoue, M.: 1980, A&AS 39, 379
Taylor, A. R., Gibson, S. J., Peracaula, M. et al.: 2003, AJ 125, 3145
Taylor, J. H., Cordes, J. M.: 1993, ApJ 411, 674
Testori, J. C., Reich, P., Reich, W.: 2004, in *The Magnetized Interstellar Medium*, ed. B. Uyanıker, W. Reich, R. Wielebinski (Katlenburg-Lindau: Copernicus), p. 57
Thomson, J. H.: 1957, Nat 180, 495
Thomson, R. C., Nelson, A. H.: 1980, MNRAS 191, 863
Tsuboi, M., Inoue, M., Handa, T. et al.: 1985, PASJ 37, 359
Turtle, A. J., Baldwin, J. E.: 1962, MNRAS 124, 459
Uyanıker, B.: 1998, PhD thesis, Bonn University
Uyanıker, B., Fürst, E., Reich, W. et al.: 1998, A&AS 132, 401
Uyanıker, B., Fürst, E., Reich, W. et al.: 1999, A&AS 138, 31
Vallée, J. P.: 1983, J. Roy. Astron. Soc. Canada 77, 177
Vallée, J. P.: 1988, AJ 95, 750
Vallée, J. P.: 1991, Ap&SS 178, 41
Vallée, J. P.: 1996, A&A 308, 433
Vallée, J. P., Kronberg, P. P.: 1975, A&A 43, 233
Vallée, J. P., Bignell, R. C.: 1983, ApJ 272, 131
Verschuur, G. L.: 1968, Phys. Rev. Lett. 21, 775
Westerhout, G.: 1958, BAN 14, 215
Westerhout, G., Seeger, C. L., Brouw, W. N., Tinbergen, J.: 1962, BAN 16, 187
Wielebinski, R., Mitra, D.: 2004, in *The Magnetized Interstellar Medium*, ed. B. Uyanıker, W. Reich, R. Wielebinski (Katlenburg-Lindau: Copernicus), p. 19
Wielebinski, R., Shakeshaft, J. R., Pauliny-Toth, I. I. K.: 1962, Observatory 82, 158
Wielebinski, R., Shakeshaft, J. R.: 1962, Nat 195, 982
Wielebinski, R., Shakeshaft, J. R.: 1964, MNRAS 128, 19
Wielebinski, R., Smith, D. H., Cardenas, X. G.: 1968, Aust. J. Phys. 21, 185
Wieringa, M. H., de Bruyn, A. G., Jansen, D. et al.: 1993, A&A 268, 215
Wilkinson, A., Smith, F. G.: 1974, MNRAS 167, 593

Wilson, A. S. Weiler, K. W.: 1976, A&A 53, 89
Wolleben, M., Reich, W.: 2004, in *The Magnetized Interstellar Medium*, ed. B. Uyanıker, W. Reich, R. Wielebinski (Katlenburg-Lindau: Copernicus), p. 99
Wolleben, M., Landecker, T. L., Reich, W., Wielebinski, R.: 2004, in *The Magnetized Interstellar Medium*, ed. B. Uyanıker, W. Reich, R. Wielebinski (Katlenburg-Lindau: Copernicus), p. 51
Yates, K. W., Wielebinski, R., Landecker, T. L.: 1967, Aust. J. Phys. 20, 595
Yusef-Zadeh, F., Morris, M., Chance, D.: 1984, Nat 310, 557

Mesoscale Magnetic Structures in Spiral Galaxies

Anvar Shukurov

School of Mathematics and Statistics, University of Newcastle, Newcastle upon Tyne, NE1 7RU, U.K.
anvar.shukurov@ncl.ac.uk

1 Introduction

Virtually all spiral galaxies host magnetic fields ordered at scales comparable to the galactic size (Beck et al., 1996; Beck, 2000, 2001). Observations of polarized radio emission at improved resolution and sensitivity have revealed details of the global magnetic structures that can shed new light on the problem of their origin. Reversals of the regular magnetic field along radius and/or azimuth and magnetic arms are such features, whose scale exceeds significantly the correlation scale of interstellar turbulence but remains smaller than the overall galactic dimension. Despite a few decades of debate, there remains doubt as to what features of the observed field could have been inherited from the pre-galactic past, and which have been formed and maintained more recently in a relatively mature galaxy. In what follows, we briefly review the current understanding of the origin of the mesoscale magnetic structures and their implications for the origin of galactic magnetic fields.

The Milky Way appears to possess a global magnetic field of unusual structure. The regular magnetic field in our Galaxy has one or more large-scale reversals, where the magnetic field coherent over a scale of order a few kiloparsecs changes its direction by about 180° along a line presumably extended along the azimuth. The number of reversals has not been firmly established, their origin has not been fully understood, and the shape of the lines along which the reversals occur is not known.

There are just a few galaxies where similar large-scale reversals cannot be excluded. The nearby galaxy M 81 might host a bisymmetric magnetic structure (Krause et al., 1989), i.e., a global structure where the regular spiral magnetic field reverses along azimuth and, perhaps, radius. However, the magnetic structure of M 81 needs to be reconsidered with observations at higher resolution and sensitivity and with more reliable interpretation techniques. A magnetic reversal between the inner and outer regions in the galaxy NGC 2997 has been suggested in Han et al. (1999). A magnetic reversal in the disc of M 51 (Berkhuijsen et al., 1997) is discussed in detail below.

The unusual structure of the Galactic magnetic field has attracted significant attention. Numerous papers have been published attempting to establish the number of reversals and their positions from observations. However, there is only a handful of papers where the origin of the magnetic reversals is addressed. In this review, we discuss some limitations of the observational evidence for the reversals and put the observational effort into a broader physical perspective of the theory of galactic magnetic fields.

Another peculiar feature of galactic magnetic fields, as yet discovered in only a few galaxies but believed to be of general significance, are magnetic arms where the large-scale magnetic field is observed to be enhanced between the gaseous spiral arms. In other words, the large-scale magnetic field in these galaxies is observed to be stronger where the gas density is smaller. This immediately implies that the strength of the regular magnetic field is not uniquely determined by gas density, and therefore that the regular magnetic field cannot be frozen into the gas at scales exceeding a few kiloparsecs (Beck, 2001). It is quite remarkable that the turbulent magnetic fields are still stronger in the gaseous arms, indicating that the turbulent and regular magnetic fields are subject to distinct physical processes. This invites explanations in terms of the mean-field dynamo theory. The origin and implications of magnetic arms are reviewed in the second half of this paper.

2 Observational Evidence for Magnetic Reversals

The first indication of a reversal of the regular magnetic field in the inner Galaxy was obtained by Simard-Normandin and Kronberg (1980) from their analysis of Faraday rotation measures of extragalactic radio sources. This reversal occurs in the inner Galaxy between the local Orion arm and the Sagittarius arm at a distance of about 0.5 kpc from the Sun. The existence of this feature has been later confirmed by most of the studies of the Faraday rotation measures of extragalactic radio sources and pulsars (Heiles, 1996, Chap. 5), and its extension to the fourth Galactic quadrant has recently been detected (Frick et al., 2001).

The only observational tracer of a large-scale magnetic field that is sensitive to its direction is the Faraday rotation measure RM (the Zeeman effect is strong enough only in relatively dense clouds that have a locally enhanced magnetic field). Thus, all the discussions of magnetic field reversals rely on the signatures of the Galactic magnetic field in the rotation measures of extragalactic radio sources and pulsars. Extragalactic polarized radio sources – radio galaxies and quasars – possess their own regular magnetic fields, so their RM contain a significant, and unknown intrinsic contribution. Polarized emission from the radio sources propagates through the turbulent magneto-ionic interstellar medium, so their Faraday rotation measures are contaminated by the strong random contribution of interstellar turbulence. The *maximum* contribution of a regular Galactic magnetic field B to the observed RM is given by

$$\begin{aligned} |\mathrm{RM}|_{\max} &= 0.81 n_{\mathrm{e}} B L_B \\ &\approx 220\,\mathrm{rad\,m}^{-2} \left(\frac{n_{\mathrm{e}}}{0.03\,\mathrm{cm}^{-3}}\right) \left(\frac{B}{3\,\mu\mathrm{G}}\right) \left(\frac{L_B}{3\,\mathrm{kpc}}\right) , \end{aligned} \tag{1}$$

when the line of sight is aligned with the field, where n_{e} is the mean number density of thermal electrons, and the path length L_B is limited to 3–6 kpc by the curvature and finite width of the spiral arms. Meanwhile, the r.m.s. contribution of the interstellar turbulence to RM along the path length L through the Milky Way is given by (Burn, 1996; Sokoloff et al., 1998)

$$\sigma_{\mathrm{RM}} = 0.81\sigma_n\sigma_B(2Ll)^{1/2}$$
$$\simeq 170\,\mathrm{rad\,m^{-2}}\left(\frac{\sigma_n}{0.03\,\mathrm{cm^{-3}}}\right)\left(\frac{\sigma_B}{5\,\mu\mathrm{G}}\right)\left(\frac{l}{0.1\,\mathrm{kpc}}\right)^{1/2}\left(\frac{L}{10\,\mathrm{kpc}}\right)^{1/2}, \quad (2)$$

where σ_n and σ_B are the standard deviations of fluctuations in thermal electron density and magnetic field, respectively, and l is the size of turbulent cells. Equations (1) and (2) are only valid if fluctuations in thermal electron density and magnetic field are statistically independent. This assumption is an oversimplification and it can affect very significantly magnetic field estimates obtained from Faraday rotation measures Beck et al. (2003).

Thus, $|\mathrm{RM}|_{\max} \simeq \sigma_{\mathrm{RM}}$, and we have data with a signal-to-noise ratio of order unity. With such a signal, we need to reveal quite delicate features of the magnetic field $\boldsymbol{B}$. In the case of pulsars, a change in the sign of the line-of-sight magnetic field, associated with a reversal, would produce a change in the slope of RM as a function of distance from the source to the Sun. For extragalactic radio sources, reversals would produce localized, relatively weak extrema of RM against a broad RM distribution produced by the local magnetic field. The situation is further complicated by nearby fluctuations in the magnetic field (e.g., associated with supernova remnants) that can occupy large regions in the sky and be stronger than the features produced by the large-scale magnetic field. H II regions can contribute significantly to the observed Faraday rotation measures, and this can affect our conclusions about the large-scale magnetic field, especially when the number of radio sources in the sample is only modest (Mitra et al., 2003). It is obvious that meaningful results can only be obtained from careful analysis based on mathematically and statistically rigorous procedures. Any 'eye-ball fitting' is dangerous and can be misleading. Equations (1) and (2) indicate that random deviations of individual RM values from a regular pattern (for both pulsars and extragalactic sources) are comparable to or exceed the magnitude of the features that might result from reversals; hence, conclusions derived using a statistically insignificant number of sources are rarely convincing. This applies to the often employed arguments relying on the dependence of pulsar RM on their dispersion measures along individual lines of sight, where just a handful of sources (often less than 10) is used.

In an attempt to reduce the noise, most authors average extragalactic RM, often with a Gaussian weight function. However, such a smoothing contaminates the data because the mean value of a Gaussian filter function differs from zero, and a few strongly deviating values of RM can distort the result. Therefore, sources with $|\mathrm{RM}| \gtrsim 100$–$300\,\mathrm{rad\,m^{-2}}$ are often excluded from analysis. It is, however, more appropriate to *filter out* small-scale fluctuations in RM rather than to smooth them. An noise filtering method used by Frick et al. (2001) involves wavelets, weight functions that have zero mean. Filtering by Fourier analysis along Galactic longitude was used by Johnston-Hollitt et al. (2004); expansion in spherical harmonics was earlier applied to a smaller data set (Seymour, 1984).

A more traditional approach involves model fitting based on well defined statistical criteria for the fit quality. Such an approach was applied to RM of both extragalactic sources and pulsars (Ruzmaikin and Sokoloff, 1977a,b; Rand and Kulkarni, 1989). It is then important to remember that the best fit is not necessarily the one that provides the minimum value of χ^2 (if this statistic is used to access the quality of the fit); instead, χ^2 must be close to an *optimal* value for the model to be acceptable (e.g., Wall and Jenkins, 2003). However, most results regarding the structure of the Galactic magnetic field are based on minimizing a certain statistic without subsequent comparison with its optimal value. As a result, models to be distinguished between are often either equally unacceptable or equally acceptable, with the data being insufficient to distinguish between them.

Since several models can provide equally good or equally bad fit to the data, it is important to explore all reasonable models before making any conclusions. For example, the dependence of RM on Galactic longitude is often used to establish the number and position of magnetic reversals from extragalactic RM, especially in the outer Galaxy. However, plausible modifications of the field configuration and strength in remote arms can produce indistinguishable longitudinal profiles of RM in models with very different magnetic configurations, including those with and without reversals (Stepanov et al., 2005).

The feature of the global magnetic field of the Milky Way confirmed with all models (including those that employ reliable statistical procedures) is the reversal in the inner Galaxy at a distance of about 0.5 kpc from the Sun. The even symmetry of the horizontal magnetic field with respect to the Galactic equator also seems to be firmly established (Frick et al., 2001). Despite rather optimistic assessments. of the reliability of the results (e.g., Han, 2004), further reversals in the inner and the outer Galaxy remain controversial, in particular because remote regions occupy small areas in the sky and are probed by a small number of sources. Recent massive determinations of Faraday rotation measures in the Canadian Galactic Plane Survey (Brown et al., 2003) have provided significant improvement in this respect. However, these data are restricted to a narrow strip extended along the Galactic equator, which complicates their analysis.

Altogether, our confident knowledge of the global magnetic structure of the Milky Way can be conservatively summarized as follows. The distance from the Sun to the reversal in the inner Galaxy is about 0.5 kpc. The reversal has been detected in the first and fourth Galactic quadrants (Frick et al., 2001). This, however, does not imply that the reversal extends over the whole Galaxy (i.e., to all azimuthal angles about the Galactic centre): this may be a relatively local phenomenon, with the reversed field extended by not more than several kiloparsecs in the azimuthal direction. Evidence for further reversals, especially in the inner Galaxy, is compelling but still not fully convincing. Forthcoming extensive RM data will hopefully help to clarify the mesoscale structure of the Galactic magnetic field.

In what follows, we discuss the theoretical understanding of the origin of magnetic field reversals in the Milky Way. There have been few attempts to explain global magnetic field reversals in the Milky Way. The dichotomy between primordial and dynamo theories of galactic magnetic fields has strongly influenced both the data interpretation and modelling. The primordial theory interprets the reversals as a global phenomenon, so that they are *assumed* to extend, in both azimuth and radius, over the whole Galaxy (Sect. 3.1). In the framework of the dynamo theory, the reversals can represent an axially symmetric magnetic configuration with alternating spiral field – then they are viewed again as a global feature (Sect. 3.2). Otherwise, a nonlinear state of the bisymmetric magnetic structure can represent a reversed magnetic field confined to a localized region near the corotation radius (Sect. 4).

3 Global Reversals

In this section we briefly review two concepts of galactic magnetic fields, the primordial and dynamo theories, with emphasis on magnetic field reversals. It is useful to draw a distinction between reversals that occur along azimuth and those occurring along radius. The azimuthal reversals, a signature of a strongly nonaxisymmetric global magnetic structure (e.g., a bisymmetric one) arise from different physical effects than reversals along radius in an (almost) axially symmetric magnetic structure. The nonaxisymmetric structures are subject to rapid wound-up by the galactic differential rotation, and so must be maintained at the global scale. Reversals in an axially symmetric magnetic field are not affected by differential rotation, and so can be supported at a smaller scale: these are genuine mesoscale structures.

3.1 Primordial Magnetic Fields

It is often (wrongly) assumed that a bisymmetric global magnetic structure is a direct indication of the primordial origin of the magnetic field. An external (extragalactic) magnetic field oriented along the plane of the galactic disc is twisted by differential rotation into a bisymmetric configuration, so magnetic field reversals arise naturally, along both radius and azimuth. This conceptual simplicity is, however, deceptive: there are no detailed models that would demonstrate that any primordial magnetic field can be twisted by differential rotation into a configuration compatible with what is known about the global magnetic field of any spiral galaxy, if only a realistic galactic model is adopted (Shukurov, 2000).

In particular, the observed pitch angle of magnetic field p (i.e., the angle between the magnetic field and the circumference) is a sensitive diagnostic of the origin of magnetic field (Shukurov, 2000). The large-scale magnetic fields observed in spiral galaxies have $p = -(10^\circ\text{–}30^\circ)$ (Ruzmaikin et al., 1988; Beck et al., 1996; Beck, 2000, 2001) (here and below, a negative value of p indicates a trailing spiral).

Consider a uniform external magnetic field of a strength B_e parallel to the disc plane and frozen into the interstellar gas (Moffatt, 1978). The differential rotation of the disc twists the field into a bisymmetric spiral. The magnetic pitch angle p of the twisted magnetic field reduces with time t as $\tan p \simeq -(|G|t)^{-1}$, where $G = r\,d\Omega/dr$ is the shear rate due to differential rotation at angular velocity Ω and r is the galactocentric radius. The winding-up proceeds until a time $t_0 \simeq 5 \times 10^9\,\mathrm{yr}$ such that $|G|t_0 \simeq |C_\omega|^{1/2}$, where $C_\omega = GR^2/\beta = 10^3\text{–}10^4$ is a dimensionless number quantifying rotational shear, with $R \simeq 10\,\mathrm{kpc}$ the representative galactic radius, and β is the turbulent magnetic diffusivity whose standard estimate is $\beta \simeq 10^{26}\,\mathrm{cm^2\,s^{-1}}$. At later times, the magnetic field decays because of diffusion and reconnection since the field direction flips over a radial scale that decreases with time as $\Delta r \simeq R/(|G|t)$ down to $\Delta r \simeq R|C_\omega|^{-1/2} \simeq 0.1\,\mathrm{kpc}$ at $t = t_0$. The wound-up magnetic field attains its maximum value $B_{\max} \simeq B_e|C_\omega|^{1/2}$ at $t = t_0$ before decaying rapidly. Even neglecting this inevitable decay, an external magnetic field of order $B_e = 10^{-8}\,\mathrm{G}$ is required to explain the observed field strength, $B_{\max} \simeq 2\,\mu\mathrm{G}$. Such an external field appears to be rather strong, but perhaps not unrealistic since an extragalactic magnetic field can be amplified by the fluctuation dynamo in the protogalaxy and by compression during its collapse (Beck et al., 1996; Kulsrud, 1999). What represents a real problem is that the corresponding value of the pitch angle is $|p| \simeq$

$|C_\omega|^{-1/2} \simeq 1^\circ$, a value much smaller than the observed values of order 10°. In order to obtain $|p| \simeq 10^\circ$ as observed (Beck et al., 1996; Beck, 2000, 2001), one would need $|C_\omega| \simeq 30$ – a value of C_ω two orders of magnitude too small – but even then the maximum field strength would be just $|C_\omega|^{1/2} \simeq 5$ times larger than the extragalactic value, which is obviously unacceptable.

A primordial magnetic field could avoid the catastrophic reduction in the radial scale due to radial twisting if it is oriented parallel to the rotation axis and is amplified by the vertical shear, $\partial\Omega/\partial z$ (e.g., Moffatt, 1978; Sofue et al., 1986). From symmetry, $\partial\Omega/\partial z \approx 0$ at the disc midplane, $z = 0$, and so this effect, if important at all, can only be significant in the galactic halo rather than in the disc. The steady state magnetic field can be strong in this case, $B_{\max} \simeq C_\omega B_e$, but this state can be reached only at a very late time, $t \gg R^2/\beta_h \simeq 10^{10}$ yr, where $\beta_h \simeq 5 \times 10^{27}\,\mathrm{cm^2\,s^{-1}}$ is a tentative estimate of the turbulent magnetic diffusivity in the galactic halo (Poezd et al., 1993), and the halo radius is assumed to be $R = 10\,\mathrm{kpc}$. At earlier times, the azimuthal field grows linearly in time as $B = rB_e(\partial\Omega/\partial z)t \simeq B_e\Omega t \simeq B_e(t/5 \times 10^7\,\mathrm{yr})$. In 10^{10} yr, an external field of a strength $B_e = 10^{-8}\,\mathrm{G}$ could be amplified to about $B \simeq 2\,\mu\mathrm{G}$, but only in the halo rather than in the disc. The field will have odd symmetry with respect to the galactic equator with its azimuthal component vanishing at the equator, contrary to what is observed in the Milky Way.

The primordial theory has never been able to resolve this difficulty (among several other problems – see Ruzmaikin et al., 1988; Shukurov, 2000), which apparently explains the lack of any quantitative comparisons of this theory with magnetic fields observed in specific galaxies, including the Milky Way.

3.2 Axisymmetric Dynamo Fields

Mean-field dynamo action is capable of maintaining both axially symmetric and bisymmetric global magnetic structures (Ruzmaikin et al., 1988; Beck et al., 1996). Axially symmetric magnetic structures are the easiest for the dynamo to produce. Bisymmetric magnetic structures can also be generated, but their *dominance* in most galaxies would be difficult to explain (Beck et al., 1996). For example, the galaxy M 81, the only candidate for the dominant bisymmetric structure, is able to support the bisymmetric dynamo mode, but there are no models that would explain convincingly its dominance throughout the galactic lifetime in the whole galaxy (cf. Moss, 1995). However, the dynamo can produce a strongly nonaxisymmetric magnetic structure with a field reversal in the neighbourhood of the corotation radius; such a structure can be described as a nonlinear state of the linear bisymmetric solution. This possibility is discussed in Sect. 4.

The dynamo action is just one of many effects that affect regular magnetic fields in galaxies, so it is natural that the perfect magnetic symmetry supported by the underlying dynamo action is distorted into the complicated observed picture by the spiral arms, Parker instability, gas outflows to the galactic halo, etc. It is therefore not surprising that recent radio polarization observations of external galaxies at enhanced sensitivity and resolution have produced radio maps where the global symmetry of the magnetic field is obscured by the wealth of details. Again, quantitative analysis of the observations is required to reveal the underlying global symmetries.

We note that strong disc-halo connections in spiral galaxies can play an important role in supporting dynamo action via advection of magnetic helicity from the disc (Shukurov, 2005). Intense exchange of gas and magnetic fields between the discs and halos of spiral galaxies has been firmly confirmed by both observations and theory (Bloemen, 1991). This makes questionable the arguments of Rafikov and Kulsrud (2000) and Kulsrud (Chap. 4) that it is unlikely that any significant quantity of magnetic flux can be expelled from the discs of the Milky Way and other galaxies.

The magnetic pitch angle of a dynamo-generated magnetic field (either bisymmetric or axisymmetric) is given by $p \simeq -\arctan(l/h) \simeq -10^\circ$ for a flat rotation curve (Shukurov, 2000), where $l \simeq 0.1\,\mathrm{kpc}$ is the turbulent scale and $h \simeq 0.5\,\mathrm{kpc}$ is the scale height of the warm interstellar gas. The magnitude of the pitch angle and the tendency of $|p|$ to decrease with galactocentric radius (because h increases with r in a flared disc) are in fair agreement with observations.

The alignment between the regular magnetic field and spiral arms is often quite tight, albeit not perfect. (We stress that the alignment of the magnetic field with the gaseous arms does not imply that magnetic field is aligned with the gas velocity which is directed along azimuth within a few degrees.) The mean-field dynamo does produce magnetic spirals with a pitch angle close to those observed. The alignment can be further improved by the refraction of magnetic lines in the gaseous arms: since the component of magnetic field normal to the arm is not affected by the gas density increase within the arm, whereas the tangential component increases in proportion to the gas density (under one-dimensional compression), the result is that the regular magnetic field becomes better aligned with the arm. If the arm-interarm density contrast is $\xi = \rho_\mathrm{a}/\rho_\mathrm{i}$ and the magnetic field between the arms makes an angle p_i to the arm, the angle between the arm axis and field within the arm follows as $\tan p_\mathrm{a} = \xi^{-1} \tan p_\mathrm{i}$, which yields $p_\mathrm{a} \approx 2.5^\circ$ for $p_\mathrm{i} = 10^\circ$ and $\xi = 4$. Velocity shear due to streaming in the galactic spiral arms can improve the alignment even further.

In the present context, it is important to appreciate that the dynamo can maintain an axisymmetric magnetic field with spiral magnetic lines and with direction alternating along radius, compatible with reversals observed in the Milky Way (Ruzmaikin et al., 1985). The dynamo mode that grows most rapidly has no reversals, the next one has one reversal, etc. Since the mode without reversals grows most rapidly, no reversals would occur at $t \to \infty$ unless nonlinear effects had halted the growth before this mode could become dominant. Since the growth rates of the different modes do not differ much in a thin disc, with the difference between the growth rates being of order $\beta/Rh \simeq (1.5 \times 10^{10}\,\mathrm{yr})^{-1}$, reversals can persist over periods of order Rh/β comparable to the galactic lifetime. We emphasize, however, that the situation is different in galaxies such as M51 where the time scales involved are an order of magnitude shorter (mainly because of stronger differential rotation), and so reversals are less plausible to survive for a long time.

This idea was further confirmed by nonlinear dynamo models (Belyanin et al., 1994; Poezd et al., 1993). Under a reasonable approximation, the signed amplitude of the axially symmetric large-scale magnetic field Q in a thin disc is governed by the equation (Poezd et al., 1993)

$$\frac{\partial Q}{\partial t} = \gamma_0 Q \left(1 - \frac{Q^2}{B_0^2}\right) + \lambda^2 \frac{\partial}{\partial r}\left(\frac{1}{r}\frac{\partial}{\partial r} rQ\right), \qquad (3)$$

where $\gamma_0(r)$ is the local growth rate of magnetic field (the same for all the modes mentioned in the previous paragraph), $B_0(r)$ is the steady-state strength of the large-scale magnetic field, and $\lambda = h_0/R = 0.5\,\mathrm{kpc}/10\,\mathrm{kpc} \ll 1$ is the aspect ratio of the galactic disc, with h_0 the typical scale height of the ionized gas layer. This equation is written in terms of dimensionless variables, where time is measured in the units of the magnetic diffusion time $h_0^2/\beta \simeq 7.5 \times 10^8\,\mathrm{yr}$, z in the units of h_0, and radius in the units of R. Different signs of $Q(r)$ correspond to magnetic fields of opposite directions. An important feature of (3) is that it admits solutions of both positive and negative sign: if $Q(r)$ is a solution, then $-Q(r)$ is also a solution; this situation is of course typical of equations with quadratic nonlinearity.

Asymptotic analysis of (3) with $\lambda \ll 1$ has shown (Belyanin et al., 1994) that, even in a nonlinear stage of the dynamo action, the reversals remain unsteady and migrate along radius at a speed of the order of the diffusion velocity $\beta/h \simeq 1\,\mathrm{km\,s^{-1}}$. However, the migration speed can be as small as $\beta h/R^2 \simeq 10^{-3}\,\mathrm{km\,s^{-1}}$ if the reversal occurs at a radius $r = R_{\mathrm{rev}}$ such that

$$U(R_{\mathrm{rev}}) = 0\,,$$

where

$$U(r) = r^2\gamma_0\left(\frac{1}{r} + 2\frac{B_0'}{B_0}\right) + \tfrac{1}{2}r^2\gamma_0' = 0\,, \tag{4}$$

and prime denotes derivative with respect to galactocentric radius. For qualitative estimates, one can use an approximate solution of the mean-field dynamo equations applicable to a quadrupole mode with $\alpha \propto \sin \pi z/h$ (Ruzmaikin et al., 1988; Shukurov, 2005):

$$\gamma_0 \simeq -\tfrac{1}{4}\pi^2 + \left(-\tfrac{1}{4}\pi D\right)^{1/2}\,, \quad D \simeq 10\frac{h^2 \Omega G}{v^2}\,, \quad G = r\frac{d\Omega}{dr}\,, \tag{5}$$

where D is the local dynamo number (i.e., the dynamo number defined at a given galactocentric radius r (Ruzmaikin et al., 1988), Ω is the angular velocity of rotation, and h is the scale height of the ionized layer. For B_0, a value corresponding to energy equipartition with turbulent kinetic energy can be adopted,

$$B_0(r) \simeq (4\pi\rho v^2)^{1/2}\,, \tag{6}$$

where $\rho(r)$ is the gas density and $v \simeq 10\,\mathrm{km\,s^{-1}}$ is the turbulent velocity. For $v = \mathrm{const}$, as observed over broad radial ranges in spiral galaxies, $B_0'/B_0 = \rho'/2\rho$. Thus, all the variables in (4) can be expressed in terms of observable quantities.

Equation (6) represents a crude heuristic estimate of the regular field strength attainable by the mean-field dynamo. Putting aside the recent controversy about the nonlinear states of the mean-field dynamo (see Brandenburg and Subramanian, 2005 and Chap. 9 for a review), this estimate can be slightly refined by invoking the balance of the Lorentz and Coriolis forces that presumably occurs in the steady state of the dynamo, to yield (Ruzmaikin et al., 1988)

$$B_0 \simeq (4\pi\rho v \Omega l)^{1/2}\,. \tag{7}$$

Note that $\Omega l \simeq 3\,\mathrm{km\,s^{-1}}$ near the Sun, so the two forms, (6) and (7), result in magnitudes of B_0 that differ by a factor of order unity.

Reversals can, but not necessarily will occur at $r = R_{rev}$. Solutions with alternating magnetic field can only arise if the initial (seed) field had reversals, and a unidirectional initial field would result in a unidirectional magnetic field in the steady state. We show in Fig. 1 $\gamma_0(r)$, $B_0(r)$ and $U(r)$ as obtained for a model of the Milky Way. Since $U(r)$ has many zeros, the occurrence of reversals in the Milky Way is quite plausible.

These results have been confirmed by a numerical solutions of the thin-disc dynamo equation (3) (Poezd et al., 1993). The numerical solutions do exhibit persistent reversals, but their number depends on the initial conditions. In particular, a reversal at $r \approx 7\,\mathrm{kpc}$ occurs for almost all configurations of the initial Galactic magnetic field; with allowance for the accuracy of the model, it can be identified with that observed between the Orion and Sagittarius arms. The other reversals, both in the inner and the outer Galaxy, do not occur for certain initial conditions. The time evolution of the signed amplitude of magnetic field, starting from a chaotic

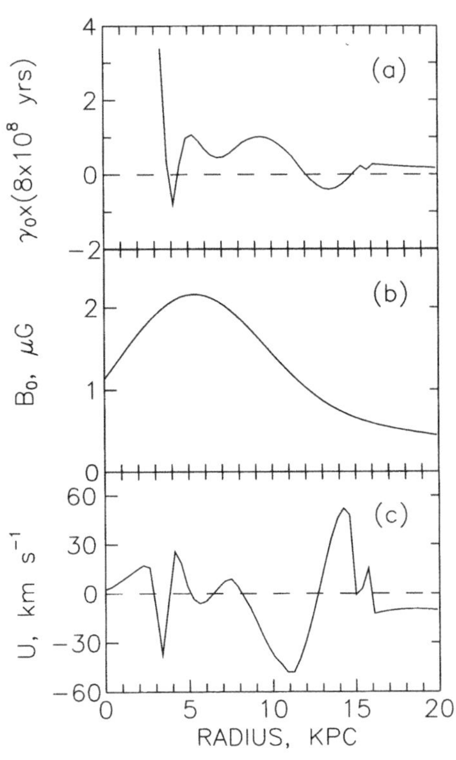

Fig. 1. Radial profiles, for the Milky Way, of (**a**) the local growth rate, (**b**) the equipartition magnetic field from (6), and (**c**) the function $U(r)$ defined in (4) Poezd et al. (1993). The model is based on the CO rotation curve of Clemens (1985) and gas density distribution of Gordon and Burton (1976), and the disc scale height $h(r) = 150\,\mathrm{pc}[1 + (r/4\,\mathrm{kpc})^2]^{1/2}$. The radius of the solar orbit was adopted to be 10 kpc

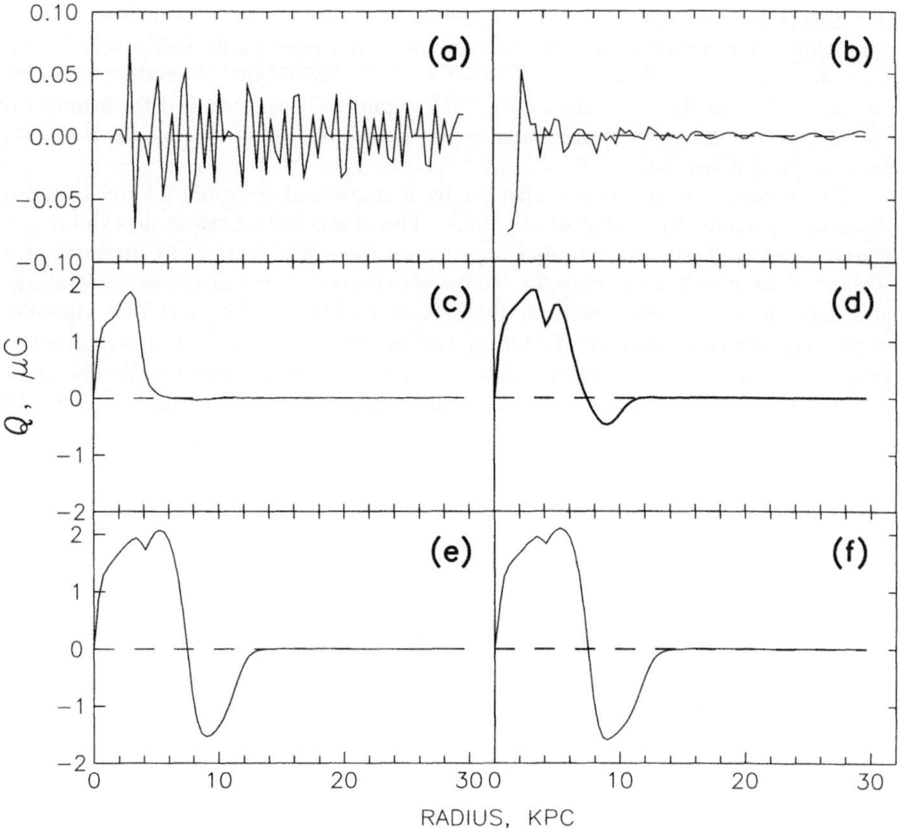

Fig. 2. Time evolution of the signed magnetic field strength in the Milky Way, according to Poezd et al. (1993), at the following times: (**a**) $t = 0$, the chaotic seed magnetic field; (**b**) 5.5×10^8 yr, (**c**) 2.2×10^9 yr, (**d**) 5.3×10^9 yr, (**e**) 8.1×10^9 yr, and (**f**) 9.6×10^9 yr. Parameters of the Milky Way model are as in Fig. 1

initial condition, is illustrated in Fig. 2. This evolution ends with just one reversal in the inner Galaxy.

The conclusion is that the regular magnetic field in the Milky Way can possess a number of reversals at rather well defined positions, but their occurrence depends on the unknown details of the initial magnetic field; as a result, the exact number of reversals is difficult to predict.

A similar nonlinear dynamo model was developed for M 31 (Poezd et al., 1993), where $U(r)$ has just one zero at $r \simeq 10$ kpc, i.e., within the synchrotron ring where we are confident that no reversals occur. The absence of the reversal can be explained by the lower growth rate of magnetic fields in M 31, so that the sign-constant (leading) mode had had enough time to become dominant before nonlinear effects have become important.

4 Localized Reversals

The above models rely on the interpretation of reversals as a global phenomenon, i.e., an *assumption* that they extend over the whole Galaxy. This is, however, not the only possibility. Since our knowledge of magnetic field of the Milky Way is limited to a relatively narrow neighbourhood of the Sun (of a size 3–5 kpc in radius and, say, 10 kpc in azimuth), it cannot be excluded that the reversed magnetic field is restricted to this neighbourhood (Shukurov, 2000). This possibility is corroborated by the magnetic structure in the disc of the galaxy M 51 (Berkhuijsen et al., 1997). As shown in Fig. 3, the regular magnetic field in the disc of M 51 is reversed in a region about 3 by 8 kpc in size elongated along the azimuth. The reversal occurs in the range of galactocentric radius 3–6 kpc centred on the corotation radius of the spiral pattern, and extends in azimuth from 280° to 20°. We suggest that the Sun can be located within a similar region with reversed magnetic field. We note that the Sun is located not far from the corotation radius of the Milky Way.

A dynamo model that clarifies the origin of such a localized region with reversed large-scale magnetic field was developed by Bykov et al. (1997) who solved numerically an equation analogous to (3), but written for nonaxisymmetric magnetic field, $Q(r,\phi)$:

$$\frac{\partial Q}{\partial t} + \Omega\frac{\partial Q}{\partial \phi} = \gamma_0 Q\left(1-\frac{Q^2}{B_0^2}\right) + \lambda^2\left[\frac{\partial}{\partial r}\left(\frac{1}{r}\frac{\partial}{\partial r}rQ\right) + \frac{1}{r^2}\frac{\partial^2 Q}{\partial \phi^2}\right] , \qquad (8)$$

where ϕ is the galactocentric azimuth and the other notation is as in (3). Now B_0 is modulated by a two-armed, logarithmic spiral pattern with the pitch angle of $p_s = -15°$. Since B_0 appears only in the nonlinear term of (8), the spiral pattern becomes important only at the nonlinear stage of magnetic field evolution in this model. Equation (8) was solved for $0 < r < 20\,\text{kpc}$, with boundary conditions $Q(0) = Q(20\,\text{kpc}) = 0$. The initial conditions represented a superposition of an axisymmetric and bisymmetric magnetic fields. In an axisymmetric disc (i.e., with the spiral modulation of B_0 neglected), the bisymmetric field rapidly decays. However, the spiral pattern can trap a bisymmetric magnetic field and preserve it near the corotation radius for a time exceeding the galactic lifetime. The radial extent of the region with reversed magnetic field is controlled by the balance of the local dynamo action and advection by the galactic differential rotation, and is estimated as

$$\delta r \simeq \frac{r_c}{|\sin p_s|^{1/2}}\left(\frac{v}{3V_0}\,\frac{l^2}{hr_c}\right)^{1/4} , \qquad (9)$$

where r_c is the corotation radius, V_0 is the rotational velocity, a constant for a flat rotation curve. For typical values of parameters, $\delta r \simeq 0.2r_c$, i.e., the trapped bisymmetric field can extend over a few kiloparsecs along radius. Equation (9) indicates that the following conditions are favourable for such a magnetic configuration to persist: smaller pitch angle of the spiral arms p_s, thinner gas disc (smaller h), weaker rotational shear (smaller V_0), and also a stronger spiral pattern. The region with reversed magnetic field would be broader in radius if the rotation curve were rising (rather than flat) near the corotation radius.

The possibility that the global magnetic structure of the Milky Way is similar to that shown in Fig. 3 has to be verified observationally. This would require a careful

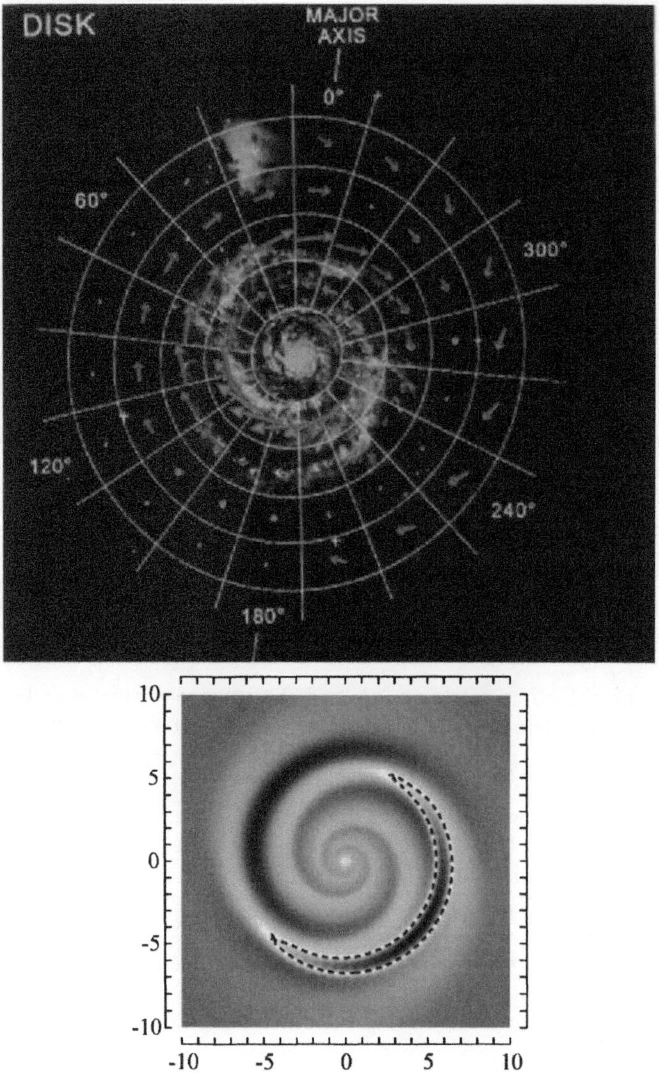

Fig. 3. *Upper panel:* The global magnetic structure in the disc of the galaxy M 51. Arrows represent the direction and strength of the regular magnetic field on a polar grid shown superimposed on an optical image of the galaxy (Berkhuijsen et al., 1997). The grid radii are 3, 6, 9, 12 and 15 kpc. The length of the arrows is proportional to $\mathcal{R} = Bn_eh$ in the inner two rings and to $3\mathcal{R}$ in the two outer rings. *Lower panel:* Magnetic field strength $|Q|$ from the dynamo model for the disc of M 51 (Bykov et al., 1997) is shown with shades of grey (darker shade means stronger field). This structure represents a bisymmetric magnetic field trapped by the spiral pattern near the corotation radius. Magnetic field is reversed within the zero-level contour shown dashed; scale is given in kpc. The magnetic structure rotates rigidly in the anticlockwise direction together with the spiral pattern visible in the shades of grey

study of Faraday rotation measures in all directions along the Galactic equator, so that the position of a magnetic field reversal(s) can be established confidently in all directions.

5 Magnetic Arms

The classical picture of the interaction of the large-scale magnetic field with galactic spiral arms was proposed by Roberts and Yuan (1970). Their two-dimensional model with magnetic field frozen into the interstellar gas predicted an enhancement of magnetic field within the arms and alignment of magnetic field with the arms, both resulting from gas compression (see Sect 3.2). This can be described as a passive behaviour of magnetic field, where it does not affect gas dynamics significantly, but rather responds to variations in gas density and velocity. However, nonthermal pressure components (including magnetic pressure) have been shown to lead, in three dimensions, to a more complicated picture with vertical motions of order $20\,\mathrm{km\,s^{-1}}$ and height-dependent displacements between maxima in gas (and magnetic field) and stellar densities (Martos and Cox, 1998; Gómez and Cox, 2002). It is not surprising that an active magnetic field produced by dynamo action is capable of an even more complicated behaviour.

The influence of the galactic spiral pattern on the dynamo can enhance the generation of nonaxisymmetric magnetic fields via parametric resonance (where with intrinsic oscillation frequency of the dynamo field is a multiple of the frequency of periodic variation in the dynamo control parameters produced by the travelling spiral pattern). This aspect of the global interaction between magnetic fields and the spiral pattern has been reviewed in Beck et al. (1996), and we refer the reader to that paper for details and references. Another, local aspect of the interaction is the position of gas ridges relative to those in the regular and total magnetic field. The fine structure of magnetized spiral arms has become accessible to observations only recently (Fletcher et al., 2005), and detailed models have not yet been developed.

Here we discuss in more detail the so-called magnetic arms, first observed in the galaxy IC 342 (Krause, 1993) and later identified as an unusual physical phenomenon in the galaxy NGC 6946 (Beck and Hoernes, 1996). We show, in the left-hand panel of Fig. 4, a map of polarized radio emission from NGC 6946 (a tracer of the large-scale magnetic field strength) superimposed on the galaxy's image in the Hα spectral line (a tracer of ionized gas). It is evident that the large-scale magnetic field is stronger *between* the gaseous spiral arms of this galaxy, i.e., where the gas density (both total and ionized) is lower. This behaviour is just opposite to what is expected of a frozen-in magnetic field that scales with a power of gas density. Spiral arm branches that may be of similar nature have been observed in NGC 2997 (Han et al., 1999). The spiral structures in gas and magnetic field in M 51 show a complicated, partially interlaced structure (Fletcher et al., 2005). It appears that the phenomenon of magnetic arms can be of general significance and some of its aspects can be common among spiral galaxies in general.

NGC 6946 remains the best studied case of magnetic arms. A quantitative morphological analysis of the spiral patterns visible in seven images in various wavelength ranges from the infrared to the radio was performed using wavelet techniques (Frick et al., 2000, 2001). Five arms and major arm segments have

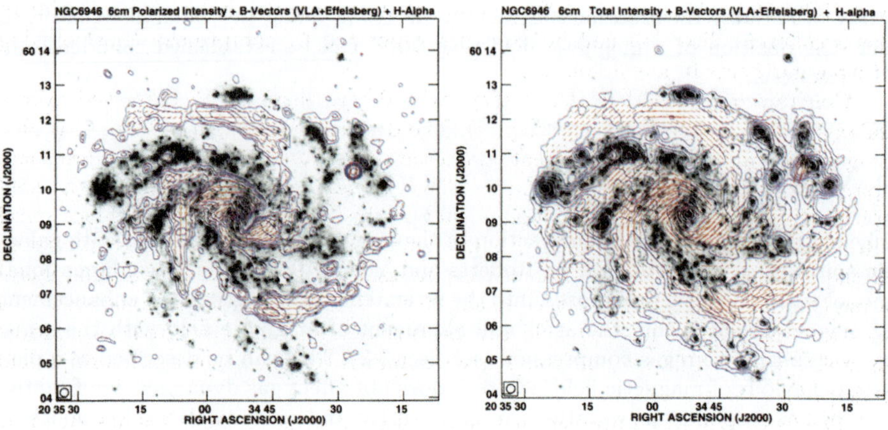

Fig. 4. *Left panel:* Magnetic arms in the galaxy NGC 6946: polarized intensity at the wavelength $\lambda = 6\,\mathrm{cm}$ (blue contours), a tracer of the large-scale magnetic field, superimposed on the galactic image in the Hα spectral line of ionized hydrogen (grey scale). Red dashes indicate the orientation of the B-vector of the polarized emission (parallel to the direction of intrinsic magnetic field if Faraday rotation is negligible), with length proportional to the fractional polarization. The spiral arms visualized by Hα are the sites where gas density is maximum. The large-scale magnetic field is evidently stronger between the arms where gas density is lower. *Right panel:* As in the left panel, but for the total synchrotron intensity, a tracer of the total magnetic field comprising both the regular and turbulent parts. The total field is enhanced in the gaseous arms. Given that the large-scale field concentrates between the arms, this means that the turbulent field is significantly stronger in the arms, a distribution very different from that of the large-scale field. The size of the beam in the radio maps is shown in the bottom left of each frame (see Chap. 3)

been identified, best visible in red light (emitted by the old stellar population) and polarized radio emissions at wavelengths 3.5 and 6.2 cm (tracers of the large-scale magnetic field). The stellar and magnetic arms are interlaced even in very fine detail, so their physical connection is evident. Each major optical arm branch has a magnetic counterpart. The arms can be reasonably approximated by logarithmic spirals, with the optical and magnetic counterparts having similar pitch angles; this implies that the phase shift between the stellar and magnetic spiral patterns is roughly independent of galactocentric radius. Total radio intensity, and H I and Hα line emissions exhibit more patchy and disordered distributions, although most features found in polarized emission and red light can also be found in the neutral and ionized hydrogen maps.

The phenomenon of magnetic arms confirms in a spectacular manner that the large-scale magnetic field is not frozen into the interstellar gas, and therefore cannot be primordial. If so, mean-field dynamo theory appears to be an appropriate framework to address the origin of magnetic spiral arms. The only alternative to the dynamo theory proposed up to date are slow magnetohydrodynamic density waves discussed in Sect. 5.2.

5.1 The Effects of Spiral Arms on the Galactic Dynamo

Two types of dynamo effects have been considered in relation to the phenomenon of magnetic arms (Moss, 1998; Shukurov, 1998; Rohde et al., 1999), discussed here and in Sect. 5.2. It was argued that dynamo number can be smaller within the gaseous spiral arms (e.g., because turbulence can be stronger there, thus providing larger turbulent magnetic diffusivity), resulting in a weaker magnetic field. In order to clarify and illustrate these arguments, consider a simple model of dynamo nonlinearity (known as α-quenching) where exponential growth of magnetic field $\boldsymbol{B}$ is saturated through the suppression of the α-effect, so that the effective dynamo number D_B decreases with B as

$$D_B = \frac{D}{1 + B^2/B_0^2} ,$$

where D is given in (5) and B_0, in (6) or (7). Replacing D by D_B in the expression for the field growth rate γ_0 given in (5), we obtain the following estimate of magnetic field strength in the steady-state, $\gamma_0 = 0$:

$$B \simeq B_0 \sqrt{\frac{D}{D_{\rm cr}} - 1} .$$

where $D_{\rm cr} \simeq -\pi^3/4 \approx -8$ is the critical dynamo number obtained from $\gamma_0 = 0$ in (5) (for $B = 0$), and the estimate for B is applicable if $D \approx D_{\rm cr}$ and $D/D_{\rm cr} > 1$. Within this simple framework, the ratio of the steady-state strengths of the regular magnetic field in the gaseous spiral arms and between them is given by (Shukurov, 1998)

$$\frac{B_{\rm a}}{B_{\rm i}} \simeq \frac{B_{0\rm a}}{B_{0\rm i}} \left(\frac{D_{\rm a}/D_{\rm cr} - 1}{D_{\rm i}/D_{\rm cr} - 1} \right)^{1/2} , \tag{10}$$

where subscripts 'a' and 'i' refer to gaseous arms and interarm regions.

It seems plausible that the turbulent kinetic energy density is larger in the arms (because gas density and star formation intensity are larger in the arms), and so $B_{0\rm a} > B_{0\rm i}$. Therefore, the regular magnetic field can be stronger between the arms, $B_{\rm a} < B_{\rm i}$, only because of the term in brackets in (10). Equation (5) shows that the local dynamo number depends on the scale height of the gas h, turbulent speed v, angular velocity Ω and the local shear rate G which may include not only shear due to differential rotation but also that arising from streaming velocities associated with the galactic spiral arms.

Therefore, we need to know how all these variables are affected by the spiral arms in order to understand the nature of magnetic arms. The available observational and theoretical knowledge of the effects of the spiral arms on the interstellar medium is yet insufficient for any firm conclusions to be made; relevant discussion can be found in Shukurov (1998) and Shukurov and Sokoloff (1998). To illustrate the nature of the problem, consider the scale height of gas disc, h. In hydrostatic equilibrium, h can be estimated as

$$h \simeq \frac{v^2 + V_A^2}{g} ,$$

where v is the turbulent speed, $V_A = B_{tot}/(4\pi\rho)^{1/2}$ is the Alfvén speed based on the total magnetic field $B_{tot} = (B^2 + \sigma_B^2)^{1/2}$ and g is the acceleration due to gravity. It might seem plausible that the disc scale height is larger in the arms because both v and V_A are larger there. An arm-interarm contrast of a factor of two in v and V_A would result in a factor of four contrast in h. Such a variation in the disc scale height seems to be unrealistically large. It can be argued (Shukurov, 1998; Shukurov and Sokoloff, 1998) that the contrast in h is reduced away from the corotation radius as the passage time of a spiral arm becomes shorter than the sound crossing time over the disc scale height. The above estimate of h can be oversimplified also because of the multi-phase nature of the interstellar medium where the filling factor of the hot gas can be significantly different within the arms and between them, thus reducing the arm-interarm contrast in total pressure in the interstellar gas, and hence in h. We discuss relevant simulations of the multi-phase interstellar medium in Sect. 5.3, which lead to an opposite conclusion that h can be slightly larger between the arms. Altogether, it seems to be reasonable to assume that h is not affected much by the spiral arms.

Thus, it is reasonable to assume that the ratio h/v is smaller in the gaseous arms, mainly because the turbulent velocity v is larger there, and then it does not seem implausible that the dynamo number can be reduced within the arms, with the arm-interarm contrast estimated in Shukurov (1998) and Shukurov and Sokoloff (1998) as

$$\frac{D_a}{D_i} \simeq \frac{1}{4} .$$

Then (10) shows that the steady-state magnetic field is stronger between the gaseous arms provided (Shukurov, 1998)

$$|D_a| < \frac{1 - B_{0i}^2/B_{0a}^2}{1 - \rho_i/\rho_a} \simeq \frac{5}{4} D_{cr} \qquad (11)$$

for $B_{0a}/B_{0i} \simeq 4$ and $\rho_a/\rho_i \simeq 4$. This indicates that magnetic arms can occur between the gaseous arms in galaxies with weak dynamos, i.e., where $|D|$ small enough to satisfy the inequality (11). A mild enhancement of the turbulent velocity suppresses significantly the dynamo action in the arms of such galaxies. In contrast, galaxies with a strong dynamo, where (11) is not satisfied, must have the strongest large-scale magnetic field in the gaseous arms. Numerical simulations of non-linear mean-field dynamos in a disc with v enhanced in the arms (Rohde and Elstner, 1997) support this explanation of interlaced magnetic arms, and confirm (10) by showing that B_a/B_i indeed decreases when the dynamo number decreases (see Fig. 4 in Rohde and Elstner, 1997).

The suppression of the dynamo number within the gaseous arms can be due to several reasons. The definition of the dynamo number from which the expression for D in (5) has been obtained is

$$D = \frac{\alpha G h^3}{\beta^2} ,$$

where $\alpha = -\frac{1}{3}\tau\langle \boldsymbol{v} \cdot \nabla \times \boldsymbol{v}\rangle$ is the so-called α-coefficient of the mean-field dynamo theory, τ is the correlation time of the random velocity field $\boldsymbol{v}$, and $\beta = \frac{1}{3}\tau\langle \boldsymbol{v}^2\rangle$

is the turbulent magnetic diffusivity. Different authors suggest different causes for the arm-interarm variation in D. A possible set of relevant estimates is presented above, following Shukurov (1998), and Shukurov and Sokoloff (1998) continued in Sect. 5.3. Rohde et al. (1999) maintain that there is no clear observational evidence for the modulation of the turbulent intensity by the spiral pattern, and therefore assume similar modulation for the turbulent correlation time. These authors assume that τ is larger within the gaseous arms than between them (unfortunately, such an assumption can hardly be verified as the correlation time is not an observable quantity; on the contrary, it is perhaps more plausible that τ is shorter in the arms because of the higher supernova rate), again resulting in a reduced D in the gaseous arms. The idea discussed in this section only relies on the appropriate modulation of the dynamo number by the spiral pattern, whatever is the eventual cause of this modulation.

The effects discussed in this section are most efficient near the corotation radius where the spiral arms do not move with respect to the gas, and so the dynamo has enough time to produce stronger large-scale magnetic field between the gaseous arms. Away from the corotation radius, the passage time of the spiral arms through a volume element can become shorter than the dynamo regeneration time $\gamma_0^{-1} = (10^8\text{–}10^9)$ yr and the azimuthal modulation of the large-scale magnetic field is averaged out. However, the interlaced magnetic and gaseous arms are expected to occur in galaxies with weak dynamos, i.e. with weak differential rotation, where the effects of azimuthal advection are minimized. Nevertheless, magnetic arms in NGC 6946 extend over a radial range broad enough for the effects of differential rotation to be potentially important. Numerical studies of the galactic mean-field dynamo model with the rotation curve of NGC 6946 and dynamo number reduced in the gaseous arms confirm that interlaced gaseous and magnetic arms persist over a broad radial range (Rohde et al., 1999). Hence, it appears that dynamo action in NGC 6946 is strong enough to balance the shearing of magnetic arms by differential rotation. Nevertheless, the situation is not completely satisfactory and we discuss in the next section an alternative explanation of magnetic arms based on travelling wave phenomena.

5.2 Dynamo Waves and Magnetohydrodynamic Density Waves

Unlike the basic axisymmetric magnetic mode, the nonaxisymmetric dynamo modes in a thin disc are oscillatory (Ruzmaikin et al., 1988), i.e., they represent dynamo waves propagating in the azimuthal direction. The spiral pattern also travels in the azimuthal direction, and so periodically modulates the dynamo parameters as described above. If this modulation is in resonance with the dynamo wave itself, the spiral pattern can facilitate the generation of this dynamo mode (Chiba and Tosa, 1990). The effect is weaker than it was first expected to be (Beck et al., 1996), but still can contribute to the support of magnetic arms (Moss, 1998). For a two-armed spiral pattern in both gas density and magnetic field strength, the dynamo mode with the azimuthal wave number $m = 1$ is to be in the resonance (then $\boldsymbol{B}$ has a bisymmetric pattern, but the field strength $|\boldsymbol{B}|$ has a two-armed structure). Numerical simulations indicate that resonance effects can indeed maintain the magnetic $m = 1$ mode interlaced with the gaseous arms if the turbulent magnetic diffusivity is enhanced within the gaseous arms (Moss, 1998) – an assumption consistent with reduced dynamo number in the arms. The resonance occurs if the oscillation frequency of the magnetic mode in the inertial frame ω is close to $2\Omega_\mathrm{p}$, where Ω_p is the angular velocity of the spiral pattern. However, the spiral pattern of NGC 6946, where the number of both gaseous and magnetic arms is different at different radii, would be difficult to explain since it seems implausible that several dynamo modes can be in resonance simultaneously (Moss, 1998). We also note that it not clear whether or not the magnetic field in the magnetic arms of NGC 6946 is consistent with the $m = 1$ symmetry.

Another theory advanced to explain magnetic arms, in NGC 6946 in particular, interprets them as the slow magnetohydrodynamic density waves in the self-gravitating galactic disc (Fan and Lou, 1996; Lou et al., 1999) (see also Lou and Fan, 2002, 2003 and references therein). This theory generalizes the density wave theory, devised to explain the galactic spiral structure, by including the large-scale magnetic field. The perturbations in gas surface density and magnetic field in the slow mode have a significant phase shift, and therefore its magnetic field is maximum away from gas density maxima, as in magnetic arms. This could be an attractive model for magnetic arms, but it appears to encounter significant difficulties. Its early versions could explain the existence of magnetic arms only in a rigidly rotating part of the galaxy, but theory has been later extended to the case of a flat rotation curve (Lou and Fan, 2002). Another limitation is that all models of magnetohydrodynamic density waves assume that the large-scale magnetic field is purely azimuthal and has a unique (and unrealistic) radial profile $B_\phi \propto r^{-1}$ or $r^{-1/2}$. Yet another difficulty is that the ratio of the amplitude of magnetic field in the magnetic arms to the mean magnetic field at a given radius, predicted by this theory, scales with galactocentric radius as $r^{-1/2}$; the amplitude of the stellar spiral arms has the same scaling (Lou and Fan, 2002). Arm strengths in magnetic field and stellar surface density in NGC 6946 have been estimated by Frick et al. (2000). Their results indicate that the mean relative intensity of magnetic spiral arms remains rather constant with galactocentric radius at a level of 0.3–0.6. On the contrary, the relative strength of the stellar arms systematically grows with radius from very small values in the inner galaxy to 0.3–0.7 at $r = 5$–$6\,\mathrm{kpc}$, and then decreases to remain at a level of 0.1–0.3 out to $r = 12\,\mathrm{kpc}$. The distinct magnitudes and radial trends in the strengths of magnetic and stellar arms in NGC 6946 do not seem to support the idea that the magnetic arms are due to MHD density waves.

However, the most important difficulty of the density wave theory of magnetic arms is of a more fundamental nature. All the existing models of magnetohydrodynamic density waves devised to explain magnetic arms are two-dimensional, with the galactic disc assumed to be infinitely thin and the perturbed magnetic field to be strictly horizontal (Lou and Fan, 2003; Lou and Zou, 2004 and references therein). A similar approximation is perfectly acceptable in theory of hydrodynamic density waves but becomes inadequate when magnetic fields are included (M. Tagger, private communication). The two-dimensional density wave models exclude the Parker instability (or magnetic buoyancy) from the analysis since this effect essentially involves vertical magnetic fields. As shown by Foglizzo and Tagger (1994, 1995), the slow branch of magnetohydrodynamic waves becomes unstable in three dimensions and transforms into a non-propagating Parker mode. This implies that the nature of the solutions applied to explain magnetic arms in this theory changes fundamentally in three dimensions; therefore, the application of this theory to magnetic arms observed in spiral galaxies is questionable.

To summarize, the nature of magnetic arms is still unclear. Galactic dynamo theory does provide mechanisms to maintain stronger large-scale magnetic field between the gaseous arms, but the development of detailed prognostic models is hampered by our insufficient knowledge of the effects of the spiral pattern on the global parameters of the interstellar medium.

5.3 Numerical Simulations of the Multi-phase Interstellar Medium

Observational evidence for the arm-interarm contrast in various parameters of the interstellar gas is still fragmentary and incomplete because of the relatively low resolution and sensitivity of the observations. However, recent numerical models of interstellar medium have become realistic enough as to shed some light on this problem (Vázquez-Semadeni et al., 2000).

The effects of the mean gas density and magnetic field on the overall parameters of the interstellar gas were recently studied with the aim to clarify the arm-interarm variation in the overall parameters of interstellar medium (Shukurov et al., 2004). These simulations are based on three-dimensional, non-ideal equations of magnetohydrodynamics with rotation, density stratification in the galactic gravity field, heat sources due to supernova explosions and UV heating, and radiative cooling (see Korpi et al., 1999a,b for details). The simulations were performed in a relatively small Cartesian box with the horizontal and vertical (z) dimensions of $0.25 \times 0.25 \times 1\,\mathrm{kpc}$ (with the midplane in the centre), modest spatial resolution of about 4 pc, and closed boundary conditions in z. The model reproduces the multi-phase structure of the interstellar medium and its stratification reasonably well. Results have been obtained for three values of the gas midplane density, assumed to model conditions within spiral arms, between them, and on average in the Solar vicinity of the Galactic disc. The three models with low, intermediate and high density are referred to as Interarm, Average and Arm. Results of the simulations are presented in Table 1.

An unexpected result of these simulations is that the density scale height is significantly larger in the Interarm model, although both thermal and turbulent pressures are a factor of about 3 larger in the Arm model. The reason for this is that the filling factor of the hot gas, together with the mean gas temperature, is significantly higher in the Interarm case, even though the SN rate is lower. An apparent reason is that the cooling rate has a stronger net dependence on gas density than the SN energy injection rate.

This conclusion is opposite to what was expected from the qualitative estimates of Sect. 5.1. An immediate implication of these simulations is that the arm-interarm contrast in gross parameters of the interstellar gas can be sensitive to quite fine details of the gas dynamics, multi-phase structure, and energy balance.

Another surprising feature of the results presented in Table 1 is that the filling factor of the hot gas is lower than expected by a factor of 2–3. This can be attributed to the geometry of the magnetic field in our models; it is uniform initially, and therefore effective in confining expanding bubbles of hot gas. The initial mid-plane field strength, $6\,\mu\mathrm{G}$, is close to that of the *total* field in the Solar vicinity, but the field

Table 1. Three models of interstellar medium driven by supernova explosions, each with a different initial mid-plane gas density, ρ_0, devised to reproduce physical conditions in the interarm regions, on average in the gas layer and within the gaseous arms (Shukurov et al., 2004). The mean temperature, pressures, root mean square velocity and filling factor are all calculated within $|z| < 0.2\,\mathrm{kpc}$, where z is the vertical distance, with the disc midplane at $z = 0$. The filling factor is for the hot gas, of a temperature $T > 10^5\,\mathrm{K}$; supernova rate and the spatial distribution of the supernovae are also obtained as the result of these simulations, and they are given in the last two lines. All the models include an imposed azimuthal magnetic field of a strength $6\,\mu\mathrm{G}$.

	Unit	Interarm	Average	Arm
Initial midplane density, ρ_0	$10^{-24}\,\mathrm{g\,cm^{-3}}$	0.7	1.4	2.9
Density Gaussian scale height	kpc	0.23	0.20	0.16
Mean temperature	$10^4\,\mathrm{K}$	35	7.8	3.4
Mean thermal pressure	$10^{-14}\,\mathrm{dyn\,cm^{-2}}$	42	68	120
Rms vertical velocity	$\mathrm{km\,s^{-1}}$	23	20	20
Mean turbulent pressure	$10^{-14}\,\mathrm{dyn\,cm^{-2}}$	39	63	110
Hot gas filling factor		0.12	0.07	0.04
SN II rate	$\mathrm{kpc^{-2}\,Myr^{-1}}$	11	38	111
SN II Gaussian scale height	kpc	0.30	0.16	0.14

is implausibly well ordered. A more realistic simulation would initialize the model with a ratio of turbulent to ordered magnetic energies of about 3. The dependence of the results on the strength of the initially uniform magnetic field is illustrated in Table 2. The filling factor of the hot gas is sensitive to the field strength and increases to 0.2 as the field becomes weaker. The density scale height marginally increases with magnetic field strength, but this effect is much less pronounced than the suppression of the hot phase; magnetic field strongly suppresses turbulence in the hot gas. Thus, magnetic field can affect the disc-halo connection and the global structure of the ISM in crucial, diverse and unexpected ways. This aspect of the ISM dynamics has not yet been fully explored. Incidentally, this implied that the problem of magnetic arms is intrinsically nonlinear, with the large-scale

Table 2. The effect of magnetic field on the multi-phase interstellar medium illustrated with three runs with varying initial magnetic field, B. All variables are as defined in Table 1. All runs have $\rho_0 = 0.7 \times 10^{-24}\,\mathrm{g\,cm^{-3}}$ (the Interarm model)

Initial Magnetic Field Strength, B	μG	0	6
Density scale height	kpc	0.20	0.23
Mean thermal pressure	$10^{-14}\,\mathrm{dyn\,cm^{-2}}$	50	42
Rms vertical velocity	$\mathrm{km\,s^{-1}}$	43	23
Mean turbulent pressure	$10^{-14}\,\mathrm{dyn\,cm^{-2}}$	54	39
Hot gas filling factor		0.19	0.12

magnetic field responding to the interstellar gas variations between gaseous arms and interarm regions, which in turn depend on the magnetic field itself.

6 Conclusions

The galactic dynamo theory has been impressively successful in explaining the gross features of galactic magnetic fields at scales exceeding a few kiloparsecs. It can be expected that the current controversy regarding the nonlinear behaviour of mean-field dynamos (Chap. 9 and references therein) will be resolved without affecting its main conclusions. The reason for this expectation is that the large-scale magnetic fields generated by the mean-field dynamo depend remarkably weakly on the detailed properties of the dynamo system (such as the poorly known α-coefficient) (Ruzmaikin et al., 1988). As argued by F. Krause and Rädler (1980), the *form* of the mean-field dynamo equations is generic: any system capable of maintaining a large-scale magnetic field independently of external electric currents must be governed by equations similar to the classical mean-field dynamo equations. The specific physical nature of the system only affects the coefficients of this system, e.g., the α-coefficient. Then, given that the form of the solutions is only weakly sensitive to the form of α, we expect the results of galactic dynamo theory to remain robust.

Improvements in the quality of radio polarization observations have revealed detailed properties of interstellar magnetic fields at scales intermediate between the global galactic scales of 10 kpc and the turbulent scale of 0.1 kpc, which can be conveniently called mesoscales. As might be expected, the details often obscure the simple and symmetric overall structure prominent in observations with lower resolution or in smoothed data. One of the outstanding mesoscale magnetic features are magnetic arms whose understanding is still far from being complete and confident.

Systematic studies of galactic magnetic structures at intermediate scales can advance our understanding of the nature of the cosmic magnetism as strongly as similar studies of the global magnetic structures. As argued above, two types of mesoscale magnetic structures, magnetic reversals and magnetic arms, are compatible with galactic dynamo theory and confirm it to a certain extent. In the framework of the dynamo theory, magnetic reversals carry information about early stages of galactic evolution and/or interaction of galactic magnetic fields with the spiral pattern. In order to understand the nature of magnetic arms, we need a much better understanding of the effects of the spiral pattern on the interstellar medium.

References

Beck, R.: 2000, Phil. Trans. R. Soc. Lond. A358, 777

Beck, R.: 2001, Sp. Sci. Rev. 99, 243

Beck, R., Hoernes, P.: 1996, Nat 379, 47

Beck, R., Brandenburg, A., Moss, D., Shukurov, A., Sokoloff, D.: 1996, ARA&A 34, 155

Beck, R., Shukurov, A., Sokoloff, D., Wielebinski, R.: 2003, A&A 411, 99

Belyanin, M. P., Sokoloff, D. D., Shukurov, A. M.: 1994, Russ. J. Math. Phys. 2, 149
Berkhuijsen, E. M., Horellou, C., Krause, M., Neininger, N., Poezd, A. D., Shukurov, A., Sokoloff, D. D.: 1997, A&A 318, 700
Bloemen, H. (ed.): 1991, *The Interstellar Disc-Halo Connection in Galaxies*, IAU Symp. 144 (Dordrecht: Kluwer)
Brandenburg, A., Subramanian, K.: 2005, Phys. Rep., in press (astro-ph/0405052)
Brown, J. C., Taylor, A. R., Jackel, B. J.: 2003, ApJS 145, 213
Burn, B. J.: 1966, MNRAS 133, 67
Bykov, A., Popov, V., Shukurov, A., Sokoloff, D.: 1997, MNRAS 292, 1
Chiba, M., Tosa, M.: 1990, MNRAS 244, 714
Clemens, D. P.: 1985, ApJ 295, 422
Fan, Z., Lou, Y.-Q.: 1996, Nat 383, 800
Fletcher, A., Beck, R., Shukurov, A. et al: 2005, in preparation
Foglizzo, T., Tagger, M.: 1994, A&A 287, 297
Foglizzo, T., Tagger, M.: 1995, A&A 301, 293
Frick, P., Beck, R., Shukurov, A., Sokoloff, D., Ehle, M., Kamphuis, J.: 2000, MNRAS 318, 925
Frick, P., Beck, R., Berkhuijsen, E. M., Patrickeyev, I.: 2001, MNRAS 327, 1145
Frick, P., Stepanov, R., Shukurov, A., Sokoloff, D.: 2001, MNRAS 325, 649
Gómez, G. C., Cox, D. P.: 2002, ApJ 580, 235
Gordon, M. A., Burton, W. B.: 1976, ApJ 208, 346
Han, J. L.: 2004, in *The Magnetized Interstellar Medium*, ed. B. Uyanıker, W. Reich, R. Wielebinski (Katlenburg-Lindau: Copernicus), p. 3
Han, J. L., Beck, R., Ehle, M., Haynes, R. F., Wielebinski, R.: 1999, A&A 348, 405
Heiles, C.: 1996, in *Polarimetry of the Interstellar Medium*, ed. W. G. Roberge, D. C. B. Whittet, ASP Conf. Ser. 97, p. 457
Johnston-Hollitt, M., Hollitt, C. P., Ekers. R. D.: 2004, in *The Magnetized Interstellar Medium*, ed. B. Uyanıker, W. Reich, R. Wielebinski (Katlenburg-Lindau: Copernicus), p. 13
Korpi, M. J., Brandenburg, A., Shukurov, A., Tuominen, I., Nordlund, Å.: 1999a, ApJL 514, L99
Korpi, M. J., Brandenburg, A., Shukurov, A., Tuominen, I.: 1999b, A&A 350, 230
Krause, F., Rädler, K.-H.: 1980, *Mean-Field Electrodynamics and Dynamo Theory* (Berlin: Akademie-Verlag, and Oxford: Pergamon)
Krause, M.: 1993, in *The Cosmic Dynamo*, IAU Symp. 157, ed. F. Krause, K.-H. Rädler, G. Rüdiger (Dordrecht: Kluwer) p. 305
Krause, M., Beck, R., Hummel, E.: 1989, A&A 217, 17
Kulsrud, R. M.: 1999, ARA&A 37, 37
Lou, Y.-Q., Fan, Z.: 2002, MNRAS 329, L62
Lou, Y.-Q., Fan, Z.: 2003, MNRAS 341, 909
Lou, Y.-Q., Zou, Y.: 2004, MNRAS 350, 1220
Lou, Y.-Q., Han, J. L., Fan, Z.: 1999, MNRAS 308, L1
Martos, M. A., Cox, D. P.: 1998, ApJ 509, 703
Mitra, D., Wielebinski, R., Kramer, M., Jessner, A.: 2003, A&A 398, 993
Moffatt, H. K.: 1978, *Magnetic Field Generation in Electrically Conducting Fluids* (Cambridge: Cambridge Univ. Press)
Moss, D.: 1995, MNRAS 275, 191

Moss, D.: 1998, MNRAS 297, 860
Poezd, A., Shukurov, A., Sokoloff, D.: 1993, MNRAS 264, 285
Rafikov, R. R., Kulsrud, R. M.: 2000, MNRAS 314, 839
Rand, R. J., Kulkarni, S. R.: 1989, ApJ 343, 760
Roberts, W. W., Yuan, C.: 1970, ApJ 161, 877
Rohde, R., Elstner, D.: 1997, A&A 333, 27
Rohde, R., Beck, R., Elstner, D.: 1999, A&A 350, 423
Ruzmaikin, A. A., Sokoloff, D. D.: 1977a, Ap&SS 52, 365
Ruzmaikin, A. A., Sokoloff, D. D.: 1977b, A&A 58, 247
Ruzmaikin, A. A., Sokoloff, D. D., Shukurov, A. M.: 1985, A&A 148, 335
Ruzmaikin, A. A., Shukurov, A. M., Sokoloff, D. D.: 1988, *Magnetic Fields of Galaxies* (Dordrecht: Kluwer)
Seymour, P. A. H.: 1984, QJRAS 25, 293
Simard-Normandin, M., Kronberg, P. P.: 1980, ApJ 242, 74
Shukurov, A.: 1998, MNRAS 299, L21
Shukurov, A.: 2000, in *The Interstellar Medium in M31 and M33*, ed. E. M. Berkhuijsen, R. Beck, R. A. M. Walterbos (Aachen: Shaker), p. 191 (astro-ph/0012460)
Shukurov, A.: 2005, in *Mathematical Aspects of Natural Dynamos*, ed. E. Dormy, B. Desjardins (Paris: EDP Sciences), in press (astro-ph/0411739)
Shukurov, A., Sokoloff, D.: 1998, Studia Geoph. et Geod. 42, 391
Shukurov, A., Sarson, G. R., Nordlund, Å., Gudiksen, B., Brandenburg, A.: 2004, Ap&SS 289, 319
Sofue, Y., Fujimoto, M., Wielebinski, R.: 1986, ARA&A 24, 459
Sokoloff, D. D., Bykov, A. A., Shukurov, A., Berkhuijsen, E. M., Beck, R., Poezd, A. D.: 1998, MNRAS 299, 189; Erratum: MNRAS 303, 207
Stepanov, R. A., Brown, J. C., Shukurov, A., Taylor, A. R.: 2005, in preparation
Vázquez-Semadeni, E., Ostriker, E. C., Passot, T., Gammie, C. F., Stone, J. M.: 2000, in *Protostars and Planets IV*, ed. V. Mannings, A. P. Boss, S.S. Russell (Tucson: Univ. Arizona Press), p. 3
Wall, J. V., Jenkins, C. R.: 2003, *Practical Statistics for Astronomers* (Cambridge: Cambridge Univ. Press)

Magnetic Fields in Diffuse H I and Molecular Clouds

Carl Heiles[1] and Richard Crutcher[2]

[1] Astronomy Department, University of California, Berkeley, USA
cheiles@astron.berkeley.edu
[2] Astronomy Department, University of Illinois, Urbana, USA
crutcher@uiuc.edu

1 Introduction

The diffuse interstellar H I is the matrix within which many molecular clouds reside and the medium that soaks up the energy injected by sources such as supernovae and stellar winds. This energy stimulates turbulence in the H I, which cascades up the turbulent wavenumber spectrum. The spectral wavelengths extend all the way down to scales most easily quoted in Astronomical Units. H I and molecular clouds enjoy a synergistic relationship, with turbulent energy, angular momentum, magnetic fields, and matter flowing across the boundaries in both directions. The molecular clouds form stars, which in turn act as energy sources to round the circle and make star formation a feedback process.

Fortunately for us who study magnetic fields, the neutral medium isn't really neutral and, as a consequence, flux freezing applies. In diffuse H I the minimum free electron fraction is, at minimum, equal to that of heavy elements that have ionization potential less than that of H I ($\gtrsim 10^{-4}$) because even in the dark reaches of space there are plenty of starlight photons available to keep any such element ionized. As a crude approximation we can model a piece of the interstellar gas as a giant inductor, for which the timescale τ for decay of a current (and its associated magnetic field) is the inductance divided by the resistance; this, in turn, goes as $\tau \propto L^2/\eta$, where L is the length scale and η the resistivity. Even with the low fractional ionization, L dominates and timescales for decay are always long in diffuse H I. In dense molecular clouds starlight is excluded and the free electrons come from cosmic-ray ionization of H; the fractional ionization is small enough that slow leakage of frozen magnetic flux allows the clouds to gradually evolve.

With flux freezing, the magnetic field becomes one of the four most important forces on the diffuse gas. The others are gas pressure, cosmic-ray pressure, and gravity. Gravity dominates on the largest scales, e.g. by keeping the gas pulled down as part of the Galactic plane; it also dominates during star formation, of course. On all other scales the gas responds only to the three pressure forces. The gas and cosmic rays are connected by the field, so they form a coupled system. The field is a – perhaps *the* – major player.

One determines the field strength in the diffuse interstellar gas in several ways. Each method has its own idiosyncrasies and provides values that are biased either up or down. Beck et al. (2003) is *required reading* to understand these biases. Synchrotron emissivity provides a volume average of $\langle B^x \rangle^{1/x}$, where $1.9 \lesssim x \lesssim 3.9$ depending on whether one assumes the electron cosmic-ray spectrum or energy equipartition (Beck, 2001). Comparing pulsar rotation and dispersion measures provides a field strength in the diffuse Warm Ionized Medium (WIM). Zeeman splitting provides the field strength in the H I.

Combining these estimates gives a typical magnetic field strength $\sim 6 \pm 2\,\mu\mathrm{G}$ (Beck, 2001), which is equivalent to a gas pressure $\tilde{P} \equiv P/k \sim 10400\,\mathrm{cm}^{-3}\,\mathrm{K}$. This is about three times the typical ISM thermal gas pressure of $\sim 3000\,\mathrm{cm}^{-3}\,\mathrm{K}$ (Jenkins and Tripp, 2001; Wolfire et al., 2003), and is comparable to the other important interstellar energy densities, namely turbulence and cosmic rays. These pressures must add to provide hydrostatic support for the gas layer, estimated to be $P_{\mathrm{tot}} \approx 28000\,\mathrm{cm}^{-3}\,\mathrm{K}$ at $z = 0$ (Boulares and Cox, 1990). Clearly, thermal pressure is a minority player; turbulence, cosmic rays, and the magnetic field dominate. One cannot hope to understand the interstellar medium without understanding the role of the magnetic field. Moreover, the crucial star formation feedback process is regulated, or stimulated, or at least greatly affected, by the magnetic field.

Magnetism makes its effects very clear in supernova shocks. These shocks compress both the gas and the field. As the gas cools behind the shock, it does so at roughly constant pressure, so its density increases. Concomitantly, the field strength increases because of flux freezing. Magnetic pressure increases as B^2, so eventually the magnetic pressure prevents the gas from condensing further. This limits the compression of gas behind the shock and over the latter stages of its evolution the magnetic field greatly increases the shell thickness relative to the idealized non-magnetic case. Moreover, on the full scale of the shell the magnetic field acts as a retarding force, increasing the deceleration of the shell and reducing its final size (Tomisaka, 1990; Ferrière et al., 1991; Slavin and Cox, 1992). Also, the strong field can inhibit the production of worms (Heiles, 1984) and chimneys (Norman and Ikeuchi, 1989).

For the study and interpretation of magnetic fields, the size scale is paramount. At the largest scales within galaxies, the global scale, the issue is field generation and maintenance, and the underlying questions are 'Primordial field or dynamo?' and 'What kind of dynamo?'. These questions are addressed by size scales ranging down to spiral arms. At smaller sizes we have the field in individual interstellar diffuse structures, which are shaped by point energy injection and condensation onto molecular clouds. At yet smaller scales we have molecular clouds, especially those that contain protostellar cores. At the smallest scales we have regions where stars have formed.

This review concentrates on the magnetic field at intermediate and small size scales, i.e. diffuse H I structures and molecular clouds and cores. See Beck (2001) for discussion of magnetic fields on larger scales.

Our chosen size range is where energy input to the ISM occurs and where energy is transferred by turbulence to smaller scales and across cloud boundaries. There are three, and only three, established[1] tracers for the field at these scales: polarization from aligned dust grains, which both absorb starlight and emit in the far-infrared, linear polarization of spectral lines, and Zeeman splitting of spectral lines. We will briefly include starlight polarization in Sect. 2.1, concentrate on Zeeman splitting

[1] Use of the difference in line widths between neutral and ionized species to infer the angle between the line of sight and the magnetic field (Houde et al., 2002) and Faraday screens in dark-cloud envelopes (Wolleben and Reich, 2004) are possible additional techniques that have not yet been fully accepted.

of the 21-cm line in Sects. 4 and 5, and discuss magnetic fields in molecular clouds starting with Sect. 6.

One major focus of this review is the magnetic field in the diffuse H I. The H I resides in two thermal phases, the Cold Neutral Medium (CNM) and the Warm Neutral Medium (WNM), each containing roughly half of the total H I. Classically, we imagine these as points of stable isobaric thermodynamic equilibrium (Field, 1965; McKee and Ostriker, 1977), with the temperatures differing by about two orders of magnitude. The CNM does, in fact, reside in the classical stable thermal equilibrium. However, the WNM is buffeted by many agents on a range of timescales, so much so that at least 50% of the WNM has temperature smaller than 5000 K, meaning that it is *not* thermally stable (Heiles and Troland, 2003). The WNM, being of much higher temperature and lower density, occupies the lion's share of the interstellar volume, roughly half the volume in the Solar vicinity (Heiles, 2000b). H I Zeeman splitting measurements refer almost exclusively to the CNM: the line widths of the WNM are large, and when combined with H I angular structure the instrumental effects have so far prohibited reliable measurements.

The other major focus is the magnetic field in molecular clouds. The most important goal is to understand the role that magnetic fields play in the fundamental astrophysical process of star formation. One view is that self-gravitating clouds are supported against collapse by magnetic fields, with ambipolar diffusion reducing support in cores and hence driving star formation (Mouschovias and Ciolek, 1999). The other view is that clouds form and disperse by the operation of compressible turbulence (e.g., Elmegreen, 2000), with clumps sometimes becoming gravitationally bound and collapsing to form stars. The issue of which (if either) of these paradigms for the evolution of molecular clouds and the formation of stars is correct is currently unresolved. We describe the state of observations of magnetic fields in molecular clouds and how these data may be used to test predictions of the two star formation paradigms.

2 Measuring the Magnetic Field in Diffuse H I and Molecular Clouds

2.1 Polarization of Starlight by Magnetically Aligned Grains

Polarization of starlight holds the enviable position of being the means by which the interstellar magnetic field was discovered (see Davis and Greenstein, 1951) for references and the original theory of grain alignment). Their alignment mechanism involves charged, spinning interstellar grains whose angular momentum vector component parallel to the field is damped by paramagnetic relaxation. The theory evolved with the introduction of superthermal spins and internal damping from Barnett relaxation (Purcell, 1979; Purcell and Spitzer, 1971). The theory continues to evolve as more exotic effects are uncovered (see Lazarian, 2003) for a comprehensive review devoted exclusively to grain alignment; also see Draine (2003) and references quoted therein). In principle, the starlight polarization can be either parallel or perpendicular to $B_\perp$, the field on the plane of the sky. However, empirically the polarization is parallel to the field, as revealed by polarization in diffuse regions near the Galactic plane: $B_\perp$ is parallel to the plane as expected for the Galactic-wide field.

Starlight polarization is produced by aligned dust that selectively absorbs one direction of linear polarization more than the orthogonal one. This makes the fractional polarization proportional to the extinction – we can't have polarization without extinction! Commonly, maps represent starlight polarization with lines whose direction is that of the polarization and whose length is proportional to the fractional polarization. The eye notices the long lines, which emphasize high extinction; these stars tend to be more distant. This is normally not the kind of bias one wants. For example, if we are interested in the nearby field structure, it is better to make all lines the same length. Accordingly, in our Fig. 1, we de-emphasize distant or high-extinction stars by placing an upper limit on the length of the lines.

The fractional starlight polarization also increases as the field becomes perpendicular to the line of sight. The dependence is $(B_\perp/B_{\rm tot})^2$. From our discussion in Sect. 3.1, for randomly oriented fields this ratio has mean value 0.67 and median 0.87. With these high numbers, most of the regions have a high ratio, so in a statistical sample the fractional polarization is relatively weakly affected by the tilt of the magnetic field. Statistically, extinction is much more important in determining the fractional polarization.

Figure 1 shows the polarization of 8662 stars from the compilation of known catalogs (Heiles, 2000a). The orientation of each star's polarization is indicated by a short line whose length L in great-circle degrees is $L = [4 < 2P]^\circ$, where P is the percentage polarization; we cap L at 4° to reduce the eye's preference for distant stars and, also, so that the lines don't become unrecognizably long. The assembly of lines is like iron filings near a bar magnet and traces out the plane-of-the-sky field lines. Note that these lines aren't vectors, because they don't indicate direction; linear polarization is defined only modulo 180°, not 360°, so it only has an orientation.

Figure 1 shows the major large-scale features in the magnetic sky:

1. In the Galactic plane, the lines tend to be parallel to the plane, showing that the large-scale field lies in the plane. This is expected, if only from the effects of differential rotation and flux freezing.
2. Near $\ell = (80^\circ, 260^\circ)$ the lines lose this tendency. Heiles (1996a) used this observed effect to determine the direction and curvature of the local magnetic field: it points towards $\ell \sim 83^\circ \pm 4.1^\circ$ and has radius of curvature 8.8 ± 1.8 kpc.
3. Figure 1 shows several small areas where the density of measurements is so high as to obliterate the individual lines. These are regions of particular interest because of their dense clouds or star formation. We label Orion, Taurus, and Perseus, but several others also stand out. In these regions the dense clouds often look filamentary.

   The observed stellar polarizations sometimes exhibit good alignment with filamentary structures, but the sense of alignment is not always the same. Three particularly good examples are Pereyra and Magalhaes (2004) and Fig. 5 in Heyer et al. (1987), where the polarizations are strikingly perpendicular to the long axis of the filaments, and Plate IX in Vrba et al. (1976), where the polarizations are parallel. The proper interpretation of these completely orthogonal senses of alignment *probably* consists of the following:

   a) Interstellar 'filaments' are edge-on sheets.

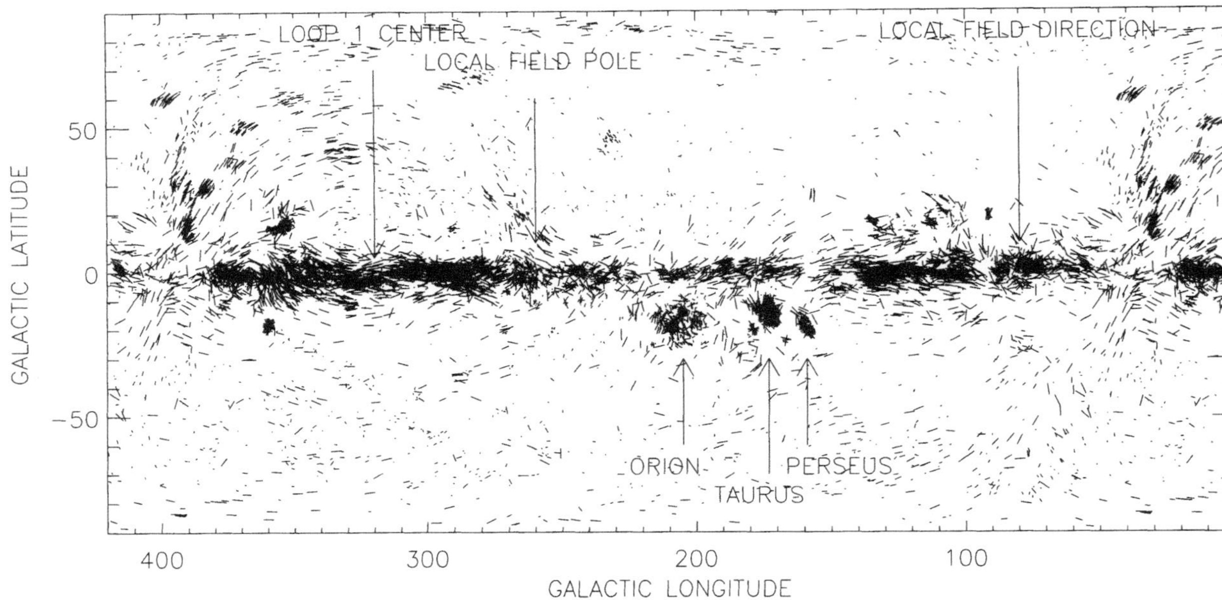

Fig. 1. Starlight polarization of 8662 stars. The orientation of each star's polarization is indicated by a short line whose length L in great-circle degrees is $L = [4 < 2P]^\circ$, where P is the percentage polarization; for L, we plot whichever of the two quantities is smaller

b) Molecular clouds are flattened triaxial ellipsoids, which are often flattened enough to be considered as slabs (Sect. 8.2 below).
c) Fat interstellar filaments are the projections of flattened ellipsoids at random angles onto the plane of the sky.
d) The apparent orientation of $B_\perp$ for such ellipsoids can adopt any position angle (call it Ψ) because of projection effects, as emphasized in the very important article by Basu (2000).
e) The only reliable way to determine the orientation of field lines with respect to the flattened ellipsoids is to compare the observed histogram Ψ for a large sample with model probability distributions for Ψ, such as Basu's. Not enough regions have been measured to accumulate sufficiently large-number statistics on Ψ. In particular, we caution that statements like 'the observed $B_\perp$ is perpendicular to the filament, i.e. perpendicular to the edge-on sheet' can be misleading when applied to a single example and can only have validity when applied to a good statistical sample.

4. Figure 1 shows the prominent distortion of the local field produced by Loop I (also known as the North Polar Spur). This distortion is also visible in the H I line and radio synchrotron continuum. It is the result of a superbubble produced by stellar winds and supernovae in the Sco/Cen association; the overall morphology of the H I, hot gas (from its X-ray emission), and magnetic field (from radio synchrotron emission) strikingly confirms the concept that the ISM is shaped by such explosions. The center of Loop I appears in different places for the radio continuum (near $(\ell, b) \sim (329°, 18°)$ (Berkhuijsen et al., 1971) and for the H I (near $(320°, 5°)$ (Heiles, 1998b). The causes for this difference are not currently understood.

   Note our discussion of the field distortion by superbubbles in Sect. 5.4. The case here, with Loop I, is clear-cut because the ambient field lies predominantly across the line of sight. Other geometries are less clear and more complicated.
5. There are other large-scale patterns in Fig. 1, which presumably trace other supernova shells or supershells. There is ample opportunity for further research here!

2.2 Polarization of Thermal Grain Emission

Starlight polarization occupies a high position, not only because of its historical importance but also because stars serve as distance markers. However, as with any tracer dependent on background sources, it is not very useful for mapping. Thermal radiation from dust is polarized, again because of the alignment of dust grains. We can look forward to the day when (1) enough stellar extinction measurements exist to determine the evolution of extinction with distance along arbitrary lines of sight, and (2) the mapping of IR emission from the diffuse interstellar gas starts in earnest. Unfortunately, (1) is in its infancy, except for particularly well defined clouds of high extinction, and regarding (2) no IR polarization data exist at all for diffuse regions.

In dense regions, however, far-infrared and millimeter wavelength observations of linearly polarized dust emission may be used to map the morphology of the magnetic field projected onto the plane of the sky, $B_\perp$ (Hildebrand, 1988). The position angle of maximum emission will be perpendicular to $B_\perp$. The mm-wavelengths sample the larger aligned grains and have the advantage that local star formation is not required because mm-wavelength emission occurs even with cold grains. These are particularly useful for places where stars have formed, because they heat the dust and provide strong emission. These regions are discussed later in this review. Other recent reviews which cover these aspects very well are Hildebrand et al. (2000), Hildebrand (2002), and Crutcher et al. (2003).

It is not possible to measure directly the strength of $B_\perp$ since fairly weak magnetic fields can align grains, so the degree of polarization is not a measure of field strength. However, in the early days of interstellar polarization studies, Chandrasekhar and Fermi (1953) suggested that analysis of the small-scale randomness of magnetic field lines could yield estimates of the field strengths. The method depends on the fact that turbulent motions will lead to irregular magnetic fields (since under interstellar conditions fields will be frozen into the matter). There will therefore be a perturbed or MHD-wave component to the field that should show up as an irregular scatter in polarization position angles relative to those that would be produced by a regular magnetic field. The stronger the regular field, the more it resists being irregularized by turbulence. They showed that the magnitude of the irregularity of field lines could yield the regular field strength in the plane of the sky:

$$B_\perp = Q\sqrt{4\pi\rho}\,\frac{\delta V}{\delta\phi} \approx 9.3\sqrt{n(H_2)}\,\frac{\Delta V}{\delta\phi}\;\mu G\,, \tag{1}$$

where $\rho = mn(H_2)$ is the gas density, δV is the velocity dispersion, $\delta\phi$ is the dispersion in polarization position angles in degrees, Q is a factor of order unity, $n(H_2)$ is the molecular hydrogen density in molecules cm^{-3}, and $\Delta V = \sqrt{8ln2}\,\delta V$ is the FWHM line width in km s^{-1}. Here we have used $Q = 0.5$, a calibration based on study of simulations of interstellar clouds by Ostriker et al. (2001), but see also Heitsch et al. (2001) and Padoan et al. (2001). These simulations found that this method could yield reliable results in molecular clouds if $\delta\phi < 25°$. One should note that while fluctuations in the field along the line of sight will be smoothed out by the polarization measurements, the calibration by the simulations referred to above include this in the Q factor. Heitsch et al. (2001) studied the effects of smoothing due to inadequate spatial resolution in the plane of the sky; although such smoothing will produce too large an estimate of $B_\perp$, the problem can be overcome so long as the region being studied, i.e. a molecular cloud or core, is adequately (a few resolution elements) resolved. The Chandrasekhar–Fermi method of estimating B is a statistical one that may be in error by ~ 2 for an individual cloud.

2.3 Spectral-line Linear Polarization

Linear polarization may also arise in radio-frequency spectral lines formed in the interstellar medium, even when Zeeman splitting is negligible. This Goldreich–Kylafis effect (Goldreich and Kylafis, 1981; Kylafis, 1983) may be used to probe magnetic field morphologies in molecular clouds. Heiles et al. (1993) provide a qualitative discussion of how the linear polarization arises. The direction of the polarization can be either parallel or perpendicular to the magnetic field, depending on the relationship between the line of sight, the direction of the magnetic field, and the direction of a velocity gradient that produces the anisotropic line optical depth that is required to produce linear polarization. Although the theory makes specific predictions for whether the field is parallel or perpendicular to the line polarization, in general the observations do not provide all of the necessary information. This ambiguity is unfortunate, but if structure in a cloud causes a flip by 90° in the polarization direction, it would easily be recognized and not confused with random magnetic fields. It therefore is a valuable tool in the measurement of magnetic field direction and in the degree of randomness of the field. As is the case for dust polarization, the Chandrasekhar–Fermi method may be applied to maps of spectral-line linear polarization to estimate field strengths.

2.4 Zeeman Splitting

Interstellar magnetic fields are very weak and in all cases except masers produce Zeeman splitting $\Delta\nu_Z$ that is much smaller than the line width $\delta\nu$, so we usually have $\Delta\nu_Z/\delta\nu \ll 1$. This makes Zeeman splitting observations sensitivity limited. Accordingly, the only hope of detecting the splitting is with an atom or molecule whose splitting is 'large', i.e. $\sim$ the Bohr magneton $e\hbar/2m_e c$; this, in turn, means that the molecule must have a large magnetic moment μ and Landé factor g. Thus, only species with electronic angular momentum are useful for Zeeman splitting observations. Other molecules have splitting $\sim$ the nuclear magneton $e\hbar/2m_n c$, which is thousands of times smaller. There is one spectacular exception, water masers, where $B_{||}$ is tens of mG in regions having volume density $n \gtrsim 10^8\,\mathrm{cm}^{-3}$ (Sarma et al., 2002).

For a given $B_{||}$, the splitting $\Delta\nu_Z$ depends on g but is independent of the line frequency itself. For species with higher line frequencies, the line widths $\delta\nu$ rise proportionally, so for a given field strength the ratio $\Delta\nu_Z/\delta\nu$ decreases proportionally. This ratio is the crucial one for sensitivity, so in the absence of other considerations it is better to use low-frequency spectral lines. Heiles et al. (1993) describe the details and provide a list of atoms and molecules having electronic angular momentum. Suitable low-frequency (<11.2 GHz) species include H I, Radio Recombination Lines, OH, CH, C_4H, and C_2S. Other molecules have much higher frequencies, but experience shows that this is not always devastating because they can exist in very dense regions where field strengths are high enough to compensate; the defining example is CN (Crutcher et al., 1999), with line frequency ~114 GHz and $B_{||}$ of several hundred μG in the Orion Molecular Cloud 1, two cores in DR21OH, and probably M17SW.

Although the Stokes parameters V, Q, and U for the Zeeman components provide in principle full information about magnetic field strength and direction, in practice full information on **B** cannot be obtained owing to the extreme weakness of Q and U. For the usual small-splitting case $\Delta\nu_Z/\delta\nu \ll 1$, Zeeman splitting is detectable in the Stokes V spectrum, which is the difference between the two circular polarizations. The V spectrum has the shape of the first derivative of the line profile (the Stokes I spectrum) with an amplitude $\propto B_{||}/\delta\nu$, where $B_{||}$ is the line-of-sight component of the field.

Why $B_{||}$ instead of $B_{\rm tot}$? Or, in colloquial terms, how do the interstellar atoms 'know' where the observer is by arranging the splitting to reveal only the particular field component that is oriented towards the observer? The answer involves the directionality associated with the circularly polarized line intensity. In contrast, when $\Delta\nu_Z/\delta\nu > 1$ the observed effect is the full splitting $\Delta\nu_Z$, which is $\propto B_{\rm tot}$, not

$B_{||}$. Crutcher et al. (1993) treat this question in detail and provide formulas for the general case.

As examples of Zeeman splitting detections, Figs. 2 and 3 illustrate Zeeman splitting for three sources from the Arecibo Millennium survey (Heiles and Troland, 2005) in order of decreasing signal/noise. The top panel of Fig. 2 shows Cas A [data from Hat Creek (HCRO)], with more than 100 hours of integration, and the bottom one shows Tau A (from Arecibo), with ~7 hours. Figure 3 shows 3C138 (from

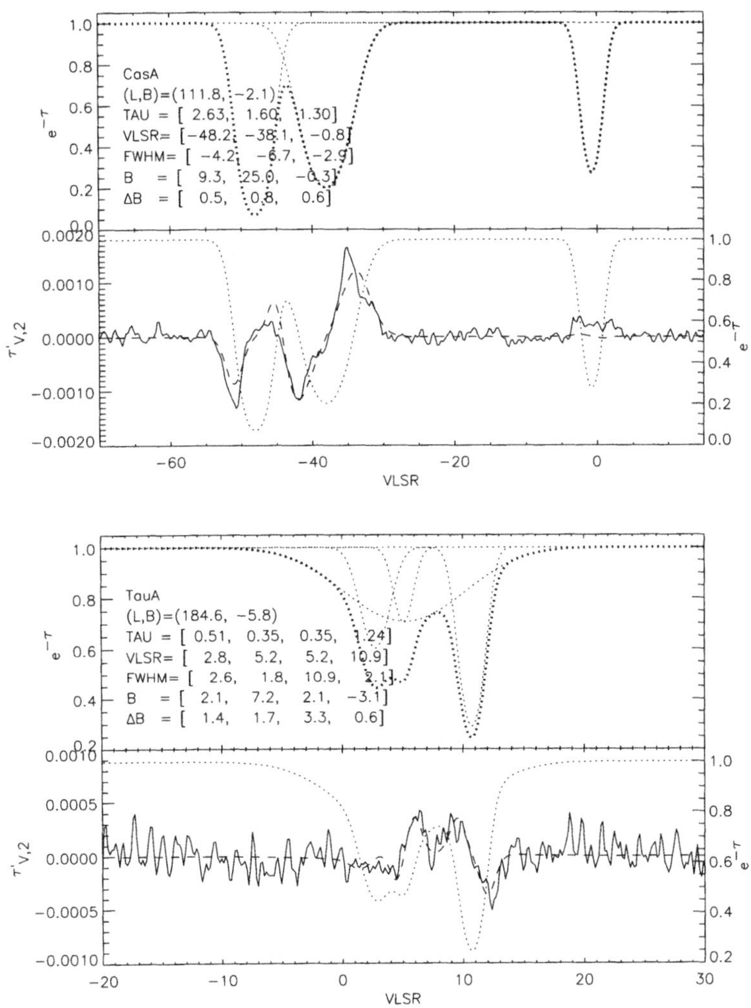

Fig. 2. Examples of H I Zeeman splitting for two sources in absorption from Heiles and Troland (2005). The *top panel* shows Cas A (data from HCRO). The *bottom panel* is Tau A. These are detections with very high signal/noise. See Sect. 4.2 for details

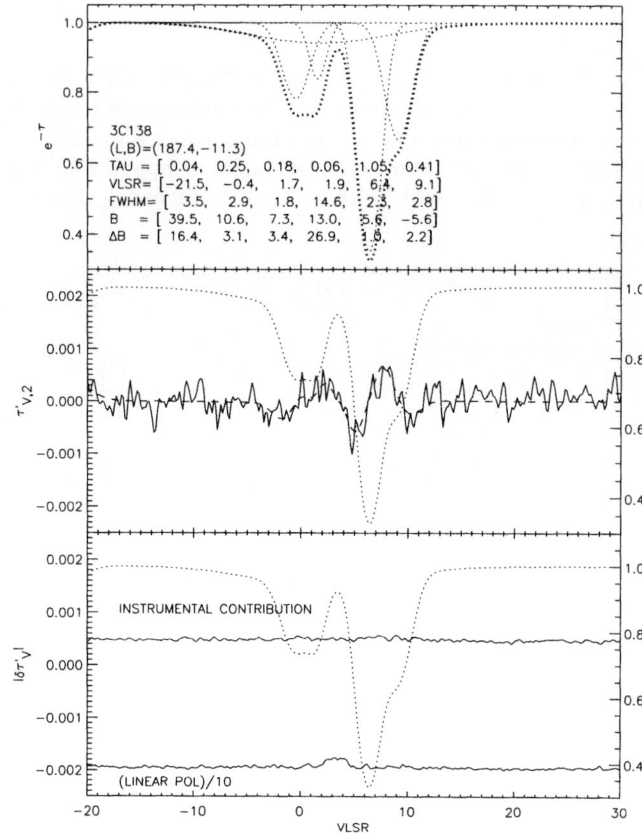

Fig. 3. An example of H I Zeeman splitting for 3C138 (Heiles and Troland, 2005). This measurement has high signal/noise relative to most other results in the Millennium survey. See Sect. 4.2 for details

Arecibo) with ∼ 17 hours. See Sect. 4 for discussion. Figure 10 shows a molecular Zeeman detection for the 3-mm CN lines toward DR 21 (OH), and Figs. 11 and 12 show a molecular Zeeman detection and $B_{||}$ map for the 18-cm line of OH toward S 106.

3 Observed vs. Intrinsic Probability Density Functions

We begin our focus on data and their interpretation with a rather technical discussion of the probability density function (pdf) of observed components of magnetic field and how they relate to the total field strength. This turns out to be surprisingly important, and because this discussion has not appeared prominently in past literature we devote considerable attention to it.

3.1 Conversion of the Intrinsic $\phi(B_{\rm tot})$ to the Observed $\psi(B_{||})$ and $\psi(B_{\perp})$

Given a field strength $B_{\rm tot}$ which can be randomly oriented to the line of sight, what is the probability of finding an observed field strength $B_{||}$? Alternatively, this is equivalent to the simple case in which all clouds have the same $B_{\rm tot}$, which is randomly oriented with respect to the observer. The line-of-sight component $B_{||}$ is

$$B_{||} = B_{\rm tot} \cos\theta \ , \tag{2}$$

where θ is the angle between the field direction and the line of sight. θ can run from 0 to π, but it's simpler and no less general to consider the smaller interval θ from 0 to $\pi/2$. In this case, the pdf of θ is the familiar

$$\phi_\theta(\theta) = \sin\theta \tag{3}$$

and we wish to know the pdf of $B_{||}$, which is given by (see Trumpler and Weaver, 1953) for a discussion of these conversions)

$$\psi(B_{||}) = \phi_\theta[\theta(B_{||})] \left| \frac{d[\theta(B_{||})]}{dB_{||}} \right| \ , \tag{4}$$

which gives

$$\psi(B_{||}) = \begin{cases} \frac{1}{B_{\rm tot}} & \text{if } 0 \leq B_{||} \leq B_{\rm tot} \\ 0 & \text{otherwise} \end{cases} \ . \tag{5}$$

In other words, $B_{||}$ is uniformly distributed between the maximum possible extremes 0 and $B_{\rm tot}$ (actually $\pm B_{\rm tot}$). This leads to the well-known results that in a large statistical sample, both the median and the mean observed field strengths are half the total field strength and also $B_{||}^2 = B_{\rm tot}^2/3$. More generally, observed fields are always smaller than the actual total fields, and with significant probability they range all the way down to zero.

Similarly, we can derive the pdf for $B_\perp$, the plane-of-the sky component; this is important for starlight polarization and synchrotron emissivity. We have

$$\psi(B_\perp) = \begin{cases} \frac{B_\perp}{{B_{\rm tot}}^2} \left[1 - \left(\frac{B_\perp}{B_{\rm tot}}\right)^2\right]^{-1/2} & \text{if } 0 \leq B_\perp \leq B_{\rm tot} \\ 0 & \text{otherwise} \end{cases} \ . \tag{6}$$

The pdf $\psi \to \infty$ as $B_\perp \to B_{\rm tot}$, but the cumulative distribution is well defined. The mean and median are $0.79\,B_{\rm tot}$ and $0.87\,B_{\rm tot}$, respectively; the high values reflect the large fraction of slabs tilted to the line of sight, where $B_\perp$ is large. The mean of ${B_\perp}^2$ is $2/3{B_{\rm tot}}^2$.

The above applies if all $B_{\rm tot}$ are the same. Now suppose $B_{\rm tot}$ has an arbitrary pdf $\phi(B_{\rm tot})$. Again, following standard techniques, we obtain

$$\psi(B_{||}) = \int_{[B_{||} > B_{\rm tot\,min}]}^{\infty} \frac{\phi(B_{\rm tot})}{B_{\rm tot}} dB_{\rm tot} \ , \tag{7}$$

where the symbol $[B_{||} > B_{\mathrm{tot\,min}}]$ means the larger of the two quantities. The presence of B_{tot} in the denominator means that smaller ranges of $B_{||}$ are emphasized. This is an obvious consequence of (5)'s uniform pdf for a single field value.

Similarly, for $B_\perp$ we obtain the more complicated

$$\psi(B_\perp) = \int_{B_\perp}^{\infty} \frac{B_\perp}{{B_{\mathrm{tot}}}^2} \left[1 - \left(\frac{B_\perp}{B_{\mathrm{tot}}}\right)^2\right]^{-1/2} \phi(B_{\mathrm{tot}}) dB_{\mathrm{tot}} \; . \tag{8}$$

It's worth illustrating these equations with some examples. Figure 4 illustrates the solution of (7) for four functional forms of $\phi(B_{\mathrm{tot}})$ plotted against $\frac{|B|}{|B_{1/2}|}$, where the subscript 1/2 denotes the median value. These forms include the following:

1. $\phi(B_{\mathrm{tot}})$ a Kronecker delta function (DELTA FCN), $\phi(B_{\mathrm{tot}}) = \delta(B_{\mathrm{tot}} - B_{\mathrm{tot},0})$, yielding ψ a flat function (as discussed immediately above, (5));

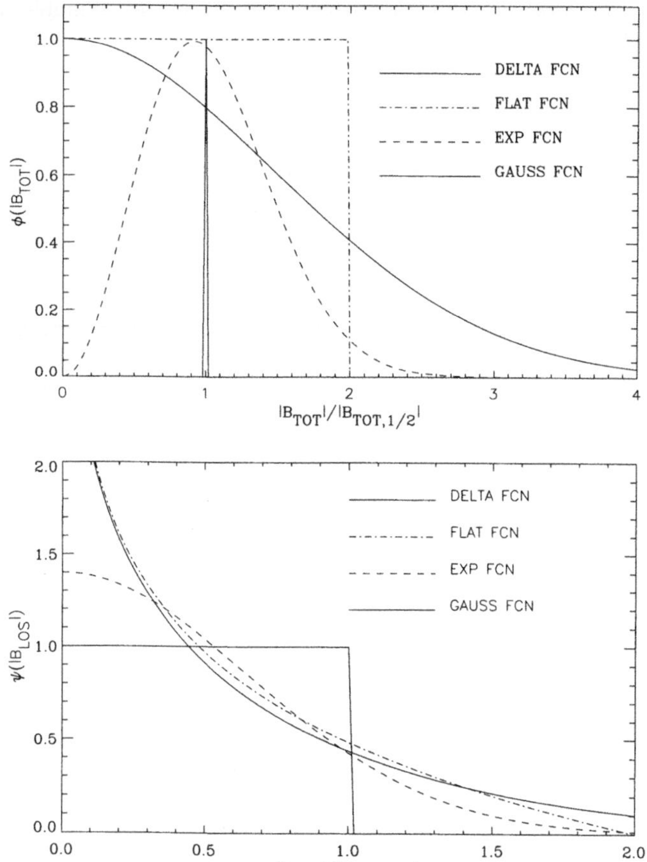

Fig. 4. *Top panel*: The intrinsic $\phi(B_{\mathrm{tot}})$ for four representative functional forms. *Bottom panel*: their line-of-sight counterparts $\psi(B_{\mathrm{los}})$. The vertical scales are arbitrary

2. $\phi(B_{\rm tot})$ a flat distribution (FLAT FCN) between $0 \leq |B_{\rm tot}| \leq B_0$, yielding $\psi \propto \ln\left(\frac{B_0}{B_{\rm los}}\right)$;
3. $\phi(B_{\rm tot})$ a weighted Gaussian (EXP FCN),

$$\phi(B_{\rm tot}) = \sqrt{\frac{2}{\pi B_0^2}\, \frac{B_{\rm tot}^2}{2B_0^2}}\; e^{-(B_{\rm tot}^2/2B_0^2)} , \qquad (9)$$

yielding ψ a Gaussian with dispersion B_0.

4. $\phi(B_{\rm tot})$ a Gaussian (GAUSS FCN) with dispersion B_0, yielding $\psi \propto E_1\left(\frac{B_{\rm los}^2}{2B_0^2}\right)$, where E_1 is the exponential integral of order 1.

All four $\phi(B_{\rm tot})$ are plotted with respect to $\frac{B_{\rm tot}}{B_{\rm tot,1/2}}$, so the medians of all lie at unity on the x-axis. However, the means differ. Similarly, the medians and means of the associated $\psi(B_{\rm los})$ differ from each other. These relationships between median and mean are summarized in Table 1. The medians and means for $\psi(B_{\rm los})$ are all about half those for $\phi(B_{\rm tot})$, which is a direct result of the weighting by $B_{\rm tot}^{-1}$ in equation 7.

Table 1. Medians and means of four representative pdfs

$\phi(B_{\rm tot})$	$B_{\rm tot,1/2}$	$\langle B_{\rm tot}\rangle$	$B_{\rm los,1/2}$	$\langle B_{\rm los}\rangle$
DELTA FCN	1.00	1.00	0.50	0.50
FLAT FCN	1.00	1.00	0.40	0.52
GAUSS FCN	1.00	1.18	0.38	0.59
EXP FCN	1.00	1.04	0.44	0.51

Figure 4 is disappointing from the observer's standpoint, because the observed distributions $\psi(B_{\rm los})$ do not differ very much. These differences become smaller – inconsequential, in fact – when there is some measurement noise. Unfortunately, given the inevitable errors in *any* observation that is sensitive to $B_{\rm los}$, it seems practically impossible to distinguish among different functional forms for $\phi(B_{\rm tot})$. Nevertheless, the average value of $B_{\rm los}$ is close to half the average value of $B_{\rm tot}$ for a wide range of intrinsic pdfs of the latter; this also applies to the medians, but less accurately. Therefore, this rule of thumb may be used to estimate the median or average $B_{\rm tot}$ from an ensemble of measurements of $B_{\rm los}$.

3.2 Conversion of the Intrinsic $\phi[\log(B_{\rm tot})]$ to $\psi[\log(B_{||})]$

Sometimes people treat $\log(B_{||})$, instead of $B_{||}$, as the important quantity. In particular, in Sect. 8.2 below, we consider least square fits of $\log(B_{||})$ for molecular clouds. The statistics for $\log(B_{||})$ differ from those of $B_{||}$. Carrying through the usual analysis, we find for the analog to (5)

$$\psi\left[\log\left(\frac{B_{||}}{B_{\rm tot}}\right)\right] = \ln(10)\; 10^{\log\left(\frac{B_{||}}{B_{\rm tot}}\right)} . \qquad (10)$$

The mean and median of $\left[\log\left(B_{||}/B_{\rm tot}\right)\right]$ are -0.434 and -0.693, which correspond to $B_{||}/B_{\rm tot} = 0.37$ and 0.21, respectively. Thus the statistics of $\log B_{\rm tot}$ favor smaller means and medians than do those of $B_{\rm tot}$, for which both numbers are 0.5.

3.3 Conversion of the Intrinsic $\phi(N_\perp)$ to the Observed $\psi(N_{\rm obs})$ for Sheets

Many interstellar morphological structures are sheets. Examples for H I include two sheets mapped in 21-cm line emission (Heiles, 1967), and an extreme sheet with aspect ratio of several hundred (Heiles and Troland, 2003). Along with Heiles and Troland (2003), we consider that all CNM structures are best considered as sheets.

As we did with the field, we discuss the pdfs of the observed column density for sheets ($N_{\rm obs}$) given the total H I column density $N_\perp$ in the direction perpendicular to the sheet, again assuming random orientations. If the normal vector to the sheet is oriented at angle θ with respect to the line of sight, then we have

$$N_{\rm obs} = \frac{N_\perp}{\cos\theta}\ . \tag{11}$$

If all sheets have the same $N_\perp$, then

$$\psi(N_{\rm obs}) = \begin{cases} \frac{N_\perp}{N_{\rm obs}^2} & \text{if } N_{\rm obs} \geq N_\perp \\ 0 & \text{otherwise} \end{cases}\ . \tag{12}$$

For a single $N_\perp$, $N_{\rm obs}$ has a long tail extending to infinity. The mean value of $N_{\rm obs}$ is not defined because, with infinite sheets, the integral diverges logarithmically; of course, this doesn't occur in the real world, where sheets don't extend to infinity. The median value of $N_{\rm obs}$ is $2N_\perp$, reflecting the increased observed column for tilted sheets. For an arbitrary pdf $\phi(N_\perp)$ we obtain

$$\psi(N_{\rm obs}) = \frac{1}{N_{\rm obs}^2}\int_0^{[N_{\rm obs}<N_{\perp\,\rm max}]} N_\perp\ \phi(N_\perp)\ dN_\perp\ . \tag{13}$$

3.4 Conversion of the Intrinsic Bivariate Distribution $\phi(B_{\rm tot}, N_\perp)$ to the Observed $\psi(B_{||}, N_{\rm obs})$ for Sheets

We can reasonably expect the magnetic field to lie either parallel or perpendicular to the sheet. If the sheet has formed by coalescence of more diffuse gas flowing more easily along the field lines, then the field should lie perpendicular to the sheet. In contrast, if the sheet is the result of a shock that has swept up both the gas and magnetic field lines, then the field should lie parallel to the sheet. Accordingly, we are led to consider the bivariate distribution of magnetic field and column density for these two cases. We assume that $B_{\rm tot}$, $N_\perp$, and of course θ are all uncorrelated. We again consider the illustrative case of delta functions for $B_{\rm tot}$ and $N_\perp$.

If $B_{\rm tot}$ is perpendicular to the sheet (the *perpendicular model*), then both $B_{||}$ and $N_\perp$ depend only on $\cos\theta$, so the bivariate pdf degenerates to the deterministic line

$$B_{||} = B_{\text{tot}} \frac{N_\perp}{N_{\text{obs}}} \tag{14}$$

which is shown in the top panel of Fig. 5. The parallel model, with B_{tot} lying in the sheet, is more complicated, with

$$\psi(B_{||}, N_{\text{obs}}) = \frac{N_\perp}{\pi N_{\text{obs}}} [(B_{\text{tot}} N_{\text{obs}})^2 - (B_{\text{tot}} N_\perp)^2 - (B_{||} N_{\text{obs}})^2]^{-1/2} . \tag{15}$$

This is illustrated by the contours in the bottom panel of Fig. 5.

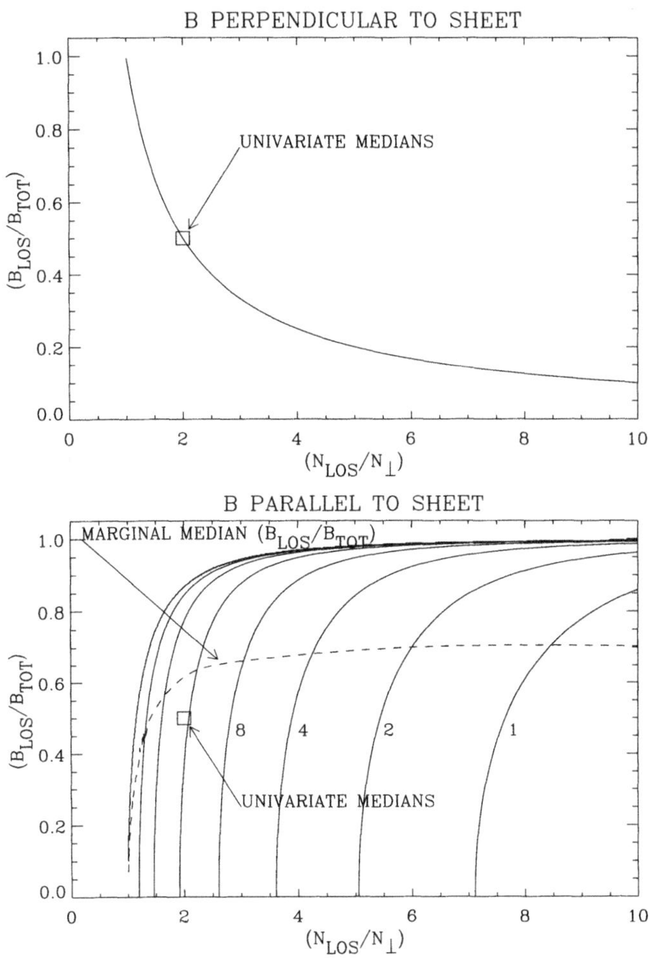

Fig. 5. The theoretical observed joint pdfs $\psi(B_{||}, N_{\text{obs}})$ for the illustrative case of δ-function distributions for $B_{||}$ and N_{obs}. The *top panel* shows the pdf for B_{tot} perpendicular to the sheets; it degenerates into a single line. The *bottom panel* is for B_{tot} parallel to the sheets; contours are spaced by factors of 2 with arbitrary scaling, and the *dashed line* shows the median $B_{||}$ versus N_{obs}

Discussion of Figure 5

The two panels of Fig. 5 exhibit the joint pdfs for the two sheet models (B_{tot} perpendicular and parallel to the sheets). The median observed column density $N_{obs1/2}$ is twice the assumed $N_\perp$ and the median observed magnetic $B_{||_{1/2}}$ is half the assumed B_{tot}; these univariate medians are indicated by squares on the top two panels. The significance of these squares is that half the observed $B_{||}$, and half the observed N_{obs}, are smaller and half larger. Finally, the dashed line in the middle panel exhibits the median $B_{||_{1/2}}$ versus N_{obs}; we calculate this by extracting the conditional pdf $\psi(B_{||}|N_{obs})$ versus N_{obs}, and calculating the medians from its cumulative distributions.

The top and middle panels illustrate a crucial observational signature at large N_{obs} that distinguishes between the two sheet models: for the perpendicular model, large N_{obs} goes with small $B_{||}$, and vice-versa for the parallel model. More quantitatively, for the perpendicular model, *all* of the datapoints having N_{obs} above its univariate median ($N_{obs} > N_{obs1/2}$, indicated by the square) have $B_{||} < B_{||_{1/2}}$. In contrast, for the parallel model *most* (66%) of the datapoints with $N_{obs} > N_{obs1/2}$ have $B_{||} > B_{||_{1/2}}$. More precisely for the parallel model, as N_{obs} gets large, the marginal pdf $\psi(B_{||}\,|\,N_{obs}) \rightarrow N_{\perp 0}/\pi N_{obs}^2\,(B_{tot0}^{\,2} - B_{||}{}^2)^{-1/2}$, which produces the median $B_{||_{1/2}} \rightarrow 0.71\,B_{tot}$; this is the asymptote of the dashed line on the middle panel of Fig. 5.

3.5 Commentary

This discussion has been rather technical, more so than is usual in a review. However, the payoff follows because we can make some powerful inferences from this discussion.

1. Consider the one-dimensional $\psi(B_\perp)$ for a given B_{tot}. $\psi(B_\perp)$ diverges as $B_\perp \rightarrow B_{tot}$; the median and mean values of $B_\perp$ are $0.79\,B_{tot}$ and $0.87\,B_{tot}$, respectively. Thus, maps of starlight polarization, or IR polarization of dust emission, tend to represent the full field strength to a considerable degree, a much higher degree than does Zeeman splitting for $B_{||}$ (see next paragraph).
2. Consider the one-dimensional $\psi(B_{||})$ for a given B_{tot}. $\psi(B_{||})$ is flat for $0 < B_{||} < B_{tot}$. Suppose we have a collection of measured $B_{||}$ and can reasonably expect the orientation to be random. Suppose we wish to fit a dependence of magnetic field on, say, volume density, as we will do below in Sect. 8.2. Then we should *not* use the standard least squares technique because it assumes that the residuals from the mean have a Gaussian distribution; in contrast, the intrinsic distribution of residuals of $B_{||}$ is flat. In particular, this means that errors derived from the distribution of residuals to the fitted points are not calculated correctly.

   Similarly, when fitting $\log B_{||}$ the distribution of residuals is asymmetric, which introduces a systematic bias into the least-squares fitted result. This must be corrected for, as we do in Sect. 8.2 below. In addition, of course, the errors are also not calculated correctly.
3. Consider an assembly of $B_{||}$ from different sources, all of which have the same B_{tot}. Then we expect some $B_{||}$ to be very small. Thus, small values of $B_{||}$ do

not necessarily mean that $B_{\rm tot}$ is small. Rather, an unbiased survey produces many small, undetectable values of $B_{||}$, which can be very frustrating for the observer but is nevertheless inevitable. A spectacular example is the local-arm ($0\,{\rm km\,s^{-1}}$) field seen against Cas A (top panel Fig. 2), $B_{||} = -0.3\pm0.6\,\mu{\rm G}$. This surprisingly small result is perfectly consistent with statistical expectation. Of course, we cannot rule out that the field actually is really small in any particular case like this, but one needs additional data to draw such a conclusion!

4. Consider the large set of magnetic fields observed in 21-cm line *emission* in morphologically obvious structures, reviewed below in Sect. 5.4. The term 'morphologically obvious' means filaments or edge-on sheets. Edge-on sheets should be edge-on shocks in which the field is parallel to the sheet, i.e. with large θ. Here, the statistics reverse and favor relatively large $B_{||}$. As explained in Sect. 3.4, as the line of sight becomes parallel to the sheet – i.e. for a morphologically obvious sheet – the median $B_{{\rm tot}\,1/2} \rightarrow 0.71\,B_{\rm tot}$. For these structures, measured fields are strong, ranging from ~ 5 to $\sim 10\,\mu{\rm G}$. This is not inconsistent with a uniform $B_{\rm tot} \sim 10\,\mu{\rm G}$, which is a factor of two above the median CNM field strength from Sect. 4. This suggests that shocks enhance the field strength, but not by large factors.

4 $B_{||}$ from H I Absorption Lines

Zeeman splitting of the H I line in absorption holds the enviable position of being the means by which the interstellar magnetic field *strength* was first measured (Verschuur, 1969). With quantitative knowledge of the magnetic field strengths came the beginning of the end of the famous theorists' refuge ('... the larger one's ignorance, the stronger the magnetic field' (Woltjer, 1967).

Zeeman splitting in absorption, instead of emission, is enviable for another important reason. It is easier to measure $B_{||}$ in the CNM than in the WNM because the H I line opacity $\propto T^{-1}$, which makes the CNM appear prominently in absorption. We detect absorption by performing $(ON - OFF)$ measurements against a radio continuum source; for such measurements the sidelobe contributions from the emission tend to cancel. This makes the CNM absorption results very much less subject to instrumental effects than emission results (Heiles and Troland, 2005). In fact, we consider the results to be statistically reliable, with Gaussian-distributed uncertainties and small systematic errors.

4.1 Early Work

Verschuur (1969) discovery of Zeeman splitting in interstellar H I, in absorption against against Cas A and Tau A, broke an earlier series of frustrating efforts focused at Jodrell Bank[2]. He continued making such measurements, but obtained

[2] Verschuur made a typographical error in labeling the sign of his Stokes V profiles (but not his derived $B_{||}$). In addition, higher sensitivity results (Fig. 2; also Heiles and Troland, 2005) reveal more Gaussian components with detected fields.

physically interesting upper limits or measurements for only five sources, which he reviewed in 1974 (Verschuur, 1974). Four of these sources had detections.

Most of Verschuur's absorption detections do not refer to diffuse H I, but rather to molecular clouds or star-forming regions. Two of the four sources (Orion A and M17; $B_{||} \sim -60$ and $+25\,\mu$G respectively) are dynamically active H II regions. One (two components in the Cas A Perseus arm with $B_{||} \sim (+9, +25)\,\mu$G; Fig. 2) is a molecular cloud probably undergoing star formation (Troland et al., 1985; Schwarz et al., 1986). None of these refer to interstellar diffuse H I. For sources that sample the diffuse H I, we are left with a single detection: Tau A, with two velocity components having $B_{||} \sim (-3, +7)\,\mu$G (Fig. 2, Heiles and Troland, 2004). Two other diffuse-cloud sources have only upper limits: Cygnus A, with $B_{||} \lesssim 3.5\,\mu$G, and Cas A Orion arm, with $B_{||} \lesssim 1\,\mu$G (Fig. 2, Heiles and Troland, 2004).

Contrary to the usual development of observational astronomy, Verschuur's discovery was not followed by the establishment of a 'cottage industry' that produced a large number of detections resulting in a significant expansion of H I absorption Zeeman splitting measurements. The reason is simply the weakness of the Zeeman splitting: typically $\Delta\nu_Z/\delta\nu \lesssim 10^{-3}$. This state of affairs lasted until the turn of the millennium (Heiles and Troland, 2005).

4.2 Recent Work: the Arecibo Millennium Survey

In our recent Arecibo Millennium survey, we (Heiles and Troland, 2005) have only 22 detections that exceed 2.5σ, out of a total of 69 measurements whose uncertainties are low enough to make them interesting. This weakness forces us to discuss the CNM Zeeman splitting results statistically. And fortunately, the statistical reliability allows us to actually carry through this statistical discussion.

Figures 2 and 3 exhibit three sources from the Millennium survey as examples of strong detections. The top two panels show Verschuur's original discovery sources Cas A and Tau A, but with higher sensitivity than his original spectra. The separate detections in two velocity components of the Perseus Arm, near $-40\,\mathrm{km\,s^{-1}}$, are very clear; the absence of a detection for the Orion arm near $0\,\mathrm{km\,s^{-1}}$ is also clear. For Taurus, there are multiple Gaussian components, more than one of which has associated features in Stokes V. The multiple-component aspect is also clear for 3C138. For these sources with multiple velocity components, we fit fields independently to each component (Heiles and Troland, 2005). The dashed lines in the three Stokes V spectra show the fits.

We emphasize that these three sources have the strongest signal/noise in Stokes V in the entire sample. Mostly we obtain upper limits instead of detections for $B_{||}$. When we include only those for which the uncertainty $\Delta B_{||} < 10\,\mu$G, the observed histogram $\psi(B_{||})$ resembles a Gaussian. Relating this to the intrinsic field B_{tot} is a complicated business requiring a Monte Carlo analysis. The end result is that the median B_{tot} is

$$B_{\mathrm{tot},1/2} = 6.0 \pm 1.8\ \mu\mathrm{G}\ . \tag{16}$$

Not surprisingly from our earlier discussion, nothing can be said about the pdf $\phi(B_{\mathrm{tot}})$.

4.3 Equipartition Between Magnetism and Turbulence in the CNM

There are no obvious correlations of $B_{||}$ with any quantity, including $N_{\rm obs}$, linewidth, or $T_{\rm k}$. However, we can compare energy densities.

Each CNM component in Heiles and Troland (2005) is characterized by measured values of not only magnetic field but also temperature, column density, and velocity dispersion. This allows us to compare energy densities. One way to do this is with the classical plasma parameter β, equal to the ratio of thermal to magnetic pressure or, alternatively, thermal to magnetic energy density. We can similarly define the ratio of turbulent to magnetic energy density (Heiles and Troland, 2005).

For comparison of turbulent and magnetic effects in the CNM, we calculate the relevant ratios for the following adopted parameter values, which are close to the medians:

$$T = 50\,{\rm K} \tag{17}$$

$$\Delta V_{\rm turb,1d} = 1.2\,{\rm km\,s^{-1}} \tag{18}$$

$$B_{\rm tot} = 6.0\,\mu{\rm G} \tag{19}$$

These values provide

$$\beta_{\rm th} = 0.29 \tag{20}$$

and

$$\beta_{\rm turb} = \frac{E_{\rm turb}}{E_{\rm mag}} = M^2_{\rm ALF,turb} = 1.3\;. \tag{21}$$

These values should be regarded as representative. Not all CNM clouds have the median values, so these parameters have a considerable spread.

4.4 Field Strengths in the CNM Versus Those in Other Phases

As mentioned in Sect. 1, Beck (2001) reviews the most recent estimate of field strength derived from synchrotron emission, minimum energy arguments, Faraday rotation, and polarization. He finds the regular component to be $\sim$4 μG and the total component to be $\sim$6 μG. The difference between regular and total components is the fluctuating component, whose scale length is probably at least tens of parsecs. Because our CNM structures are physically small, it is more appropriate to compare their field strengths with the total component. The CNM median of $\sim$6 μG is nominally identical to Beck's local Galactic total component of $\sim$6 μG (see also Chap. 5).

All of the other diffuse ISM phases are less dense than the CNM. For example, both the WNM and the WIM are nearly two orders of magnitude less dense. Thus the ISM field strength does not depend very sensitively on volume density. In contrast, for the larger densities associated with molecular clouds, in which gravity plays a significant role, the field strength does increase with density, roughly $B_{\rm tot} \propto n^{1/2}$ (Crutcher, 1999). The density independence for diffuse gas is well known from past studies (Crutcher et al., 2003), so this is hardly news; nevertheless, we tend to forget these things and, moreover, from an observer's standpoint the paucity of detectable fields is disappointing.

4.5 Astrophilosophical Discussion

These numbers indicate that magnetism and turbulence are in approximate equipartition. The approximate equipartition suggests that turbulence and magnetism are intimately related by mutual exchange of energy. Magnetic energies do not dissipate because the magnetic field cannot decay on short time scales. On the contrary, supersonic turbulence does dissipate rapidly: numerical simulations of turbulence suggest that the magnetic field does not mitigate turbulent dissipation (MacLow et al., 1998). Thus, the equipartition between the dissipative turbulent energy and nondissipative magnetic energy must arise from a mechanism other than energy decay.

We suspect the answer is that the CNM components result from the transient nature of turbulent flow: the CNM occupies regions where densities are high, produced by converging flows, and the density rise is limited by pressure forces. This idea is discussed and reviewed thoroughly by Mac Low and Klessen (2004). These limiting pressures are magnetic because the gas has small $\beta_{\rm th}$, meaning that thermal pressure is negligible and the dynamical equality makes the magnetic pressure comparable to the converging ram pressure. The equipartition looks like a steady-state equilibrium, but it is really a snapshot of time-varying density fields and our immediate observational view is a statistical result over a large sample. In other words, our current observational snapshot shows an ensemble at a given time. Against this we compare the numerical simulations, which are stationary in the sense that they have been allowed to run long enough that the statistical properties become time-independent. Such simulations are also ergodic, with statistical properties over time being equivalent to those over space. With this view, the ISM dynamically evolves through turbulence and its properties are governed by statistical equilibrium of energy inputs and dissipation.

An alternative picture is based on the classical model of *static equilibrium* in which all forces balance. Static clouds are formed and evolve by gas moving adiabatically from one equilibrium state to another as ambipolar diffusion allows magnetic flux to slowly unfreeze. These slow adjustments in morphology occur primarily along the field lines. At each stage there is a well-defined morphological structure in quasistatic equilibrium. This idea was originated by Mouschovias (1976) and has been well-developed by the 'Mouschovias school' of students and collaborators, consisting of Ciolek, Fiedler, and Basu (see Ciolek and Basu, 2000) and references quoted therein), and by Shu and collaborators (see Shu et al., 1999). The picture of static equilibrium predicts the linear relationship between $B_{\rm tot}$ and $\sigma_{\rm v} n^{1/2}$, which is found for molecular clouds (Sect. 8.2 below), which is equivalent to the energy equipartition found in (21) above.

Both models predict the same result, namely approximate equipartition between turbulent and magnetic energy densities. However, the concepts on which they are based are in direct opposition. Which one is correct for diffuse clouds? The role of gravity in diffuse clouds is negligible. Given this, the static equilibrium models, for which gravity is a major player, cannot apply to diffuse clouds. Thus, for diffuse gas (but not for molecular clouds) we favor the concept of statistical equilibrium as briefly outlined above. Analytical and numerical research is being intently pursued on this topic; an excellent review is Mac Low and Klessen (2004).

5 $B_{||}$ from H I Emission Lines

Zeeman splitting of the H I line in emission holds the enviable position of not requiring a background source: one can look anywhere, so that the field in interesting regions can be measured and mapped. However, nothing comes for free: emission measurements are prone to instrumental error from polarized sidelobes. These errors have been the subject of much controversy and here we will devote considerable attention to explaining these matters. We will conclude that most published Zeeman detections in H I emission are fairly reliable. We begin our examination of this question with a discussion of instrumental effects arising from polarized structure in the telescope beam.

5.1 Instrumental Effects from Polarized Sidelobes and their Description by a Taylor Series

The instrumental effects in H I Zeeman splitting measurements arise from angular structure in the Stokes V beam interacting with H I structure on the sky. The V beam has angular structure, even to the extent of having sign changes. Troland and Heiles (1982) used both their empirical investigations of the HCRO telescope and theoretical investigations published by others to classify this V structure into three primary categories; here we split one, the sidelobe component, into two subcomponents, near and far sidelobes. This gives:

1. *Beam squint*, in which the two circular polarizations point in slightly different directions with typical separation (Ψ_{BS}) of a few arcseconds. This angular separation doesn't seem like much, but given a small velocity gradient with position the two beams see different frequencies, and this mimics the tiny splitting resulting from the Zeeman effect.
2. *Beam squash*, in which the Stokes V beam has slightly different beamwidths in orthogonal directions. These 'four-lobed' polarized beams, in which two lobes on opposite sides of beam center have the same sign and two lobes rotated 90° in position angle have the opposite sign, are sometimes described as 'cloverleafs'. This four-lobed structure responds to the second derivative of the 21-cm line on the sky. Theoretically, beam squash occurs only for the linearly polarized Stokes parameters Q and U, but in practice it can also for Stokes V (e.g. Heiles et al., 2001, 2003).
3. *Near-in sidelobes*, which can be considered as standard diffraction effects and have polarization structure similar to that of the main beam described above.
4. *Far-out sidelobes*. For most telescopes the total power in these 'distant sidelobes' is nontrivial: even though the sidelobes are weak, they cover very large solid angles and tend to be elliptically polarized. Troland and Heiles (1982) present one of the very few, perhaps the only, map of the circular polarization of far-out sidelobes; the pattern looks like a windmill and obviously results from feed legs. These distant sidelobes are a result of telescope surface roughness and the feed leg structure, so their structure is impossible to predict and can be time variable.

The classification is useful because it allows one to parameterize the beam polarization effects. These parameters can be measured and corrections applied.

Nearly all H I emission Zeeman splitting measurements have made these corrections in one form or another.

The appropriateness of this fourfold classification applies to all telescopes that have been used for emission Zeeman splitting observations: HCRO (Heiles, 1996b), the Green Bank 140-foot telescope (Verschuur, 1969, 1989), Arecibo (Heiles et al., 2001), and the Green Bank Telescope (GBT) (Heiles et al., 2003). For example, Verschuur (1969) Fig. 2 presents the V beam pattern for the 140-foot telescope as it was in the late 1960's. At that time, it was very well described by beam squint with a peak-to-peak amplitude of about 1.4%; this corresponds to a beam squint $\Psi_{\rm BS} \approx 7''$. Our maps of the complete polarized sidelobe structure of the HCRO telescope always produced similar results, although with much smaller beam squint. Verschuur (1989) Fig. 1 presents the 140-foot polarized beam structure as it was in the late 1980's, and shows a drastic difference: the newer map shows primarily the four-lobed pattern of our category (2) with little beam squint. (The feed system had been changed between the two epochs.) The 1960's version of the beam pattern made the 140-foot telescope unsuitable for Zeeman-splitting measurements of H I in emission because the beam squint contribution to instrumental error would have been excessive. However, the 1980's version, with its small beam squint but higher second-derivative component, was satisfactory – as shown by the fact that Verschuur reobserved four positions that had previously been observed with the HCRO telescope and found excellent agreement in three.

5.2 Verschuur's Bombshell

Measurements of Zeeman splitting of H I emission lines have been made by Troland, Heiles and other collaborators, and Verschuur. Until 1993, the agreement was quite good.

Despite the apparent agreement of the measurements, in 1993 Verschuur became highly suspicious of all emission results and dropped a bombshell. He asserted that '... claims of Zeeman effect detections in H I emission features ... based on observations made with presently available single-dish radio telescopes cannot be regarded as reliable.' At the time of his paper, the HCRO telescope had already been destroyed, but he meant his claim to apply to that telescope as well as other telescopes that were then available. This is a strong statement and it has had a dampening effect on the field, making many astronomers highly suspicious of the published results. Accordingly, we believe a thorough discussion is in order. This discussion is excerpted from Heiles (1998a), a reference which is difficult to find.

We believe Verschuur's claim to be incorrect. His claim is based on his estimates of the instrumental effects, which in turn are based *solely* on measurements of the velocity gradient of the H I line (Verschuur, 1995a,b). In particular, his estimates of the instrumental effects are not based at all on the *properties of the polarized beam.* To clarify his procedure and its inadequacy, we describe its six steps:

1. Observe V and I spectra at the central position P; denote these $V_{\rm obs}(v)$ and $I_{\rm obs}(v)$.
2. Make an 8-point map of I spectra around P. Each map position is displaced from P by $15'$; in position angle the 8 points are equally spaced ($45°$), with the displacements of 4 points towards the cardinal directions in equatorial coordinates.

3. Find the pair of profiles whose difference spectrum $\Delta(v)$ is strongest and mimics the shape of $V_{obs}(v)$.
4. Find the coefficient R that scales the $\Delta(v)$ spectrum to the V_{obs} spectrum, i.e. the best fit for $R\Delta(v) = V_{obs}(v)$.
5. Produce the 'corrected' V spectrum $V_{corr}(v) = V_{obs}(v) - R\Delta(v)$.
6. Derive the Zeeman splitting from $V_{corr}(v)$.

The fatal flaw is that R, which represents the beam squint, is not measured directly for the *telescope*. Rather, it is given the particular value that minimizes the observed V spectrum $V_{obs}(v)$.

As explained above, the beam squint samples the first derivative of the 21-cm line on the sky, which must contain a velocity gradient at some level. Steps 2 and 3 of the above procedure measure the velocity gradient. Step 4 fits this velocity gradient to the observed V spectrum and derives the coefficient R. Then, no matter how large R is, it is used to subtract away the scaled Δ profile from the observed V spectrum. With this step, R implicitly represents the amplitude of the beam squint in units of $30'$.

But the amplitude of the beam squint can be independently measured for a telescope. The proper procedure would be to measure the beam squint and velocity gradient, multiply the two vectorially, and subtract the result from the observed V spectrum.

Consider one particular entry in Verschuur's (1995b) Table 2 as an example: NCPShell.4. For this position he obtains $R = 0.0052$. This corresponds to a beam squint of $(30' \times 0.0052) = 9''.4$. He uses this value of R to subtract away a velocity derivative from the V_{obs} profile that amounts to $10.8\,\mu$G, obtaining a 'corrected' field strength $2.1 \pm 1.0\,\mu$G. In doing this he has removed the contribution to V_{obs} that arises from the magnetic field – he has removed the 'signal'. In colloquial English, this is known as 'throwing out the baby with the bathwater'.

The data in Verschuur's papers (1995a,b) could be reanalyzed taking account of the fact that the beam squint of the 140-foot telescope is limited to some maximum value. Unfortunately, this is not discussed by Verschuur, but judging from his earlier paper in this field (Verschuur, 1989) the upper limit on 140-foot beam squint is probably $\sim 3''$, which corresponds to $R = 0.0017$ (0.17%). Many entries in Verschuur's table have $R > 0.0017$ and these probably represent real measurements of Zeeman splitting.

5.3 Reliability of the HCRO H I Emission Results

Nearly all published results in H I emission are from the HCRO telescope. Verschuur's bombshell was directed primarily at those results. Having dealt with Verschuur's criticisms, it remains to show that our HCRO emission measurements are, in fact, correct. Heiles (1996b) discussed his correction procedures for the HCRO data. He also tested these correction procedures on the North Celestial Pole, which is the one point on the sky where, for the HCRO equatorially mounted telescope, the telescope beam could rotate in a complete circle.

Heiles divides the data into 12 time ('Right Ascension' or RA) bins and measures the magnetic field strength $B_{\|}$ separately and independently for each. He then Fourier analyzes the 12 results. The Fourier terms respond differently to the beam components listed above. Beam squint, with a two-lobed pattern on the sky, works with the first derivative of the H I emission to produce one cycle of variation per 24 hours. Beam squash produces two cycles per 24 hours, and higher order terms can come from the sidelobes.

These Fourier coefficients constitute *empirically determined* squint and squash contributions for the NCP. He also *predicted* the squint contribution by measuring the first derivatives of the H I emission and applying the previously-measured beam squint. The two methods gave comparable results, which shows that one can, indeed, apply measured beam squint and squash to measured angular derivatives of H I emission to derive – and subtract out – the instrumental contribution.

Averaging over all 24 hours zeros out the contributions from beam squint and squash, because their structure in the azimuthal direction around beam center averages to zero. It also eliminates some, and probably nearly all, of the sidelobe contributions. For the average of all RAs the V spectrum is an excellent fit to the derivative of the I spectrum, with $B_{||} = 8.8 \pm 0.4\,\mu\mathrm{G}$ (Heiles, 1996b); this is in excellent agreement with the measurements nearby in the sky (Heiles, 1989). He also found a systematic variation of $B_{||}$ with RA from ~ 7 to $12\,\mu\mathrm{G}$, indicating the contribution of instrumental errors. The amplitude of the first Fourier component $\sim 2.0\,\mu\mathrm{G}$ and of the second $\sim 0.58\,\mu\mathrm{G}$. The additional uncertainty produced by this variation, calculated as an r.m.s., is $1.4\,\mu\mathrm{G}$. The first Fourier component is significantly higher than the others, while the second is comparable to them and is probably not significant with respect to noise.

Heiles et al. (2003) have performed a similar analysis of the North Celestial Pole using the Green Bank Telescope. The analysis is not yet complete because the data were taken recently. Nevertheless, the 24-hour average for the GBT is in excellent agreement with the above HCRO results, yielding $B_{||} = 8.5 \pm 0.8\,\mu\mathrm{G}$. Figure 6 compares the results for the two telescopes; recall that the beam areas differ by a factor of 16! If anything, sidelobe effects in the line wings seem higher for the GBT spectrum.

Most of the published HCRO results did not, in fact, go through the procedure of subtracting out the instrumental contribution. Rather, any position having a significant instrumental contribution, i.e. one that exceeded about one third of the measured results, was not published. Quoted errors on the published results do not include the instrumental contribution, so they are too small; a conservative estimate of the instrumental error in quoted results depends on circumstances, but is typically of order 30% of the derived value. This is relatively high, and a few quoted values may be incorrect and even of the wrong sign. Nevertheless, the published results should be relatively reliable given these caveats.

All this means that HCRO reliably measured strong fields in H I emission, but not weak fields. Thus, those measurements cannot be used statistically, as the absorption measurements of Sect. 4 can be.

5.4 Overview of the HCRO H I Emission Results

The HCRO telescope was devoted almost exclusively to Zeeman splitting during the years before its catastrophic demise in 1993 (Heiles, 1993). It made many Zeeman splitting detections in H I emission. Figure 7 shows a global map of these detections, which are presented in five publications (Heiles, 1988, 1989; Goodman and Heiles, 1994; Myers et al., 1995; Heiles, 1997). Below we present the briefest of brief summaries of each.

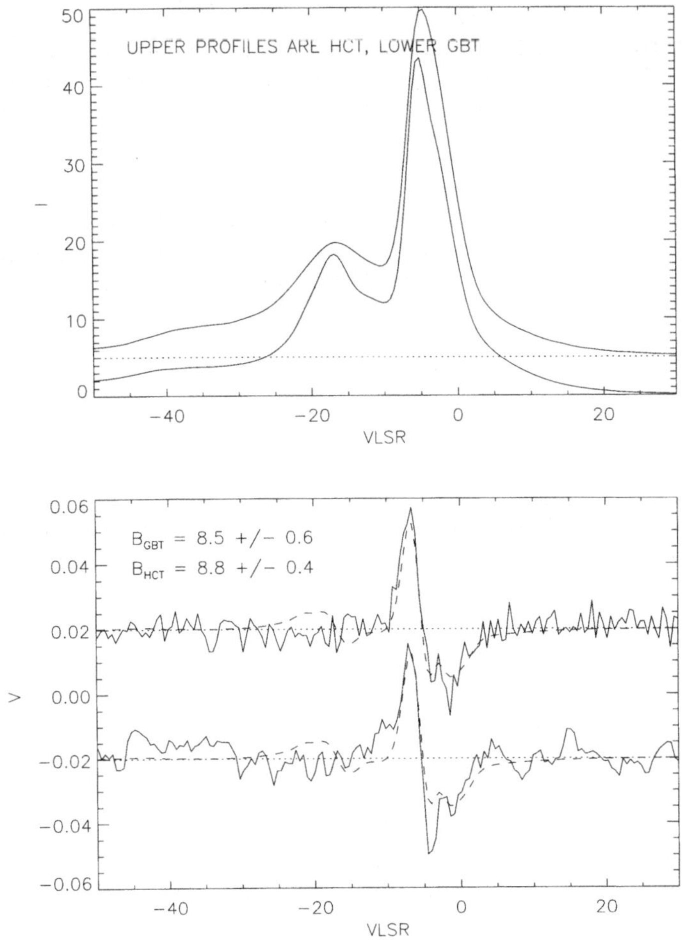

Fig. 6. Stokes I (*top*) and V (*bottom*) for H I emission towards the NCP for two telescopes, the HCRO and the GBT. The upper profile in each panel is from HCRO

- Heiles (1988) mapped $B_{||}$ for 27 positions in the vicinity of the filamentary dark cloud L204, detecting Zeeman splitting in H I emission for all 27 and also H I self-absorption for 12 positions. This remains the best $B_{||}$-mapped example of a well-defined, isolated dark cloud. The $B_{||}$ exhibits correlation with starlight polarization, CO velocities V_{CO}, and the shape of the curvy filament, implying that projection effects are responsible for much of the structure and allowing an estimate $B_{tot} = 12\,\mu G$. The field dominates ram pressure from systematic flows and also dominates the self-gravity of the molecular gas. This cloud seems worth further study because it is well-defined with interesting correlations, and would benefit from redoing the correlations with better angular resolutions.
- Heiles (1989) mapped $B_{||}$ in a number of morphologically obvious regions, meaning high-contrast filaments. These included several supernova or superbubble shells such as Eridanus, the North Polar Spur, and the North Celestial

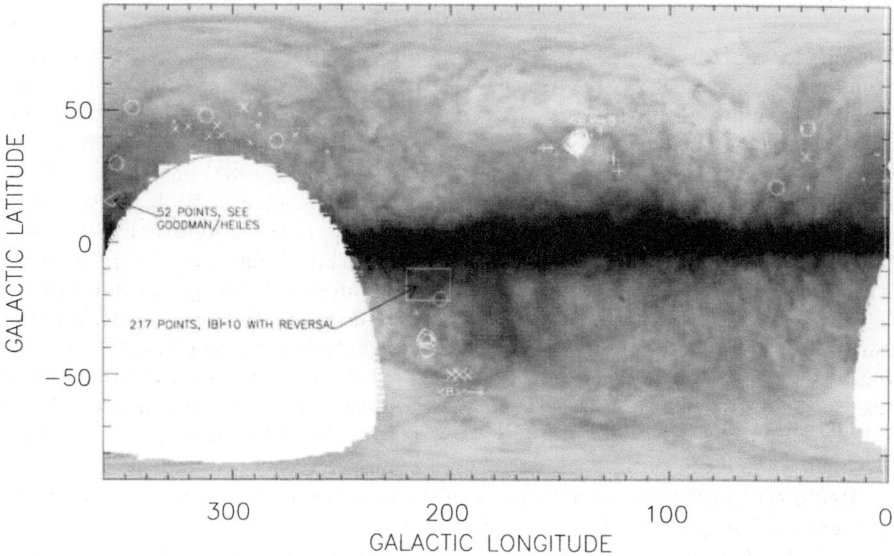

Fig. 7. Detections of magnetic fields in emission from Heiles (1988, 1989), Goodman and Heiles (1994), Myers et al. (1995) and Heiles (1997), superposed on a map of H I in which blacker means more H I

Pole Loop. In every morphologically obvious structure, the fields were strong ($|B_{||}| \gtrsim 5\,\mu$G) and the field retained the same sign over the feature. Magnetic pressure overwhelmingly dominates thermal pressure, and it even dominates turbulent pressure. The paper considers the filaments to be true filaments instead of edge-on sheets, but we wonder if this is correct; this is an important question and needs to be resolved. If the structures are edge-on sheets, then the observed values $|B_{||}| \gtrsim 5\,\mu$G imply $B_{\rm tot} \sim 10\,\mu$G from our discussion in Sect. 3.4, meaning that the field is mildly amplified in old supernova shocks.

- Goodman and Heiles (1994) mapped $B_{||}$ for 52 positions in Ophiuchus, detecting it for 43 Gaussian components in 29 positions. 16 of the 43 components were in self-absorption having the same velocity as, and therefore associated with, molecular gas. Combining the Zeeman-splitting results with optical polarization data allows them to determine not only $B_{||}$ but also $B_\perp$ and, consequently, $B_{\rm tot}$; it is 10.6 μG, with the field inclined to the line of sight by 32°. About half the magnetic energy is associated with the random field component, and the magnetic and kinetic energy densities are comparable.
- Myers et al. (1995) detected $B_{||}$ for 1 position in the Draco dark cloud and 31 positions in the Ursa Major (North Celestial Pole) loop. Magnetic and kinetic energy densities are comparable.

One HCRO detection, at $(\ell, b) = (141.^\circ1, 38.^\circ8)$, is remarkably strong, with $B_{||} = 18.9 \pm 1.8\,\mu$G. However, the same position observed with the Effelsberg 100-m telescope yields the completely discrepant $B_{||} = 3.5 \pm 3.7\,\mu$G. This is a real problem and not simply a difficulty with one of the telescopes, because two other HCRO positions observed with Effelsberg yielded consistent results. Given the factor 16 difference in beam area, it would seem that there is much angular structure in $B_{||}$ at this position! But this needs to be checked by mapping the locale with, say, the GBT.

– Heiles (1997) mapped $B_{||}$ for 217 positions covering ~100 deg^2 in the Orion/-Eridanus loop region. The goal was to develop a holistic interpretation of the magnetic field structure on small and large size scales. The observations were interpreted as a large-scale ambient field distorted by the superbubble's shock, together with smaller-scale structure produced by local perturbations. But the match to the data is sketchy and vague, at least in part because of the geometrical situation described in the next paragraph, so the goal was realized only in part. Nearly all of the area mapped is permeated by a negative field (pointing towards the observer); a small (~10 deg^2) region has a uniformly positive field, which is associated with a unique velocity component, different from those associated with the negative field. The reversal in sign had been previously interpreted as a toroidal field, but this may not be correct because of the different velocity components; an alternative interpretation involves field lines wrapped around a molecular filament by the shock front produced by the superbubble explosions.

  As part of the analysis, Heiles (1997) develops a simple geometrical model of field lines distorted by the Eridanus superbubble shock front. For individuals who are interested in studying the magnetic field perturbations produced by shocks, this model is worth some study as an illustrative example of the general case. The patterns of $B_\perp$ and $B_{||}$, revealed by observations of starlight polarization and of Zeeman splitting, are very complicated, more than one naively imagines. They depend, firstly, on the direction of the ambient field relative to the line of sight. They also depend on the position within the structure. Most importantly, they also depend on which wall of the superbubble – the near or the far wall – produces most of the extinction or H I column density. The North Polar Spur, with its easily recognizable starlight polarization effect, is a very unusual and deceptively simple case because we see the ambient field nearly in the plane of the sky.

6 Importance of Magnetic Fields in Molecular Clouds

Here we will both review the observational data and focus on one of the main reasons for observing magnetic fields in molecular clouds – to try to understand their role in the evolution of dense clouds and in the star formation process. Understanding star formation is one of the outstanding challenges of modern astrophysics. However, in spite of significant progress in recent years, there remain unanswered fundamental questions about the basic physics of star formation. In particular, what drives the star formation process? The prevailing view has been that self-gravitating clouds are supported against collapse by magnetic fields, with ambipolar diffusion reducing support in cores and hence driving star formation (e.g., Mouschovias and Ciolek, 1999). The other extreme is that molecular clouds are intermittent phenomena in an interstellar medium dominated by turbulence, and the problem of cloud support for long time periods is irrelevant (e.g., Elmegreen, 2000). In this paradigm, clouds form and disperse by the operation of compressible turbulence (Mac Low and Klessen, 2004), with clumps sometimes becoming gravitationally bound. Turbulence then dissipates rapidly, and the cores collapse to form stars. Hence, there are two competing models for driving the star formation process. The fundamental issue of what drives star formation is far from settled, on either observational or theoretical grounds. Since the main difference between the two star-formation scenarios listed above is the role of magnetic fields, observations of magnetic fields in star formation regions are crucial.

Observations of magnetic fields in molecular clouds have now become a fairly routine procedure. Great progress has been made in mapping polarized emission from dust, and the first detections of linearly polarized spectral lines have been made. Only the Zeeman technique has been used for both diffuse H I and dense molecular clouds. Measuring Zeeman splitting in molecular clouds is both easier and harder than in the H I. Instrumental effects are less important because the sources are confined in angle so that polarized sidelobes often lie off of the source; this makes it easier. However, molecular lines are typically much weaker than the H I line, the frequencies are all higher, and the Landé g factors are somewhat smaller; although this makes it harder, there is compensation in the form of narrower line widths and higher field strengths in the denser molecular clouds. So progress in molecular Zeeman measurements has been possible.

7 Molecular Cloud Observational Results

There has been a remarkable explosion in the observational data on magnetic fields in molecular clouds in the last few years. Hildebrand and collaborators have mapped warm molecular clouds in the far infrared; that work is reviewed by Hildebrand (2002, 2003). The JCMT SCUBA polarimeter has been used by multiple investigators (Matthews et al., 2001; Chrysostomou et al., 2002; Wolf et al., 2003; Crutcher et al., 2004) to map polarized dust emission at 850 μm in both warm clouds and cool cores. The BIMA millimeter array has been used to map linearly polarized dust and spectral line emission at 3 and 1.3 mm at $2'' - 6''$ resolution (Lai et al., 2003). Crutcher (1999) reviewed all molecular Zeeman observations made at that time and analyzed in detail the 15 positive detections. Since then, two major surveys of OH Zeeman have been carried out (Bourke et al., 2001; Troland and Crutcher, 2005) that have added to the total. Finally, Zeeman measurements in OH (Fish et al., 2003; Caswell, 2003, 2004) and H_2O (Sarma et al., 2002) masers, which probably probe magnetic fields in shocked molecular regions, have been made. See references to additional results in the above papers.

Space precludes discussion of all the results. Instead, we discuss magnetic field results for a small number of molecular clouds, chosen to illustrate the range of the data available and the astrophysical conclusions that may be inferred. These are a starless, low-mass core (L 183), a region of low-mass star formation with a CO bipolar outflow (NGC 1333 IRAS4A), a region with evidence of high-mass star formation but no H II region (DR 21 OH), and a region with high-mass star formation and an H II region (S 106).

7.1 The Starless Core L 183

L 183 is a dark cloud that contains a starless core – a dense concentration of a few solar masses with no evidence that a protostar or star has yet formed. Figure 8 shows observational results for the magnetic field; the left panel shows the SCUBA dust emission and polarization map at 850 µm (Crutcher et al., 2004), while the right panel shows the NRAO 43-m telescope observation of Stokes I and V spectra of 18-cm OH lines (Crutcher et al., 1993). The dust polarization map has an angular resolution of 21″ and covers 3′; the observed dust polarization position angles have been rotated by 90° so the line segments are in the direction of $B_\perp$. The OH spectra were obtained with a telescope beam diameter of 18′.

The dust polarization map samples the core of L 183, with a density of $n(H_2) \approx 3 \times 10^5$ cm^3. The magnetic field is fairly regular, in agreement with the field being strong enough to resist turbulent twisting. But the dispersion in position angles of 14° is significant, implying that some turbulent twisting is present. The angle between the projected minor axis of the core and the mean direction of $B_\perp$ is $\sim 30°$. Applying the Chandrasekhar–Fermi technique yields $B_\perp \approx 80$ µG. The OH Zeeman spectra sample a much larger area – the extended envelope of the L 183 core, for which $n(H_2) \approx 1 \times 10^3$ cm^3. The Zeeman effect is not detected to a 3-σ upper limit of $B_{||} < 16$ µG.

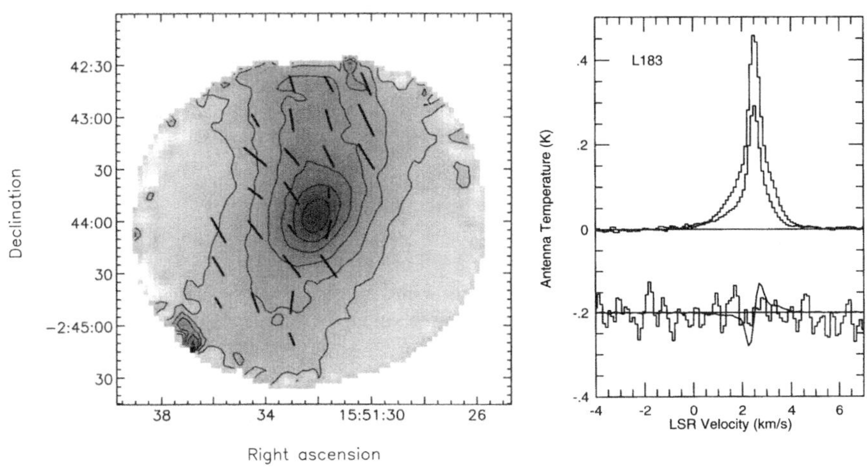

Fig. 8. *Left*: Dust polarization map of the starless core L 183. Grey-scale and contours show the dust emission at 850 µm. *Thick line* segments show the direction of the magnetic field projected on the sky; lengths are proportional to the polarized flux. *Right*: OH 1665 and 1667 MHz line profiles toward L 183. Observed data are histogram plots; the fit to Stokes V in the lower panel is a line. *Top panel* shows the two Stokes I spectra. *Bottom panel* shows the mean Stokes V spectrum for the two lines with a 3-σ upper limit fit

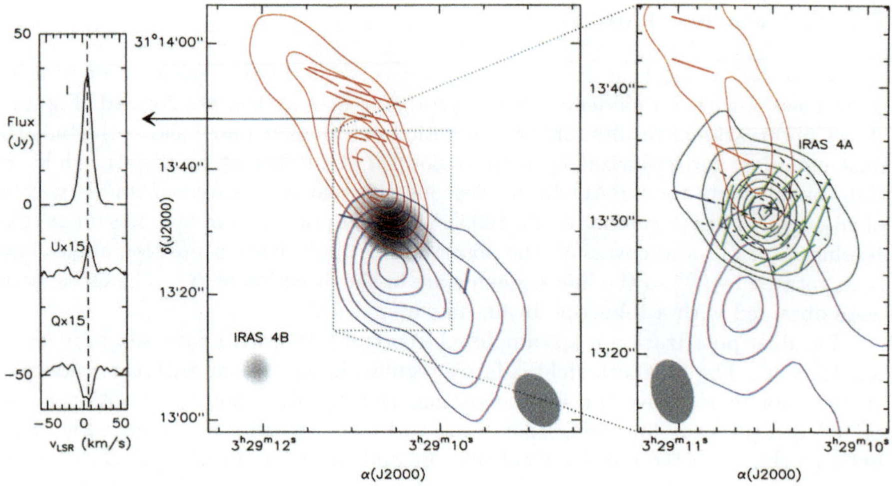

Fig. 9. BIMA observations of NGC 1333 IRAS4A. The *middle panel* shows dust emission (greyscale) and CO 2–1 emission from the bipolar outflow (contours). Line segments superposed on the outflow show the polarization of the line emission. The mean Stokes I, U, and Q profiles for the northern lobe are shown in the *left panel.* The *right panel* shows the central region dust emission (thick contours), CO outflow (thin contours), CO polarization (black line segments), and dust polarization (grey line segments). *Dotted lines* show a possible hourglass morphology for (**B**)

7.2 NGC 1333 IRAS4A

NGC 1333 IRAS4A is a later stage in star formation than L 183 – a very young low-mass star formation region with multiple young stellar systems and an associated molecular outflow. Figure 9 shows BIMA observations (Girart et al., 1999) of the dust and CO outflow emission and polarization at 1.3 mm. The line polarization is perpendicular to the dust polarization. In the outflow, where the direction of the velocity gradient is known, it is possible to predict theoretically (Kylafis, 1983) that the line polarization should be parallel to $B_\perp$ and therefore perpendicular to the dust polarization, as observed. The outflow is initially north-south, at about a 50° angle to $B_\perp$. A successful theory of molecular outflows must account for such a difference between **B** and the outflow. However, about 25″ from the center the difference is only 15°, suggesting that the field has deflected the outflow. The morphology of the dust polarization is again smooth and suggestive of a pinched or hourglass morphology.

7.3 DR 21 (OH)

Figure 10 shows results for the high-mass star formation region DR 21 (OH); the left panel shows the BIMA dust and CO emission and polarization map at 1.3 mm (Lai et al., 2003), while the right panel shows IRAM 30-m telescope Stokes I and V spectra of the 3-mm CN lines (Crutcher et al., 1999). In millimeter-wave dust emission the main component of DR 21 (OH) consists of two compact cores (Woody et al., 1989) with a total mass of $\sim 100\,M_\odot$. The two CN velocity components are each centered on a different one of the two compact cores. The region has associated masers of OH (Norris et al., 1982), H_2O (Genzel and Downes, 1977), and CH_3OH (Batrla and Menten, 1988), and high-velocity outflows powered by the two compact cores (Lai et al., 2003). The results from the dust and CO 2–1 linear polarization maps suggest that the magnetic field direction in DR 21 (OH) is parallel to the CO polarization and therefore parallel to the major axis of DR 21 (OH). This could be explained by a toroidal field produced by rotation of the double core. The strong correlation between the CO and dust polarization suggests that magnetic fields are remarkably uniform throughout the envelope and the cores. Both the dust emission and the CN lines sample a density $n(H_2) \approx 1 \times 10^6\,\mathrm{cm}^3$. The Chandrasekhar–Fermi technique yields $B_\perp \approx 1\,\mathrm{mG}$, compared with $B_{||} = -0.4 \pm 0.1\,\mathrm{mG}$ and $B_{||} = -0.7 \pm 0.1\,\mathrm{mG}$ inferred from the CN Zeeman detections shown in Fig. 10. Combining these results, the total field strength $B_{\mathrm{tot}} \approx 1.1\,\mathrm{mG}$ and $\mathbf{B}$ is at an angle $\theta \sim 60°$ to the line of sight. However, uncertainties in $B_\perp$ and in $B_{||}$ are sufficiently large that θ is quite uncertain.

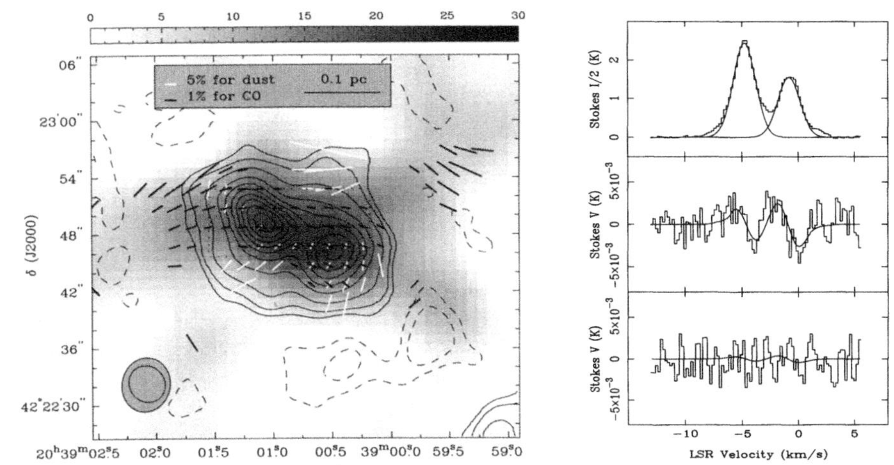

Fig. 10. *Left*: BIMA map of the high-mass star formation region DR 21 (OH). Contours show the 1.3-mm dust emission, grey scale shows the CO 2–1 line emission integrated over velocity, white line segments show the dust polarization, and black line segments show the CO linear polarization. *Right*: CN 1–0 line profiles toward DR 21 (OH). Observed data are histogram plots, fits are lines. *Top panel* shows the Stokes I spectrum with two Gaussians fitted. *Middle panel* shows the mean Stokes V spectrum for the four hyperfine components that have strong Zeeman splitting coefficients Z; the bottom panel shows the three components with weak Z. $B_{||}$ was fitted independently for the two Gaussian lines. The fields derived from these data are $B_{||} = -0.4 \pm 0.1\,\mathrm{mG}$ and $B_{||} = -0.7 \pm 0.1\,\mathrm{mG}$ for the velocity components at $-4.7\,\mathrm{km\,s^{-1}}$ and $-1.0\,\mathrm{km\,s^{-1}}$, respectively

7.4 S 106

S 106 is a bipolar H II region $\sim$ 0.5 pc in length embedded in an $\sim$ 4 pc diameter molecular cloud with $\bar{n}(H_2) \approx 1.4 \times 10^3\ \mathrm{cm}^{-3}$ and $M \approx 2000\, M_\odot$ (Schneider et al., 2002). Roberts et al. (1995) mapped $B_{||}$ in OH absorption lines with the VLA. Figure 11 shows the line optical depth profile, to which three Gaussian components have been fit. Component B is a narrow component that corresponds with the CO emission seen over the entire molecular cloud; this is gas undisturbed by the H II region. The broader component A arises in gas that has been shocked by the expansion of the H II region. The Zeeman effect is seen (Fig. 11) in component B, so the $B_{||}$ map is of the undisturbed molecular gas and not material that has been compressed into a shell surrounding the H II region. Figure 12 shows maps of N(OH) and $B_{||}$. The component B gas has a strong peak to the east of H II region, which is seen as a high-density clump in the molecular emission line maps; Schneider et al. (2002) find $N(H_2) \approx 3 \times 10^{22}\ \mathrm{cm}^{-2}$ for this clump.

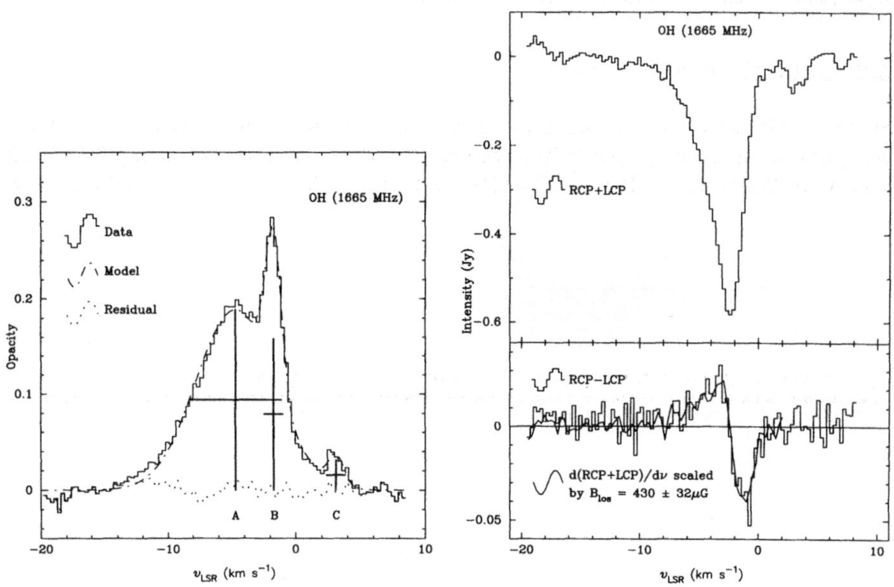

Fig. 11. *Left*: Optical depth profile for the 1665 MHz line toward S 106. *Right*: Stokes I and V spectra toward the position of maximum $B_{||}$ toward S 106

7.5 Maser Zeeman Observations

OH masers are found associated with the early stage of massive star formation, with maser spots coming from the dense ($\sim 10^7\ \mathrm{cm}^{-3}$) molecular envelope surrounding the massive star. Because of their brightness, they serve as signposts identifying sites of recently formed massive stars, and can be used to study kinematic and physical conditions in the dense molecular material. The ground state $^2\Pi_{3/2}, J = 3/2$ OH

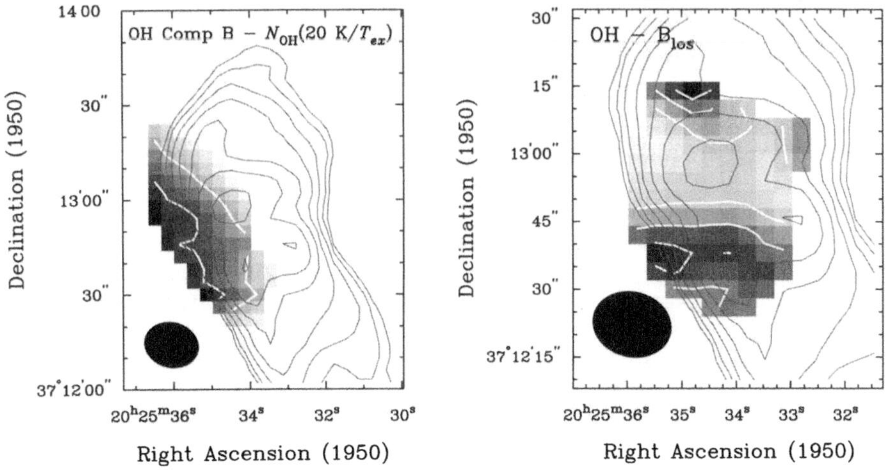

Fig. 12. *Left*: Map of N(OH) for the narrow 'B' line component toward S 106. For $T_{ex} = 50$ K (Schneider et al., 2002), contours are 1 and $2 \times 10^{15}\,cm^{-2}$. *Right*: Map of $B_{||}$ toward S 106. Contours are at 200, 300, and 400 μG

masers sometimes have clearly identifiable Zeeman pairs, that imply milligauss magnetic field strengths. Here B_{tot} is measured since the two Zeeman pairs are (generally) separated. Argon et al. (2000) surveyed 91 regions with the VLA A-array in both senses of circular polarization simultaneously, in order to identify Zeeman pairs.

Fish et al. (2003) analyzed this sample and found more than 100 Zeeman pairs in more than 50 regions. Field strengths range from ~ 0.1 mG to ~ 10 mG. They derived a magnetic field direction for each massive star formation region and looked for correlations, such as the correlations between maser field directions and the large-scale Galactic field suggested by Davies (1974) based on a much smaller data set. The more complete data did not show this correlation, which if present would have required a preservation in field direction between the very diffuse and the very dense gas.

Excited state OH ($^2\Pi_{3/2}, J = 5/2$ and $J = 7/2$) maser lines were observed by Caswell (2003, 2004). The excited-state masers tend to have fewer components and 'cleaner' Zeeman pairs than the ground-state masers. Field strengths are similar to those found in the ground-state maser lines.

Fiebig and Güsten (1989) detected Zeeman splitting in the ($6_{16} - 5_{23}$) H_2O maser lines toward W 3, Orion KL, W49N, and S 140 and inferred field strengths up to 50 mG. H_2O masers probe densities $\sim 10^{8-9}\,cm^{-3}$. Because H_2O does not have an unpaired electron, the Zeeman splitting is proportional to the nuclear magneton, and only $B_{||}$ could be measured. Sarma et al. (2002) used the VLA to continue these studies, finding $B_{||} \approx 13$–49 mG in four massive star formation regions. They argued that the masers arise in C-shock regions, and that the magnetic and turbulent energies are close to equilibrium. Sarma et al. (2001) used the VLBA to map four H_2O maser spots in W3 IRS5, finding that $B_{||}$ varied by a factor of three over

150 au but did not change sign. This might be expected if the masers and magnetic field are entrained in a coherent outflow.

8 Model Predictions and Observational Tests

Crutcher (1999)'s review of the molecular Zeeman-splitting measurements available at that time included a detailed discussion of physical conditions and an astrophysical discussion of the implications of the data. He found that magnetic fields play an important role in molecular clouds, as they do in the diffuse H I reviewed above. Typically $\beta_{th} \sim 0.04$ and $\beta_{turb} \sim 1$, so the turbulent and magnetic energy densities are comparable. He also discussed the 'mass to magnetic flux' ratio and the scaling of $B_{||}$ with density ρ. These topics will be considered in more detail below.

8.1 Mass-to-flux Ratio

In contrast to the diffuse H I, gravity plays an important role in molecular clouds. From the virial theorem and assuming flux freezing, one can straightforwardly derive the result that the ratio of gravitational to magnetic energy is independent of size. This, in turn, means that the relative importance of gravity and magnetism is maintained. This relative importance is measured by the 'mass to magnetic flux' ratio M/Φ, which is proportional to the ratio $N_\perp/B_{tot}$ (where $N_\perp$ is the column density perpendicular to the sheet or disk of matter, i.e., along the magnetic field direction for a magnetically supported cloud). We use the symbol $\mu_{intrinsic}$ to denote M/Φ in units of the critical value for a slab, $\mu_{intrinsic} = (2\pi G^{1/2})^{-1}$ (Nakano and Nakamura, 1978). Then

$$\mu_{intrinsic} = 7.6 \times 10^{-21} \frac{N_\perp(H_2)}{B_{tot}} . \tag{22}$$

In the ambipolar diffusion model clouds are initially subcritical, $\mu_{intrinsic} < 1$. Ambipolar diffusion is fastest in shielded, high-density cores, so cores become supercritical, and rapid collapse ensues. The envelope continues to be supported by the magnetic field. Hence, the prediction is that $\mu_{intrinsic}$ must be < 1 in cloud envelopes, while in collapsing cores $\mu_{intrinsic}$ becomes slightly > 1. Hence, this model tightly constrains $\mu_{intrinsic}$. On the other hand, the turbulent model imposes no direct constraints on $\mu_{intrinsic}$, although strong magnetic fields would resist the formation of gravitationally bound clouds by compressible turbulence. Also, if magnetic support is to be insufficient to prevent collapse of self-gravitating clumps that are formed by compressible turbulence, the field must be supercritical, $\mu_{intrinsic} > 1$. $\mu_{intrinsic}$ may take any value > 1, although of course for turbulence models that happen to have weak magnetic fields, clouds will be highly supercritical, $\mu_{intrinsic} \gg 1$ (Mac Low and Klessen, 2004).

If B_{tot} is strong, clouds will have a disk morphology with **B** along the minor axis. To properly measure $\mu_{intrinsic}$, one needs B and N along a flux tube, i.e., B_{tot} and $N_\perp$. We use our discussion in Sect. 3.4 to relate μ_{obs} to $\mu_{intrinsic}$, which is $\propto N_\perp/B_{tot}$. For a randomly oriented assembly of sheets all having the same $N_\perp$, the median N_{obs} is $2N_\perp$. For a randomly oriented set of uniformly strong magnetic fields, the median $B_{||} = B_{tot}/2$. Thus, the median value of the ratio $N_{obs}/B_{||}$ is

$4N_\perp/B_{\rm tot}$. However, it may be more appropriate to use the mean rather than the median value:

$$\left\langle \frac{M}{\Phi} \right\rangle = \int_0^{\pi/2} \frac{M_{\rm obs}\cos\theta}{\Phi_{\rm obs}/\cos\theta} \sin\theta d\theta$$
$$= \int_0^{\pi/2} \left(\frac{M}{\Phi}\right)_{\rm obs} \cos^2\theta \sin\theta d\theta = \frac{1}{3}\left\langle \frac{M}{\Phi} \right\rangle_{\rm obs} . \quad (23)$$

Thus, the mean value of the observed ratio is three times the intrinsic ratio, i.e. $\langle N_{\rm obs}/B_{||}\rangle = 3\langle N_\perp/B_{\rm tot}\rangle$.

Crutcher (1999) listed values of $\mu_{\rm obs} \propto N_{\rm obs}/2B_{||}$, which are derived from observed values instead of the intrinsic ones $N_\perp$ and $B_{\rm tot}$. He included the factor of 2 for the magnetic field, but not the additional correction factor for the column density. He noted that such a correction would be necessary for magnetically supported clouds that would have a disk morphology, but preferred not to apply an additional geometry factor since the morphology of the molecular clouds was not known directly from the observations. However, the prediction of the magnetic support model is a disk morphology, so one must apply the column density correction to test this model.

Crutcher reported the median $\mu_{\rm obs,1/2} = 2.2 \pm 0.3$. We conclude that for that sample of molecular clouds, the intrinsic and observed μ are related by $\mu_{\rm intrinsic,1/2} = \mu_{\rm obs,1/2}/2$ if we choose the median and by $\mu_{\rm intrinsic} = \mu_{\rm obs,1/2}/1.5$ for the mean. Therefore, $\mu_{\rm intrinsic,1/2} \sim 1.1$ (median) or 1.5 (mean). This puts these clouds into the regime in which magnetism is closely comparable to gravity. Presumably they are in general not currently suffering gravitational collapse, because they appear to be stable entities. (Once a core becomes supercritical, the time scale for collapse is very short, so few cores can be at this stage.) They are on the verge of becoming supercritical: in the absence of external perturbations, they will gradually evolve by ambipolar diffusion to the point where gravitational collapse can occur. Estimates of $\mu_{\rm obs}$ for additional clouds may be obtained from the OH Zeeman surveys of Bourke et al. (2001) and Troland and Crutcher (2005), and from estimates of $B_\perp$ with the Chandrasekhar–Fermi method applied to linear polarization maps of cores (Crutcher et al., 2004). Figure 13 shows all of the $\mu_{\rm intrinsic}$ now available, where the mean value correction of 1/3 has been used. That is, the plotted $\mu_{\rm intrinsic} = \mu_{\rm obs}/3$. The observations are distributed roughly equally above and below the $\mu_{\rm intrinsic} = 1$ line that divides subcritical and supercritical M/Φ ratios for disk geometries. Therefore, the data suggest that $\overline{\mu}_{\rm intrinsic} \approx 1$; that is, the typical mass to magnetic flux ratio is approximately critical. There is a slight indication that for large column densities, $\overline{\mu}_{\rm intrinsic}$ may be supercritical, and for small column densities, subcritical.

It is also relevant to consider mass-to-flux ratios in H I clouds, from which molecular clouds presumably form. Results from the Arecibo Millennium Survey showed that for all of the detections, the $\mu_{\rm obs}$ were significantly subcritical. Moreover, almost all of the non-detections were also consistent with $\mu_{\rm obs} < 1$. If these points were to be plotted on Fig. 13, they would lie to the left of and below the $\mu_{\rm intrinsic} = 1$ line. Hence, the H I data suggest that the precursors to molecular clouds are subcritical, as required by the magnetic support model.

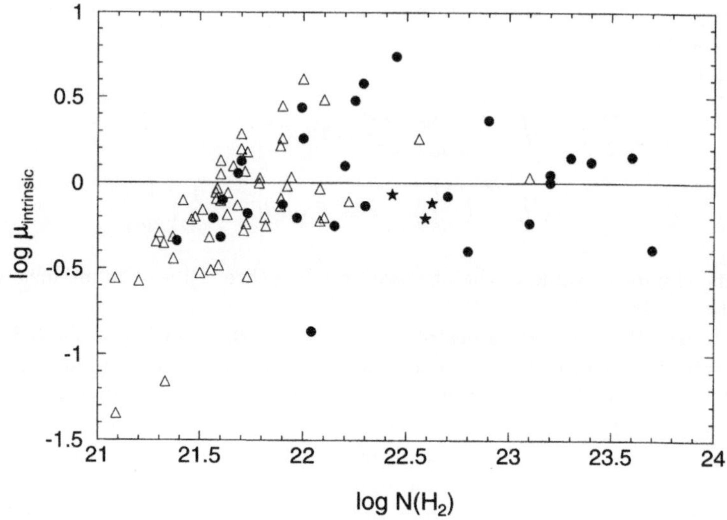

Fig. 13. $\mu_{\mathrm{intrinsic}}$ are the observed mass to magnetic flux ratios, divided by 3 to correct for projection bias, in units of the critical values. $\mu_{\mathrm{intrinsic}} > 1$ is supercritical, $\mu_{\mathrm{intrinsic}} < 1$ is subcritical. Dots are for Zeeman data with $B_{||} > 3\sigma(B_{||})$, stars are for Chandrasekhar–Fermi estimates of $B_{\perp}$, and triangles are lower limits plotted at $B_{||} = 3\sigma(B_{||})$. Although the statistical correction of 1/3 for geometrical bias has been applied to each point, so that statistically this plot should be valid, for any individual point the true μ could be higher or lower than the plotted $\mu_{\mathrm{intrinsic}}$. Some of the scatter is therefore still due to geometrical projection effects

In the ambipolar diffusion model the envelopes of dark clouds are the regions where M/Φ remains essentially unchanged while ambipolar diffusion drives M/Φ supercritical in the core. Hence, envelopes of dark clouds provide a crucial test of magnetic support models – M/Φ *must* be subcritical in these regions. Observations of dark-cloud cores were carried out by Crutcher et al. (1993), but the 18′ telescope beam size meant that the cores occupied a small fraction of the beam; mainly, the envelope regions were sampled. The result was $\overline{\mu}_{\mathrm{intrinsic}} \gtrsim 1$, rather than the $\mu_{\mathrm{intrinsic}} < 1$ required by magnetic support. However, the geometrical correction to the column density was not applied; with this correction, $\overline{\mu}_{\mathrm{intrinsic}}$ would be slightly subcritical, as required by the magnetic support model.

8.2 Scaling

The scaling of B_{tot} with density ρ is usually parameterized as $B_{\mathrm{tot}} \propto \rho^{\kappa}$, so our discussion will be in terms of κ. For strong magnetic fields, a cloud may be supported perpendicular to the field, but the field provides no support along the field. Then clouds will be disks rather than spheres. With the assumption that self-gravity is balanced only by internal thermal pressure along the symmetry axis z, $2\pi G\rho z^2 = c^2$ (this expression was derived for the plane-parallel or infinite thin disk case and first applied in astrophysics by Spitzer (1942) to the structure of the

Galaxy perpendicular to the plane). Then the expression for magnetic flux freezing ($\frac{M}{\Phi} \propto 2\pi\rho R^2 z/\pi R^2 B$) makes it possible to eliminate z from Spitzer's expression, yielding $B \propto \sqrt{\rho T}$. For an isothermal core, $\kappa = 1/2$. Detailed calculations of the evolution of a cloud collapsing due to ambibolar diffusion show that since the ambipolar diffusion timescale is much shorter in a core than in an envelope, the core will become supercritical and collapse while the envelope remains subcritical and supported by the field. Hence, B_{tot} in cloud envelopes remains virtually unchanged, so at lower densities no strong correlation between B_{tot} and density ρ is predicted, and $\kappa \sim 0$. As ambipolar diffusion increases M/Φ in a core, ρ increases faster than B_{tot} and κ increases rapidly. After the core becomes supercritical, it will collapse much more rapidly than the ambipolar diffusion rate, and κ continues to increase and approaches a limit of 0.5 (Ciolek and Basu, 2000).

Once a self-gravitating clump is formed by turbulence, if gravity exceeds both turbulent and magnetic support, the clump will collapse rapidly, at near the free-fall rate. Mestel and Spitzer (1956) considered the case of a spherically contracting cloud, for which the magnetic field was too weak to affect the collapse morphology; they showed that $\kappa = 2/3$ for this case. Hence, this would be the prediction for a core formed by turbulence with no significant magnetic support against gravity. On the other hand, if virial equilibrium is achieved between gravity and turbulence ($3GM^2/5R = 3M\sigma^2/2$), then $\rho R^2 \propto \sigma^2$. Flux freezing ($M \propto \Phi$) gives $\rho R \propto B_{\text{tot}}$, so $B_{\text{tot}} \propto \sigma\rho^{1/2}$ is predicted.

Determining κ observationally can distinguish between the various scenarios. $\kappa = 2/3$ implies a collapsing core with no significant magnetic or kinetic support. $\kappa < 0.5$ suggests a magnetically supported cloud, with $\kappa \rightarrow 0.5$ as M/Φ goes from subcritical to supercritical. Finally, $\kappa = 1/2$ but with an additional scaling of B_{tot} with the turbulent velocity dispersion σ is predicted for a core in virial equilibrium, with magnetic fields and turbulence (or thermal motions) providing support.

At low densities $n \sim 0.1$–$100\,\text{cm}^{-3}$, it has been clear for some time that there is no correlation of B_{tot} with ρ (Troland and Heiles, 1986). Crutcher's analysis of the higher density, molecular cloud data used the observed parameters N_{obs} and $B_{||}$ (not the intrinsic ones $N_\perp$ and B_{tot}). A least squares fit showed that $\log B_{||} \propto [\log n(H_2)]^{0.47}$, which is consistent with ambipolar diffusion driven contraction of clouds (Fiedler and Mouschovias, 1993) or, alternatively, with a constant Alfvénic Mach number M_{ALF}.

One year later, Basu (2000) extended Crutcher's analysis by including the velocity dispersion in the correlation. For slablike clouds, the combination of hydrostatic pressure equilibrium and the mass to flux ratio yields the expected relationship from Basu's equation (3),

$$B_{\text{tot}} = (8\pi)^{1/2}\sigma_v \rho^{1/2} \frac{c_1^{1/2}}{\mu_{\text{intrinsic}}} \tag{24}$$

where σ_v is the velocity dispersion and ρ the mean mass density across the slab. The parameter c_1 relates the midplane volume density to the mean density ($c_1 \geq 1$). Basu replotted Crutcher's points, with the remarkable result shown in Fig. 14: the rms scatter in $\log B_{||}$ dropped by nearly a factor of two, from Crutcher's fit with $\Delta(\log B_{||}) \sim 0.40$, to Basu's with $\Delta(\log B_{||}) \sim 0.23$. The data and Basu's fit are shown in Fig. 14 as the diamonds and solid line. The dashed line is the theoretical

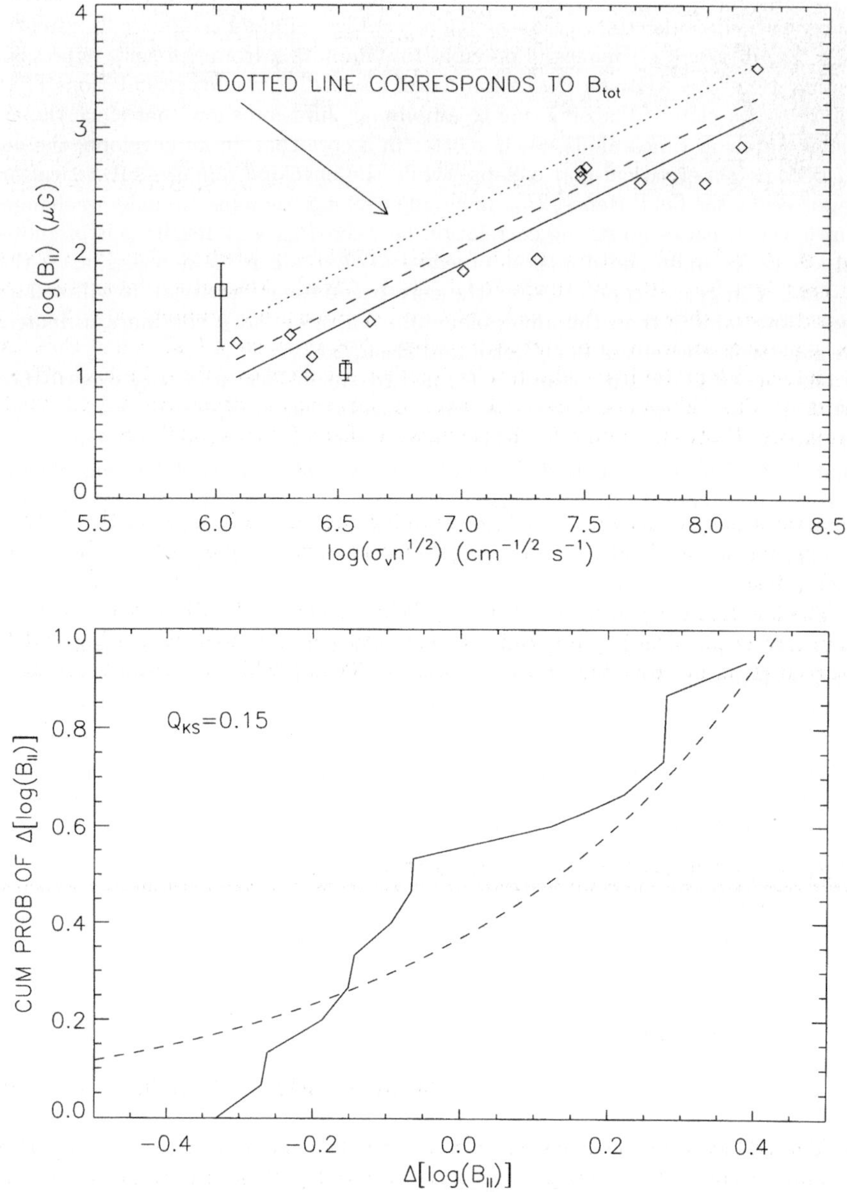

Fig. 14. The *top panel* shows molecular cloud data from Crutcher (1999), together with the least-squares fit by Basu (2000) (*solid line*), the correction to $B_{\rm tot}$ (*dotted line*), and the line for $\mu_{\rm intrinsic} = 1$ (*dashed line*). The *bottom panel* is the cumulative distribution of the residuals from the fit; the *dashed line* is the theoretical cumulative distribution from (26)

prediction from (24) for $c_1^{1/2}/\mu_{\text{intrinsic}} = 1$, which is parallel to and just little larger than the solid-line fit to the data.

The logarithmic rms dispersion $\Delta(\log B_{||}) \sim 0.23$ is remarkably small. This corresponds to dispersion of a factor of only 1.7 in magnetic field $B_{||}$; alternatively, because the slope is one, it also corresponds to a factor 1.7 in $\sigma_v n^{1/2}$. We expect large variations in $B_{||}$ because of the projection factor $\cos\theta$. We expect considerable uncertainty in the volume density n, because it is estimated using a variety of rather imprecise methods. And we also expect some cosmic scatter! The small residuals $\Delta(\log B_{||}) \sim 0.23$ show that this fit has physical meaning.

Basu's result is robust with respect to the addition of new data. The two squares with errorbars in Fig. 14 are new datapoints, published after his analysis. The one with small errorbars is from OH Zeeman splitting in L 1544 (Crutcher and Troland, 2000). The one with large errorbars is not regarded as a detection (Levin et al., 2001). Both are consistent with Basu's fit. Although there are additional Zeeman detections in the Bourke et al. (2001) and Troland and Crutcher (2005) surveys, data on ρ for these clouds are not yet available; these will provide an additional test of the robustness of the Basu result.

Basu's result convincingly shows that his model of the molecular clouds, which is slabs in which pressure, gravity, and magnetism all play important roles, is correct. The straightforward interpretation from comparing the solid and dashed lines in Fig. 14 is that the parameter $c_1^{1/2}/\mu_{\text{obs}}$ is close to unity, which implies both that there isn't much variation in density within the slab and also that the mass to flux ratio is close to the critical value.

We can go further by using the statistical discussion of Sect. 3 to relate the observed field to the total one. We consider two results where this extension is relevant.

We now return to Basu's correlation shown in Fig. 14. The scatter of the datapoints is small, and we must ask whether it is consistent with the statistical distribution of Sect. 3.2 for $\Delta \log B_{||}$. In particular, is the scatter too small to be consistent with a random distribution of orientation of magnetic field?

A least squares fit, such as done by Basu, selects the mean value of datapoints with respect to the fitted function. The residuals of the measured points are $\Delta(\log B_{||}) = \log B_{||} - \langle \log B_{||} \rangle$, where $\langle \log B_{||} \rangle$ is the mean of the distribution. As discussed in Sect. 3.2, the mean of $\log(B_{||}/B_{\text{tot}}) = -0.43$. The distribution of the residuals $\Delta \log(B_{||}/B_{\text{tot}})$ should follow

$$\psi\left(\Delta \log \frac{B_{||}}{B_{\text{tot}}}\right) = 0.85\ 10^{\Delta \log(B_{||}/B_{\text{tot}})} \tag{25}$$

We wish to compare this predicted distribution with the observed one. Such comparisons are best done on the cumulative distribution using the Kolmogorov–Smirnov (K-S) test. The cumulative distribution that corresponds to (25) is

$$\text{cum}\left(\Delta \log \frac{B_{||}}{B_{\text{tot}}}\right) = 0.368\ 10^{\Delta \log(B_{||}/B_{\text{tot}})} \tag{26}$$

The bottom panel of Fig. 14 shows the cumulative distribution of the residuals as the solid curve together with the predicted one as the dashed curve. The K-S test gives the probability P_{KS} that the two distributions are not dissimilar; here we have $P_{KS} = 0.15$, which although it seems small does indeed indicate that the distributions are consistent with being identical.

We conclude that Basu's fit to Crutcher's data is statistically consistent with a randomly oriented set of slabs. Being a least squares fit, Basu's result provides a value $\langle \log(B_{||}/B_{tot}) \rangle = -0.43$, meaning that it gives $B_{||}/B_{tot} = 0.37$. To obtain B_{tot} from this fit we should raise the fitted line by the factor $1/0.37 = 2.72$ (which is the base of Naperian logarithms e). The dotted line in the top panel of Fig. 14 shows this correction, which a factor 1.9 times higher than the dashed curve, which represents $\mu_{intrinsic} = 1$.

In (24), this means that the factor $c_1^{1/2}/\mu_{intrinsic} = 1.9$. Above we corrected Crutcher's observed mass-to-flux ratios to give $\mu_{intrinsic} \sim 1.1$. If this is accurate, then the molecular clouds are magnetically dominated subcritical slabs with density contrast of ~ 4. However, the uncertainties are such that a more appropriate summary statement is as follows: the molecular clouds are close to the cusp of being supercritical and have some density structure within the slab.

8.3 Morphology

In the magnetic support model, the dominant magnetic field means field lines should be smooth, without irregular structure. Clouds will be thin disks or oblate spheroids, since thermal pressure provides the only support along field lines. The field lines should be parallel to the minor axes of clouds. Finally, an original morphology with parallel magnetic field lines will be transformed into an hourglass morphology since it is the tension of the bent field lines that provides support. In the turbulent model, the magnetic field will be too weak to resist twisting by the dominant turbulence, and field lines will not be smooth but chaotic, with small-scale irregular structure. No correlation with cloud morphology is expected.

Maps of dust and spectral-line linear polarization and of the Zeeman effect generally show a regular field morphology (e.g., Figs. 8, 9, and 10), and an hourglass morphology is sometimes seen (e.g., Fig. 9; see also Schleuning, 1998). A regular field dominating a random field and an hourglass morphology toward cores are predictions of the strong magnetic field model. However, the magnetic field vector projected onto the sky is not observed to be parallel to the minor axes of starless cores as predicted by magnetic support (e.g., Fig. 8). Finally, even though fairly small, the dispersion in polarization position angles is often greater than observational errors (e.g., Fig. 8), implying that turbulence is producing an irregular component to the magnetic field.

9 Magnetic Field Observations, Present and Future

The field is currently in excellent health, with an unbiased survey of absorption lines that provide statistically reliable (if noisy) magnetic field strengths in the CNM, and a host of statistically biased measurements with some instrumental errors in emission regions. There are a number of molecular clouds with measured field strengths or sensitive limits, and study of the field morphology in the plane of the sky from dust and spectral-line linear polarization mapping is rapidly advancing. From all these measurements we conclude that the magnetic energy density is comparable to turbulence, or larger in some regions, and that molecular clouds are well-defined by models that incorporate both gravity and magnetism. These results are hard-won: they require much telescope time and, for the emission measurements, careful evaluation and correction of instrumental contributions.

What does the future hold? In particular, what can we expect from new instruments?

9.1 H I Zeeman in Absorption

Current Telescopes

The Arecibo Millennium survey, discussed in Sect. 4, has provided much useful statistical quantitative information about magnetic fields in the CNM. It used nearly 1000 hours of Arecibo telescope time to survey 79 sources in H I absorption, of which 40 (plus Cas A from HCRO) had useful sensitivity for Zeeman-splitting analysis. The survey was sensitivity limited. To significantly improve the statistics, one would want, say, four times as many sources. As we go for more sources we inevitably go for weaker sources, so a significant improvement would cost perhaps 10000 hours of Arecibo time. In our opinion, getting such a time block for Zeeman splitting measurements – indeed, for any single scientific project – is unlikely. And using any other telescope, with its necessarily lower sensitivity, takes even longer. Except for special purpose projects, we see no useful future for H I absorption Zeeman splitting measurements using existing telescopes[3].

The SKA

The Square Kilometer Array (SKA) will have sensitivity about 40 times larger than Arecibo. However, this doesn't mean that the sensitivity-limited results go $40^2 = 1600$ times faster. The reason is that any set of reasonable sources would all be stronger than the SKA's system noise so integration time would be independent of source flux or system sensitivity. In other words, 10 hours on the SKA would provide the same limiting magnetic field strength for both a 100 mJy source and a much stronger 1 Jy source. If a new Millennium survey were performed using 1000 hours of SKA time, then about the same number of sources could be covered as in the original Millennium survey. This would be nice, but would probably not represent a major scientific advance. We conclude that H I Zeeman-splitting absorption line survey work using the SKA is unlikely to prosper.

9.2 H I in Emission

Current and Future Telescopes

For H I emission, minimizing sidelobes, with their concomitant instrumental contribution to Zeeman splitting, is paramount. This rules out Arecibo (Heiles and Troland, 2005). It makes two telescopes very attractive:

[3] This statement applies only to diffuse H I. The excellent set of Zeeman-splitting measurements in H I associated with H II regions and supernova remnants, made with the VLA (e.g., Brogan and Troland, 2001) can be extended to many more sources.

1. The Green Bank Telescope. The GBT is totally unique as a single dish because, with its clear aperture, it should have no significant distant sidelobes. While its sidelobes are indeed low, nevertheless we see their effects, both in ordinary H I profiles (Stokes I) and also in Zeeman splitting (Stokes V). We have measured these sidelobes with complete sampling to $\sim 7^\circ$ from beam center and with incomplete sampling out to $\sim 24^\circ$. This larger field shows, surprisingly, that there seems to be little spillover from over-illumination of the secondary. Rather, most of the Stokes V effects come from within the smaller angular field. This is good news, because it means that it might be possible to correct for their instrumental contributions.

   We are currently studying the details of these sidelobes and expect to understand them well enough to subtract out their contribution to H I emission Stokes V spectra. The degree to which we can correct the GBT's sidelobes will determine what projects in H I emission are feasible. Projects for which the corrections should be easy include external galaxies other than M 31 (because emission is restricted in angle) and the CNM in the Milky Way (because lines are narrow). Projects for which success should depend more seriously on corrections include M 31 (emission is extended, with large velocity gradients) and the WNM in the Milky Way (lines are weak and broad). Time will tell which projects are feasible.
2. The Allen Telescope Array. The ATA is unique among arrays in having plenty of small baselines, which helps to provide good brightness temperature sensitivity. At the 21-cm line the angular resolution will about ten arcsec and the field of view some $2^\circ\!.5$; a long integration on one field of view will produce a map with 10^6 pixels. Moreover, the sidelobe properties of synthesis arrays are very well understood, so their effects should be removable with rather good accuracy. This will be an exciting instrument and has the potential of revolutionizing our understanding of magnetic fields in the ISM!

9.3 Molecular Clouds

Current Telescopes

The major telescopes used for Zeeman studies of molecular clouds are the VLA, Arecibo, the IRAM 30-m, and the GBT. Including the recently completed but unpublished survey of OH Zeeman toward dark clouds at Arecibo by Troland and Crutcher, there are 27 detections toward 81 positions or clouds. Because of the very large amount of telescope time that has been expended in the OH surveys, further advances with single-dish telescopes will probably come from Zeeman detections in CN and other species (excited OH, SO, C_2S, C_2H, ...) that sample high-density gas rather than from additional surveys in H I and the ground-state OH lines. The improvements to the VLA (including especially the new correlator) that will result in the EVLA will improve H I and OH absorption-line Zeeman mapping of clouds.

Current telescopes that have been actively used for mapping polarized dust emission include the CSO, JCMT, and BIMA. The upgrade of the SCUBA array on the JCMT and the combination of the BIMA and OVRO arrays into CARMA will lead to significant improvements in sensitivity that will allow many more clouds to be mapped with higher sensitivity. Similarly, CARMA should extend studies of linearly polarized line emission to additional clouds. And the SMA will complement CARMA with access to higher frequencies, although with a smaller number of antennas.

Future Telescopes

ALMA will very significantly improve the sensitivity available for dust polarization and spectral-line linear polarization observations. With its single-dish and compact array components, very large number of antennas, and high site, ALMA should routinely allow high fidelity polarization mapping over extended areas of molecular clouds. For Zeeman observations of millimeter-wave spectral lines, the improvement in sensitivity will be more modest, but should make possible mapping of $B_{||}$ in (for example) CN in a limited number of clouds.

Although as noted above the SKA will not make it possible to significantly improve the astrophysical results that were obtained from the Millennium Survey, its high sensitivity will greatly increase the surface density of background continuum sources that are strong enough for H I and OH Zeeman-splitting measurements, making it possible to measure and map magnetic field strengths in just about any specific cloud of interest.

It is a pleasure to acknowledge the pleasurable collaborations with many Zeeman-splitting friends over the years, especially Tom Troland. Tim Robishaw was indispensable for the GBT data. Mordecai-Mark MacLow made the important suggestion regarding equipartition of turbulence and magnetism, which we discussed in Sect. 4.5. This work was partially supported by NSF grants AST 02-05810 and AST 04-06987.

References

Argon, A. L., Reid, M. J., Menten, K. M.: 2000, ApJS 129, 159
Basu, S.: 2000, ApJ 540, L103
Batrla, W., Menten, K. M.: 1988, ApJ 329, L117
Beck, R.: 2001, Sp. Sci. Rev. 99, 243
Beck, R., Shukurov, A., Sokoloff, D., Wielebinski, R.: 2003, A&A 411, 98
Berkhuijsen, E. M., Haslam, C. G. T., Salter, C. J.: 1971, A&A 14, 252
Boulares, A., Cox, D. P.: 1990, ApJ 365, 544
Bourke, T. L., Myers, P. C., Robinson, G., Hyland, A. R.: 2001, ApJ 554, 916
Brogan, C. L., Troland, T. H.: 2001, ApJ 560, 821
Caswell, J. L.: 2003, MNRAS 341, 551
Caswell, J. L.: 2004, MNRAS 352, 101
Chandrasekhar, S., Fermi, E.: 1953, ApJ 118, 113
Chrysostomou, A., Aitken, D. K., Jenness, T., Davis, C. J., Hough, J. H., Curran, R., Tamura, M.: 2002, A&A 385, 1014
Ciolek, G. E., Basu, S.: 2000, ApJ 529, 925
Crutcher, R. M.: 1999, ApJ 520, 706
Crutcher, R. M., Troland, T. H.: 2000, ApJ 537, L139

Crutcher, R. M., Heiles, C., Troland, T.: 2003, in *Turbulence and Magnetic Fields in Astrophysics*, ed. E. Falgarone, T. Passot, Lecture Notes in Physics 614 (Berlin: Springer), p. 155

Crutcher, R. M., Nutter, D., Ward-Thompson, D., Kirk, J. M.: 2004, ApJ 600, 279

Crutcher, R. M., Troland, T. H., Goodman, A. A., Heiles, C., Kazès, I., Myers, P. C.: 1993, ApJ 407, 175

Crutcher, R. M., Troland, T. H., Lazareff, B., Paubert, G., Kazès, I.: 1999, ApJ 514, L121

Davies, R. D.: 1974, in *Galactic Radio Astronomy*, IAU Symp. 60, ed. F. J. Kerr, S. C. Simonson (Dordrecht: Reidel), p. 275

Davis, L., Greenstein, J. L.: 1951, ApJ 114, 206

Draine, B. T.: 2003, ARA&A 41, 241

Elmegreen, B. G.: 2000, ApJ 530, 277

Ferrière, K., Mac Low, M.-M., Zweibel, E.: 1991, ApJ 375, 239

Fiebig, D., Güsten, R.: 1989, A&A 214, 333

Fiedler, R. A., Mouschovias, T. Ch.: 1993, ApJ 415, 680

Field, G. B.: 1965, ApJ 142, 531

Fish, V. L., Reid, M. J., Argon, A. L., Menten, K. M.: 2003, ApJ 596, 328

Genzel, R., Downes, D.: 1977, A&AS 30, 145

Girart, J. M., Crutcher, R. M., Rao, R.: 1999, ApJ 525, L109

Goldreich, P., Kylafis, N. D.: 1981, ApJ 243, L75

Goodman, A. A., Heiles, C.: 1994, ApJ 424, 208

Heiles, C.: 1967, ApJS 15, 97

Heiles, C.: 1984, ApJS 55, 585

Heiles, C.: 1988, ApJ 324, 321

Heiles, C.: 1989, ApJ 336, 808

Heiles, C.: 1993, BAAS 25, 829

Heiles, C.: 1996a, ApJ 462, 316

Heiles, C.: 1996b, ApJ 466, 224

Heiles, C.: 1997, ApJS 111, 245

Heiles, C.: 1998a, Astrophys. Lett. and Comm. 37, 85

Heiles, C.: 1998b, in *The Local Bubble and Beyond*, ed. D. Breitschwerdt, M. J. Freyberg, J. Trümper (Berlin: Springer), p. 229

Heiles, C.: 2000a, AJ 119, 923

Heiles, C.: 2000b, in *Tetons 4: Galactic Structure, Stars, and the Interstellar Medium*, ed. C. E. Woodward, M. D. Bicay, J. M. Shull, ASP Conf. Ser. 231, p. 294

Heiles, C., Troland, T. H.: 2003, ApJ 586, 106

Heiles, C., Troland, T. H.: 2005, ApJ, in preparation

Heiles, C., Goodman, A. A., McKee, C. F., Zweibel, E. G.: 1993, in *Protostars and Planets III*, ed. E. H. Levy, J. I. Lunine (Tucson: Univ. of Arizona Press), p. 279

Heiles, C., Perillat, P., Nolan, M. et al.: 2001, PASP 113, 1247

Heiles, C., Robishaw, T., Troland, T. H.: 2003, Calibrating the GBT at L, C, and X Bands, GBT Commissioning Memo 23, available at `wwwlocal.gb.nrao.edu/~rmaddale/GBT/Commissioning/memolist.html`

Heitsch, F., Zweibel, E. G., MacLow, M.-M., Li, P. S., Norman, M. L.: 2001, ApJ 561, 800

Heyer, M. H., Vrba, F. J., Snell, R. L. et al.: 1987, ApJ 321, 855

Hildebrand, R. H.: 1988, QJRAS 29, 327
Hildebrand, R. H.: 2002, in *Astrophysical Spectropolarimetry*, Proc. XII Canary Islands Winter School of Astrophysics, ed. J. Trujillo-Bueno, F. Moreno-Insertis, F. Sánchez (Cambridge: Cambridge University Press), p. 265
Hildebrand, R.: 2003, New Astro. Rev. 47, 1009
Hildebrand, R. H., Davidson, J. A., Dotson, J. L., Dowell, C. D., Novak, G. F., Vaillancourt, J. E.: 2000, PASP 112, 1215
Jenkins, E. B., Tripp, T. M.: 2001, ApJS 137, 297
Kylafis, N. D.: 1983, ApJ 275, 135
Lai, S.-P., Girart, J. M., Crutcher, R. M.: 2003, ApJ 598, 392
Lazarian, A.: 2003, J. Quant. Spectrosc. Radiat. Trans. 79, 881 (also astro-ph/0208487)
Levin, S. M., Langer, W. D., Velusamy, T., Kuiper, T. B. H., Crutcher, R. M.: 2001, ApJ 555, 850
MacLow, M., Klessen, R. S., Burkert, A., Smith, M. D.: 1998, Phys. Rev. Lett. 80, 2754
MacLow, M., Klessen, R. S.: 2004, Rev. Mod. Phys. 76, 125
Matthews, B. C., Wilson, C. D., Fiege, J. D.: 2001, ApJ 562, 400
McKee, C. F., Ostriker, J. P.: 1977, ApJ 218, 148
Mestel, L., Spitzer, L., Jr.: 1956, MNRAS 116, 505
Mouschovias, T. Ch.: 1976, ApJ 206, 753
Mouschovias, T. Ch., Ciolek, G. E.: 1999, in *The Origin of Stars and Planetary Systems*, ed. C. J. Lada, N. D. Kylafis (Dordrecht: Kluwer), p. 305
Myers, P. C., Goodman, A. A., Güsten, R., Heiles, C.: 1995, ApJ 442, 177
Nakano, T., Nakamura, T.: 1978, PASJ 30, 671
Norman, C., Ikeuchi, S.: 1989, ApJ 345, 372
Norris, R. P., Booth, R. S., Diamond, P. J., Porter, N. D.: 1982, MNRAS 201, 191
Ostriker, E. C., Stone, J. M., Gammie, C. F.: 2001, ApJ 546, 980
Padoan, P., Goodman, A., Draine, B. T., Juvela, M., Nordlund, Å., Rögnvaldsson, Ö. E.: 2001, ApJ 559, 1005
Pereyra, A., Magalhaes, A. M.: 2004, ApJ 603, 584
Purcell, E. M.: 1979, ApJ 231, 404
Purcell, E. M., Spitzer, L.: 1971, ApJ 167, 31
Roberts, D. A., Crutcher, R. M., Troland, T. H.: 1995, ApJ 442, 208
Sarma, A. P., Troland, T. H., Romney, J. D.: 2001, ApJ 554, L217
Sarma, A. P., Troland, T. H., Crutcher, R. M., Roberts, D. A.: 2002, ApJ 580, 928
Schleuning, D. A.: 1998, ApJ 493, 811
Schneider, N., Simon, R., Kramer, C., Stutzki, J., Bontemps, S.: 2002, A&A 384, 225
Schwarz, U. J., Troland, T. H., Albinson, J. S., Bregman, J. D., Goss, W. M., Heiles, C.: 1986, ApJ 301, 320
Slavin, J. D., Cox, D. P.: 1992, ApJ 392, 131
Spitzer, L., Jr.: 1942, ApJ 95, 329
Tomisaka, K.: 1990, ApJ 361, L5
Troland, T. H.: 2004, in *The Magnetized Interstellar Medium*, ed. B. Uyanıker, W. Reich, R. Wielebinski (Katlenburg-Lindau: Copernicus), p. 105
Troland, T. H., Crutcher, R. M.: 2005, ApJ, to be submitted
Troland, T. H., Heiles, C.: 1982, ApJ 372, 179

Troland, T. H., Heiles, C.: 1986, ApJ 301, 339
Troland, T. H., Crutcher, R. M., Heiles, C.: 1985, ApJ 298, 808
Trumpler, R. J., Weaver, H. F.: 1953, *Statistical Astronomy* (New York: Dover)
Verschuur, G. L.: 1969, ApJ 156, 861
Verschuur, G. L.: 1974, in *Galactic and Extra-Galactic Radio Astronomy*, ed. G. L. Verschuur, K. I. Kellermann (New York: Springer), p. 179
Verschuur, G. L.: 1989, ApJ 339, 163
Verschuur, G. L.: 1993, BAAS 25, 1466
Verschuur, G. L.: 1995a, ApJ 451, 624
Verschuur, G. L.: 1995b, ApJ 451, 645
Vrba, F. J., Strom, S. E., Strom, K. M.: 1976, AJ 81, 958
Wolf, S., Launhardt, R., Henning, T.: 2003, ApJ 592, 233
Wolfire, M. G., McKee, C. F., Hollenbach, D., Tielens, A. G. G. M.: 2003, ApJ 587, 278
Wolleben, M., Reich, W.: 2004, in *The Magnetized Interstellar Medium*, ed. B. Uyanıker, W. Reich, R. Wielebinski (Katlenburg-Lindau: Copernicus), p. 99
Woltjer, L.: 1967, in *Radio Astronomy and the Galactic System*, IAU Symp. 31, ed. H. van Woerden (London: Academic Press), p. 479
Woody, D. P., Scott, S. L., Scoville, N. Z., Mundy, L. G., Sargent, A. I., Padin, S., Tinney, C. G., Wilson, C. D.: 1989, ApJ 337, L41

Stellar Magnetic Fields*

Leon Mestel[1] and John D. Landstreet[2]

[1] Astronomy Centre, University of Sussex, Falmer, Brighton, BN1 9QH, England
lmestel@sussex.ac.uk
[2] Department of Physics and Astronomy, University of Western Ontario, London, ON N6A 3K7, Canada
jlandstr@astro.uwo.ca

Abstract. Stellar magnetic fields are directly detected or inferred across the whole Hertzsprung–Russell Diagram. Attention in this chapter is concentrated on pre-main sequence and on late- and early-type main sequence stars, with some discussion also of red giants and white dwarfs, and with a brief reference to neutron stars. The current observational situation is reviewed, and some of the consequences for our ideas on star formation, on stellar structure and evolution and on stellar rotation are summarized.

1 Stellar Magnetism

1.1 General Theoretical Considerations

The gross dynamical effect of a large-scale magnetic field on a self-gravitating mass can be estimated formally from the Chandrasekhar–Fermi virial theorem, which compares the total magnetic and gravitational energies, yielding as a convenient parameter the non-dimensional flux-to-mass ratio

$$f \equiv \frac{F}{G^{1/2}M} , \tag{1}$$

where F is the magnetic flux threading the mass M. The strongest field-strength observed on a non-collapsed star is that of Babcock's Ap star HD215441, which has a surface polar field of $\simeq 3.4 \times 10^4$ G. If this field strength were to increase inwards by as much as a factor $\approx 10^2$, the total magnetic energy would still be only one-thousandth of the gravitational; and for most stars, it would be very much smaller. Thus over the bulk of a star, the parameter $f \ll 1$: the star is magnetically 'weak', with the Lorentz force making a negligible contribution to the balance against gravity.

It is possible or even likely that the parent clouds from which stars form are by contrast magnetically 'strong', so that the first proto-stars begin their lives with a much larger flux/mass ratio; but the excess magnetic flux may very well have been destroyed either in the late pre-opaque phases of star formation or in the optically thick pre-main sequence phases (cf. Sect. 2.2). However, suppose that a star has reached the main sequence while retaining a markedly higher flux than is inferred from observation. As the simplest example, suppose a stellar field within a

* **This chapter is dedicated to the memory of Horace Babcock.**

sub-adiabatic, radiative zone were to consist of a toroidal flux tube, in pressure equilibrium with its non-magnetic surroundings, and thin enough to be approximated as a cylinder, so that

$$p_{\rm i} + B^2/8\pi = p_{\rm e} \ , \tag{2}$$

where the suffices i, e stand respectively for 'internal' and 'external'. If the internal and external temperatures are the same, the thermal pressure deficit requires a lower internal density, and by Archimedes' principle the tube will rise and expand adiabatically, approaching a state in which the internal and external densities are the same, and the pressure deficit is made up rather by a lower internal temperature ($T_{\rm i} < T_{\rm e}$). The consequent heat flow down the local temperature gradient into the tube will then cause the tube to rise in a thermal time-scale, with a compensating down-flow in the domain outside the tube. The process is analogous to the Eddington–Sweet-type circulation that occurs in a stellar radiative zone where centrifugal acceleration is the dominant perturbation (e.g. Mestel, 1999). In the low-temperature surface regions, the rising tubes will be destroyed by the much higher Ohmic resistivity.

A realistic stellar field will almost certainly consist of mutually linked poloidal and toroidal components, if only to satisfy the requirements (necessary but not sufficient) for dynamical stability Wright (1973); Wilson (1978); Markey and Tayler (1973, 1974); Tayler (1980); Braithwaite and Spruit (2004). Nevertheless, extrapolation from this over-simple case suggests that, in general, magnetic flux will tend to float to the surface in a time of order

$$\frac{\tau_{\rm KH}}{(F/G^{1/2}M)^2} \ , \tag{3}$$

where $\tau_{\rm KH}$ is the global Kelvin–Helmholtz time. For this to exceed the nuclear lifetime of the star, $F/G^{1/2}M$ must be small: even if a star had managed to reach the main sequence as magnetically 'strong', it would inevitably have become magnetically 'weak' over the bulk of the main sequence lifetime and later.

The most striking effect of a large-scale **B**-field is on the star's rotation field. Consider first the simplest case of a radiative domain in an axisymmetric star – the aligned rotator – with a purely poloidal field $\mathbf{B_p}$ having its axis coinciding with the axis of an initial non-uniform rotation. With the field frozen into the gas, the shear then generates a toroidal component $\mathbf{B_t}$, and the consequent toroidal Lorentz force component $(\nabla \times \mathbf{B_t}) \times \mathbf{B_p}/4\pi$ generates torsional Alfvén waves which redistribute angular momentum along individual field lines via the Maxwell tensions. The smallness of the ratio $F/G^{1/2}M$ implies that the Alfvén travel time along a field line traversing the bulk of the star is much longer than the gravitational free-fall time, but it is still much less than the Kelvin–Helmholtz time, and a fortiori than the nuclear evolution time, even for very low field strengths. When the initial shear is large, strict flux-freezing will yield large values of $|B_{\rm t}/B_{\rm p}|$. Hydromagnetic instabilities may in fact set in and act as an effective macro-resistivity, preventing the generation of a toroidal component $|B_{\rm t}|$ much greater than $|B_{\rm p}|$, and so limiting the growth in the toroidal force, but the estimated times for shear reversal remain short compared with the K–H time (Mestel and Weiss, 1987).

With axial symmetry still assumed, damping of the torsional oscillations may in principle allow a weakly viscous radiative domain to settle into a steady state of *isorotation*, with each field line rotating with its own angular velocity (Ferraro,

1937), virtually uncoupled from its neighbours. Axial symmetry is however an oversimplification, and in fact a plausible phenomenological model for the early-type main sequence magnetic stars is the essentially non-axisymmetric *oblique rotator*, with the axis of the large-scale magnetic field inclined to the rotation axis (cf. Sect. 6). Any relative shearing of individual field lines will now generate azimuthal magnetic pressure gradients which distribute angular momentum between different flux tubes, so that the only likely asymptotic steady state is that of near uniform rotation.

Further, whereas in the axisymmetric problem, rotational shearing generates a toroidal field but has only a weak effect on the basic poloidal field, rotational distortion of a highly oblique field can lead to the juxtaposition of oppositely directed lines of the basic field, and so can cause accelerated Ohmic decay (Rädler, 1986). A weakish field that is initially highly oblique could thus be converted into a more nearly aligned field, simply through the accelerated decay of the component perpendicular to the rotation axis. By contrast, a strong enough field should be able to reverse the initially imposed non-uniform rotation before the accelerated Ohmic decay gets under way; dissipative processes will again act to destroy the non-uniform rotation, and the star will settle down as an oblique rotator. An idealized illustrative model (Moss et al., 1990; Moss, 1992) does indeed find a critical value for the ratio of Alfvén speed to rotation velocity.

It is clear that any discussion of the rotation law in a radiative domain which ignores magnetic effects is implicitly putting a very severe constraint on the strength of any magnetic field present. In a convective zone there is an ongoing spontaneous conversion of local gravitational energy into turbulent energy, and both the structure of an imposed magnetic field and the distribution of angular momentum will be affected strongly by the Reynolds stresses of the turbulence. Many computations demonstrate how magnetic flux, prescribed at the boundaries, tends to be concentrated into thin tubes within which the field is dynamically dominant (reminiscent of sunspot models), surrounded by domains which are nearly flux-free and in which normal convection persists (cf. Proctor and Weiss, 1982, and references therein, and Mestel, 1999). Parametrization of the convection by generalized mixing length theory, incorporating both anisotropic macroscopic viscosity and heat conductivity, appears to have some success in yielding a rotation law for the Sun's convective envelope that is consistent with helioseismological data (Kitchatinov and Rüdiger, 1995).

The phenomenal development of supercomputers is enabling detailed numerical study of non-magnetic and magneto-convection in rotating media, and associated dynamo models, over a wide domain of Rossby, Taylor and Reynolds numbers. The computations enable one to assess both the usefulness and the limitations of the classical mixing length formulation, and the standard 'mean-field' dynamo equations.

A full discussion of the rotation field in a stellar radiative zone – such as the radiative core of the Sun – must clearly take account of the boundary conditions on Ω and $\mathbf{B}$ set by the dynamics and electrodynamics of the contiguous convective envelope. In the low-density surface regions, the magnetic energy density can be comparable with the thermal; but in many model calculations, the magnetic tension and pressure terms tend to combine to yield a field that is nearly curl-free or force-free in the mean, so that the magnetic perturbing forces are weaker than would be anticipated from a naive comparison of energy densities. Well above the photosphere, in the surrounding magnetosphere, the mean thermal pressure will fall off with height more rapidly than the magnetic pressure, so that the field will tend to be nearly force-free except in local pinched zones. Coupling of the field emanating from a rotating star with external gas will lead to outward transport of angular momentum by the Maxwell stresses exerted by the twisted field (cf. Sects. 3–7 below).

We now look in more detail at the structure, evolution, and effects of magnetic fields in specific stellar settings. As we proceed with this examination, we will include in each section summaries of the most important information and constraints available from observations. As a preliminary to this aspect of the chapter, we summarize briefly the principal methods of field measurement.

1.2 Stellar Magnetic Fields: Measurement Techniques

The Zeeman effect (and the closely related Paschen–Back effect) makes possible the observation of magnetic fields in settings ranging from the interstellar medium, where the observed field strengths are of the order of some μG ($1\,\mathrm{G} = 10^{-4}\,\mathrm{T}$), to main sequence stars, many of which have fields of the order of 10^3 G, to a few percent of white dwarfs, some with fields well in excess of 1 MG. This physical effect thus makes possible the detection of magnetic fields over a range of more than 12 orders of magnitude in field strength. Briefly, in the presence of a magnetic field, an atomic or molecular energy level E_i of total angular momentum $J\hbar$ splits into $2J+1$ magnetic sublevels. Transitions between such levels, which generally must respect the selection rule $\Delta M_J = 0, \pm 1$ where M_J is the magnetic quantum number, then lead to spectral lines which are also split into several components at slightly different wavelengths. For fields of less than a few tens of kG, the components with $\Delta M_J = 0$ (the π components) are distributed symmetrically about the wavelength of the unperturbed line, while the σ_+ components of $\Delta M_J = +1$ are displaced to one side of the unperturbed line position and the σ_- components are displaced to the other side of the unperturbed position. Up to a field of the order of 10^6 G, the displacement of the mean position of the $\sigma_\pm$ components is proportional to the field strength. The typical separation between the π and one of the σ groups is

$$\Delta\lambda_B = \bar{g} e B \lambda^2 / 4\pi m c^2 \,, \tag{4}$$

where $\bar{g}$ is a number of order 1 which varies from one transition to another. In familiar units, the splitting is

$$\Delta\lambda_B(\mathrm{nm}) = 4.67 \times 10^{-3} \bar{g} B(\mathrm{kilogauss}) \lambda(\mathrm{microns})^2 \,. \tag{5}$$

Thus, if a stellar field is of the order of $2\,10^3$ G, this splitting may exceed both the local thermal width and the rotational Doppler broadening of the line, making possible a direct measurement of the magnitude of the field strength $\langle B \rangle$ (the 'mean surface field') averaged over the visible hemisphere. However, in many magnetic stars, the stellar rotation is large enough to mask the Zeeman splitting, and the field must be detected in some other way.

In the presence of a longitudinal field, the two σ components are circularly polarized with opposite handedness, and the π components vanish. If the field is transverse, the π components (in emission) are linearly polarized parallel to the field, while the σ components are linearly polarized normal to the field. The fact that the mean wavelength of the two σ components is different when a longitudinal field component is present is the basis of the most sensitive method of measuring a stellar field. This difference in mean wavelength gives rise to circular polarization in the wings of the local line profile (the profile at one point on the stellar disk) because at a given wavelength in the line wing the depth in the σ_+ and in the σ_- profiles is not the same. Since the circular polarization in the continuum is extremely small (except for fields far in excess of 1 MG), this polarization can be detected when it is as small as a few parts in 10^5, and stellar fields as small as a few G are measurable.

In the Sun, measurement of this line wing circular polarization makes it possible to deduce the strength of the local line-of-sight component of the field at the point of observation. However, in a star, the resolved disk is not observable. In this case, the simplest measure that such an observation provides is the value of the 'mean longitudinal field' $\langle B_z \rangle$, the line-of-sight component averaged over the visible disk. Clearly this quantity will be non-zero only if the field has a relatively simple structure. In practice, non-zero values of $\langle B_z \rangle$ seem to occur largely in stars whose overall field structure is topologically dipolar (i.e. much of the stellar magnetic flux emerges from one hemisphere of the star and re-enters in the opposite hemisphere). If the stellar magnetic field is topologically sufficiently complex so that $\langle B_z \rangle$ vanishes, but is not too complex, the local field may still be detectable if the stellar disk is resolved in the line profile via the Doppler shift of rotation. Detailed discussions of various methods of field measurement may be found, for example, in Landstreet (1982) and Mathys (1989).

The first detection of any cosmic magnetic field was the detection by Hale (1908) of the circular polarization due to the Zeeman effect in the spectral lines of sunspots. Study of the magnetic field of the Sun became truly systematic with the invention of the solar magnetograph by Horace and Howard Babcock (1952); with this device, a map of the distribution of B_z over the visible solar hemisphere could be obtained, and soon such observations were being obtained on a routine basis. The first magnetic field in a star other than the Sun was discovered by Horace Babcock (1947) after he built a circular polarization analyzer for the coudé spectrograph of the Mount Wilson 100-inch telescope and used it to observe the undistinguished A star 78 Vir.

Instruments for measuring fields in both the Sun and in other stars have undergone steady improvement since. One of the most important advances has been the development of spectropolarimeters capable of measuring all four Stokes parameters (i.e. intensity, circular and both independent components of linear polarization) in spectral lines. Circular polarization measurements have been obtained since the first experiments by Hale, and half a century later, by Babcock. In contrast, linear polarization measurements require detection of polarization that is typically at least an order of magnitude smaller than the circular polarization signal, and are consequently far more difficult to make. Such four-Stokes parameter measurements (using 'vector magnetographs') provide enough information to recover the vector direction and strength of the local magnetic field from solar observations, thus making possible true field mapping over the visible hemisphere (although this is still largely restricted to the level of the photosphere).

An exciting recent measurement advance in stellar magnetic measurements has been the development of the capability by the MuSiCoS spectropolarimeter (Donati et al., 1999a; Wade et al., 2000) to obtain routine observations of polarization profiles of many spectral lines of a magnetic Ap stars in all four Stokes components. This has finally made possible accurate reconstruction of surface magnetic field distributions in magnetic middle main sequence stars.

Another effect, in addition to the high-field form of the Zeeman effect in spectral lines, becomes measurable for fields of a few MG or more. As pointed out by Landstreet (1967) and Kemp (1970), the fact that electrons spiral around field lines with a definite handedness means that the absorption of right circularly polarized continuum radiation is different from the absorption of left circularly polarized radiation, and the net continuum radiation of the star becomes circularly polarized at a detectable level. For fields of order 10^2 MG, continuum linear polarization is also found. This effect has provided a valuable method of detecting really large fields in white dwarfs.

2 Magnetism and Star Formation

2.1 Star Formation: the Early Phases

Magnetic fields have been observationally detected in a number of settings in the interstellar medium via the Zeeman effect in the 21-cm line of neutral H and in lines of the OH molecule. In diffuse clouds (regions which do not appear at present to be gravitationally unstable), the observed field line-of-sight components $\langle B_z \rangle$ are typically of the order of a few µG (e.g. Myers et al., 1995; Heiles, 1997). A number of observations are also available of regions in molecular clouds, in which it appears that the earliest stages of star formation are occurring. The observed $\langle B_z \rangle$ values in such molecular clouds range from 10 or 20 µG in regions with inferred particle densities of the order of $n \sim 10^3$ H_2 molecules cm^{-3}, to about 1 mG in regions with $n \sim 10^5 - 10^6\,\mathrm{cm}^{-3}$ (Crutcher, 1999, see Chap. 7).

In considering the dynamics of star formation from the magnetized galactic gas, there are two related but distinct questions: (a) how do self-gravitating bodies manage to contract to high densities; and (b) why are stars magnetically 'weak', in the sense of having the ratio (1) much below unity. Let us suppose that the interstellar gas that ultimately forms the star has accumulated from the mean galactic background into a roughly spherical blob, *subject to strict freezing of the galactic field.* Then in terms of the canonical background values $B_0 \simeq 3 \times 10^{-6}$ G in domains with number density $n_0 = 1$, to form a body that is just gravitationally bound – i.e. with the ratio (1) no more than about 2 – the accumulating column must have a length greater than $L_0 \simeq B_0/G^{1/2}\rho_0 \simeq 10^3(B_0/3 \times 10^{-6})/(n_0/1)$ parsec. As a corollary: to produce a body with the ratio (1) as small as in Babcock's star, the accumulation length would have to be absurdly long – 40 kpc is a conservative estimate; and a fortiori for other magnetic stars.

The conclusion is that at some stage, there must be a radical departure from flux-freezing. In the first discussion, (Mestel and Spitzer, 1956), it was suggested that in a lightly ionized gas cloud, sufficiently rapid flux loss would occur by the process of plasma drift, later referred to as 'ambipolar diffusion', and discussed in many papers since (e.g. Nakano, 1976; Mouschovias, 1987; Shu et al., 1987; Barker and Mestel, 1996). In essence, it is the small fraction of gas in the form of the heavy ions and associated electrons (the 'plasma') which is both inductively coupled to the field and which also feels the Lorentz force. There is thus a steady outward leakage of flux-plus-plasma with respect to the neutral bulk as the gravitationally distorted field tries to straighten itself, resisted by the friction due to collisions between the ions and the neutral particles.

In the original treatment, it was thought that in dusty H I clouds, the plasma density could become low enough for the leakage time-scale to be shorter than the free-fall time, so that the 'magnetic flux problem' would be resolved early in the star formation process. It was later recognized that because of cosmic ray ionization, ambipolar diffusion would not be important until the molecular cloud phase. Within such a cloud, differential flux-diffusion – with a time-scale longer than the free-fall time – may lead to the separation-out of 'fragments' which are gravitationally-bound, with a mildly sub-critical flux-to-mass ratio, and which can contract as proto-stars. The persisting very strong magnetic field would in fact assist the contraction process through its efficient transport of excess angular momentum.

There is still controversy as to how and at what stage these very strongly magnetic bodies lose the bulk of their remnant flux. Nishi et al. (1991) have argued that in the final pre-opaque phases of contemporary star formation, the currents maintaining the field are carried by ions, which through collisions with the negatively charged grains suffer a large Ohmic dissipation, leaving fragments with flux much below the virial limit, i.e. with f defined by (1) much below unity. But this conclusion is challenged by Mouschovias (2001), who argues that ambipolar diffusion becomes fast enough in the late molecular cloud phases. In a recent study, Heitsch et al. (2003) present the results of numerical experiments that illustrate how turbulence in a weakly ionized gas can yield a markedly enhanced rate of ambipolar diffusion. And one can plausibly argue that a fragment which has retained an excess magnetic flux will lose it by buoyancy once the fragment has become an optically thick 'proto-star'.

2.2 Star Formation: the Later Phases

A newly formed star is believed to approach the main sequence through a Hayashi-type phase, in which stars of solar mass or less will be fully convective, while more massive stars will have an extensive outer convective zone surrounding a radiative core. In a convective domain that is even slightly superadiabatic, a field of energy density greater than that of the turbulence will again be subject to magnetic buoyancy, with flux tubes rising at a dynamical rather than a thermal rate. Dissipation of this excess magnetic energy has in fact been suggested as at least a contributory power source for the T Tauri phenomenon.

There has been a variety of suggestions for the effect of a dynamically dominant turbulence on a primeval magnetic field. Tangling of the field reduces the length scale and so decreases the local Ohmic decay time to much below the global Cowling decay time; however, this need not imply accelerated flux destruction, but rather expulsion of the flux from the bulk of the turbulent domain. An early study by Spitzer (1957) showed that flux expelled from the bulk of the zone may just be compressed against the boundaries, without any significant reduction in the decay time of the dipole moment. A similar picture emerges from the cited studies by Weiss and colleagues, which show demarcation into respective zones that are nearly flux-free and those with the Maxwell and Reynolds stresses locally comparable. However, for a long while, workers in dynamo theory argued that toroidal flux tubes, generated within or just below a superadiabatic, convective envelope domain, would again be driven upwards by magnetic buoyancy, leading to an embarrassingly rapid loss of flux at the surface. More recently a new consensus appears to have arisen: at least for convective domains surrounding a radiative core, three-dimensional numerical simulations (e.g. Nordlund et al., 1992) confirm earlier suggestions that there is strong downward turbulent pumping of flux into the stably stratified region below.

Provisionally, one can argue as follows. A star that has managed to reach the Hayashi phase with flux anything like the virial maximum will indeed suffer spontaneous loss, until the magnetic energy is less than the turbulent energy. A star of mass $\simeq 2\,M_{\odot}$ will retain during the Hayashi phase a sizeable radiative core which can retain remnant flux pumped down into the core by the turbulence. This need not prevent some primeval field lines emanating from the star along the magnetic axis, to couple with external gas. As such a star approaches the main sequence, its surface temperature reaches 10^4 K or more, the photospheric hydrogen becomes almost fully ionized, and the envelope convection dies out. Simultaneously, the onset of the C-N cycle generates a convective core, and any remnant primeval flux may very well be expelled into a layer between the core and the radiative envelope. Subsequent diffusion or advection by laminar circulation can yield an observable flux emanating from the photosphere. A less massive star such as the Sun, fully convective during the Hayashi phase, retains a convective envelope on the main sequence, but energy generation by the pp-chain dominates and allows a radiative core to develop. It is perhaps less clear that a significant primeval flux will have survived the epoch of full convection, but if so, it will again be largely confined to the radiative interior.

In the original Hayashi work, stars are supposed born with radii $R \simeq 50 - 100R_{\rm ms}$. The surface temperature is prevented from falling below $T_{\rm Hay} \simeq 4000\,{\rm K}$, by the requirement that the opacity of the H^--ion should yield a surface optical depth near unity. The consequent surface radiation loss far exceeds the Eddington estimate for radiative transport from within, so that all masses are found to begin their pre-main sequence lives as fully convective. More recent models (e.g. Palla and Stahler, 1993) have proto-star formation by accretion on to a core, with very much lower initial radii and luminosities. Low mass stars, with $M < 2\,M_{\odot}$, are found still to begin as fully convective. Intermediate masses have radiative cores: if $M = 2.5\,M_{\odot}$, the convective outer zone has mass $\simeq .7M$; if $M = 3\,M_{\odot}$, only $(1/3)M$; while for $M = (4-5)\,M_{\odot}$, the outermost layers are fully ionized, $T_{\rm s} > T_{\rm Hay}$, and the energy transport through the star is given essentially by the Eddington theory, as in the earlier Henyey calculations of pre-main sequence contraction. As the main sequence is approached, low mass stars retain their convective envelopes but develop radiative cores. Intermediate mass stars lose their outer convective zones, but the onset of the highly temperature-sensitive CN-cycle yields convective cores in intermediate and high mass stars.

This revised picture is particularly relevant to our attempts to understand the magnetic properties of main sequence stars. Low mass stars pass through a fully convective Hayashi phase. Although not rigorously proven, it is at least plausible that any fossil magnetic flux will have been tangled up by the turbulence and Ohmically destroyed. (For an alternative view, see Moss, 2003) A relic d-c dynamo-built field will be trapped in the main sequence radiative core, and can be of importance for study of the internal solar rotation, but the observable flux will be that maintained by the periodic or chaotic contemporary dynamo. By contrast, during the Hayashi phase of an intermediate mass star, any primeval flux expelled from the convective envelope can persist in the radiative core. In the approach to the main sequence, the envelope convection dies out; and on the main sequence, the flux threading about the central fifth of the mass is expelled from the developing convective core. The field structure evolves steadily towards that of the slowest decaying Cowling mode, subject to the diamagnetic convective core condition. The process of diffusion-plus-floating towards a state with observable flux penetrating the surface is estimated to take 10^8 yr, indeed a sizeable fraction of the main sequence lifetime. In the more massive stars, the absence of a Hayashi convective envelope allows any primeval flux to penetrate the surface during the approach to the main sequence, so that such stars should appear magnetic throughout their main sequence phase.

3 Pre-main Sequence Stars: Observation and Theory

An important development in recent years has been the direct detection and measurement of magnetic fields in a number of pre-main sequence (PMS) stars. The presence of fields in PMS stars has been suspected for some time on the grounds that they are needed to transfer angular momentum to accreting material so as to preserve the rather low rotation velocities of these objects in spite of accretion and contraction. Now the expected fields are being observed.

One fruitful method of field detection in T Tauri stars, the progenitors of low-mass (solar-type) stars has been to observe red spectral regions at high enough dispersion to resolve spectral lines. It is typically found that the magnetically sensitive spectral lines (lines of large g values – see (4) – which are usually lines of atoms such as Ti I and Fe I) are significantly wider than insensitive lines (most molecular lines, such as those of CO, are insensitive to a magnetic field). Modelling such spectra allows one to deduce the approximate field strength, or even a field strength distribution, and to estimate the fraction of the stellar surface on which fields are found. The deduced fields tend to be of the order of 2–3 kG, and they seem to cover a large fraction of the T Tauri stars on which they are observed (e.g. Johns-Krull et al., 1999, 2001; Valenti and Johns-Krull, 2001). From such data it has not yet been possible to deduce much about the geometric structure of the observed fields.

Another very promising development has been the detection of circular polarization in spectral line profiles of a number of PMS stars. Such polarization is almost certainly the signature of a magnetic field. The size of the polarization detected is very small, of the order of 0.01% of the continuum intensity. Detection of such small levels of polarization in single spectral lines of stars as faint as PMS stars is generally not yet possible, but non-zero values of Stokes V can be detected by combining polarization measurements of many spectral lines at once. Using a method of combining data from many lines called Least Squares Deconvolution (see Donati et al., 1997), magnetic fields have been detected by Donati et al. (1997) in two T Tauri stars, a PMS binary system, and a Herbig Ae star (the progenitor of a middle or upper main sequence star of several $M_{\odot}$). The few spectra available for each star so far have not been modelled yet to permit the derivation of surface magnetic field strengths, but in principle when a series of spectra are obtained through the rotation period of one of these stars, it will be possible to derive a map of the stellar field geometry, a type of data that is already available for several main sequence stars (see below).

Attention should be called to a major gap in observational data concerning magnetic fields during star formation. Fields are observed in gas clouds in star-formation regions (and elsewhere in the interstellar medium) with strengths which range from a few μG up to a little over 1 mG (e.g. Crutcher, 1999). Fields of at least 1 kG are observed in PMS stars. In between, six decades of field strength are not constrained at all by field observations. This is an important missing category of observation which it would be very valuable to obtain.

The pre-main sequence stars which have been most studied are the classical (CTTS) and the associated weak-line (WTTS) T Tauri stars, both of which show strong surface activity – star-spots, X-ray emission – qualitatively similar to solar magnetic activity (cf. Sect. 4), but orders of magnitude more intense. Bouvier (in Lynden-Bell, 1994b) and others cited therein report an overall correlation between X-ray emission and measured rotations as evidence for pure dynamo maintenance of the magnetic fields, analogous to what we see in the Sun and other solar-type stars, whereas Montmerle et al. (in Lynden-Bell, 1994b) argue that at least part of the T Tauri magnetic flux is primeval. Earlier, Tayler (1987) suggested that the excess activity for a given rotation period P observed by Gilliland is due to the presence also of a fossil field, of strength unrelated either to P or to the structure of the star, but dependent on its pre-history. By the time the star has reached the main sequence, either the fossil field must have decayed during the later Hayashi phase, or it is pumped down into the radiative core where it will make little if any contribution to surface magnetic activity.

The differences between the CTTS and the WTTS appear due to a massive rotating accretion disc surrounding the CTTS, yielding a strong non-stellar contribution to the standard chromospheric indicators. The evidence that the CTTS typically rotate half as rapidly as the WTTS forces one to extend the canonical models of accretion disks (e.g. Lynden-Bell and Pringle, 1974) to allow for magnetic coupling between disc and star. Magnetic torques exerted by field lines that reach the disc beyond the Keplerian corotation radius have a sign opposite to the accretion torque and so may account for the slower rotation. Although a complete consensus appears to be lacking (e.g. Ostriker and Shu, 1995), we are impressed particularly by the calculations by Collier Cameron et al. (1995) and by Armitage and Clarke (1996) on pre-main sequence evolution. They find that coupling with a very massive disc yields a fairly low zero-age main sequence rotation, independent of the disc mass; whereas stars with lower mass discs can reach the main sequence with a rotation anywhere up to the centrifugal limit (cf. Sect. 4).

There is now a veritable industry on magnetic phenomena in accretion discs. The long-standing question of the physical cause of the 'turbulent viscosity' has stimulated renewed study of magnetically driven instability in a rotating disc, especially by Balbus and Hawley (1991) and their collaborators. Again, one wants to know whether significant dynamo action can occur not only in the central star but also in the disc, analogously to dynamo action in spiral galaxies. The first papers applied the now classical 'mean-field' formalism, but later there has been at least a partial return to first principles, linking up with the Balbus–Hawley work, and avoiding the semi-phenomenological α-effect (Tout and Pringle, 1992; Brandenburg et al., 1995, 1996). Magnetic field lines emanating from a rotating disc can drive a centrifugal wind, which may contribute to the evolution of the disc through angular momentum transfer (e.g. Blandford, 1989); and there is the related somewhat controversial question of the possible magnetic collimation of the wind (Blandford, in (Lynden-Bell, 1994a), p. 171, Spruit, in (Lynden-Bell, 1994a), p. 33, Spruit et al., 1997; Okamoto, 1999, 2003). A resumé of much of the literature is given especially by Campbell (1997).

4 The Main-sequence: Late-type Stars

The transition between early- and late-type main sequence stars takes place near effective temperature $T_e = 10^4$ K, the temperature at which hydrogen ionization occurs. An early-type star has a mass M above $\simeq 2\,M_\odot$, with a consequent central temperature T_c high enough for energy generation to occur via the temperature-sensitive C-N cycle, yielding a structure with a convective core and an essentially radiative envelope. By contrast, in a late-type star energy generation is via the proton-proton chain, yielding a radiative core. However, with T_e below 10^4 K, convective instability is triggered a little way below the photosphere by the low γ resulting from the ionization of hydrogen and helium, and by the extremely high opacity of hydrogen; in fact the high opacity ensures that the resulting convective envelope extends to depths with temperatures of 10^5 K or more. Helioseismological observations (Christensen-Dalsgaard et al., 1991, and several later studies) have confirmed the depth of the solar convective zone, as predicted originally by Biermann (1935). This sub-photospheric convection leads to dynamo activity, and generation of a magnetic field in the Sun. The field is responsible for 'solar activity' – sunspots, prominences, surges, flares, the solar cycle and its long-period modulation, the X-ray emitting hot corona and the consequent solar wind.

The line-of-sight component of the solar field is directly measured on a routine basis, and the full vector field in the photosphere can be mapped as well. The observed photospheric field structures are observed to be extremely complex, and to occur on a variety of scales. Remarkably, however, over at least three orders of magnitude of diameter, flux appears to emerge into the photosphere in the form of flux tubes, and except for the smallest tubes, the field strength in these tubes seems to average to about 1–2 kG. In between the flux tubes (i.e. over the vast majority of the solar surface) the field strength is considerably smaller Solanki (2001).

The largest magnetic structures are the sunspots, where flux tubes with typical diameters of the order of 10^4 km are prominently visible as dark spots on white light solar images. Sunspots frequently occur in groups. Sunspot fields are usually found to show a structure in which most of the flux that emerges in one spot then re-enters the photosphere in one or more nearby spots. We thus expect to find fields above the spots to have the form of loops, and in fact loops of dense gas are observed in the corona above sunspots. Spot pairs with emerging and re-entering flux tend to be oriented east-west on the solar surface, with one polarity leading in one hemisphere and the other polarity leading in the the other hemisphere.

Smaller flux tubes are found in active regions with sizes of the order of 10^5 km surrounding prominent sunspots. Individual tubes have diameters of hundreds or thousands of km. Magnetic flux in active regions tends to repeat on a larger scale the bipolar distribution of sunspots; emerging and re-entering flux is found in two irregular regions oriented roughly east-west, with the same polarity leading as is the case for the embedded sunspots. These active regions of concentrated flux tubes are also distinguished by higher temperatures in the overlying chromosphere (faintly visible as plage regions in white light images), but at the level of the photosphere the flux tubes themselves are slightly darker than the surrounding continuum (Schrijver and Zwaan, 2000).

Outside the active regions, still smaller flux tubes, but still with kG fields, are found as small bipoles, particularly at the cell boundaries of the solar supergranulation (the 'network' field). These weaker fields have been described as a kind of magnetic carpet, in which small loops appear and then drift towards the edges of the convective blobs known as granules, where they cancel or merge; this flux is apparently renewed on a time scale of a few days (Schrijver et al., 1997).

Pressure balance between a flux tube and the surrounding gas is apparently maintained, and the flux confined, by having a lower gas pressure inside the flux tube. In a large flux tube (such as that of a sunspot) the reduction of convective heat transport, together with the inhibition of inward radiative transfer by the opacity of the gas, ensures that at the photospheric level the central region of tube is less luminous than the nearby photosphere. In smaller tubes, convective transport is still inhibited, but radiation keeps the tubes at temperatures near those of their surroundings (Solanki, 2001).

Above the photosphere level, declining gas pressure allows flux tubes to expand, and their diameter increases roughly exponentially with the exponentially declining gas pressure, until in the low corona most of the volume is filled with magnetic field lines (Solanki, 2001).

The solar field clearly dominates the structure of the chromosphere and corona. The photosphere, in which 99% of the area has weak or insignificant fields, presents a bland and uniform appearance marked only by the regions where the largest flux tubes emerge as sunspots, and by the homogeneous granulation which reveals the top of the convection zone. However, above this level the higher atmosphere is strikingly structured. The chromosphere (observed in the cores and wings of strong spectral lines such as those of Ca II) is substantially brighter in the active regions around sunspots than elsewhere. The corona, as revealed by soft X-ray images, is even more strongly structured, with bright regions of gas confined by loops of the solar magnetic field, and other regions (coronal holes) from which the weak, predominantly radial field allows gas to flow outwards into the solar wind. Figure 1 shows a magnetogram of the solar disc and a simultaneous photograph of the Sun taken in X-rays.

The solar field changes continuously. The most obvious manifestation of this is the appearance and disappearance of sunspots and sunspot groups in a time interval of weeks. Another, sometimes spectacular, change is the occurrence of a sudden (minutes to hours in length) flare in a sunspot region, apparently due to sudden reorganization of the local field, with a consequent large release of energy. Variation on a longer timescale is evident in the sunspot cycle, a roughly regular

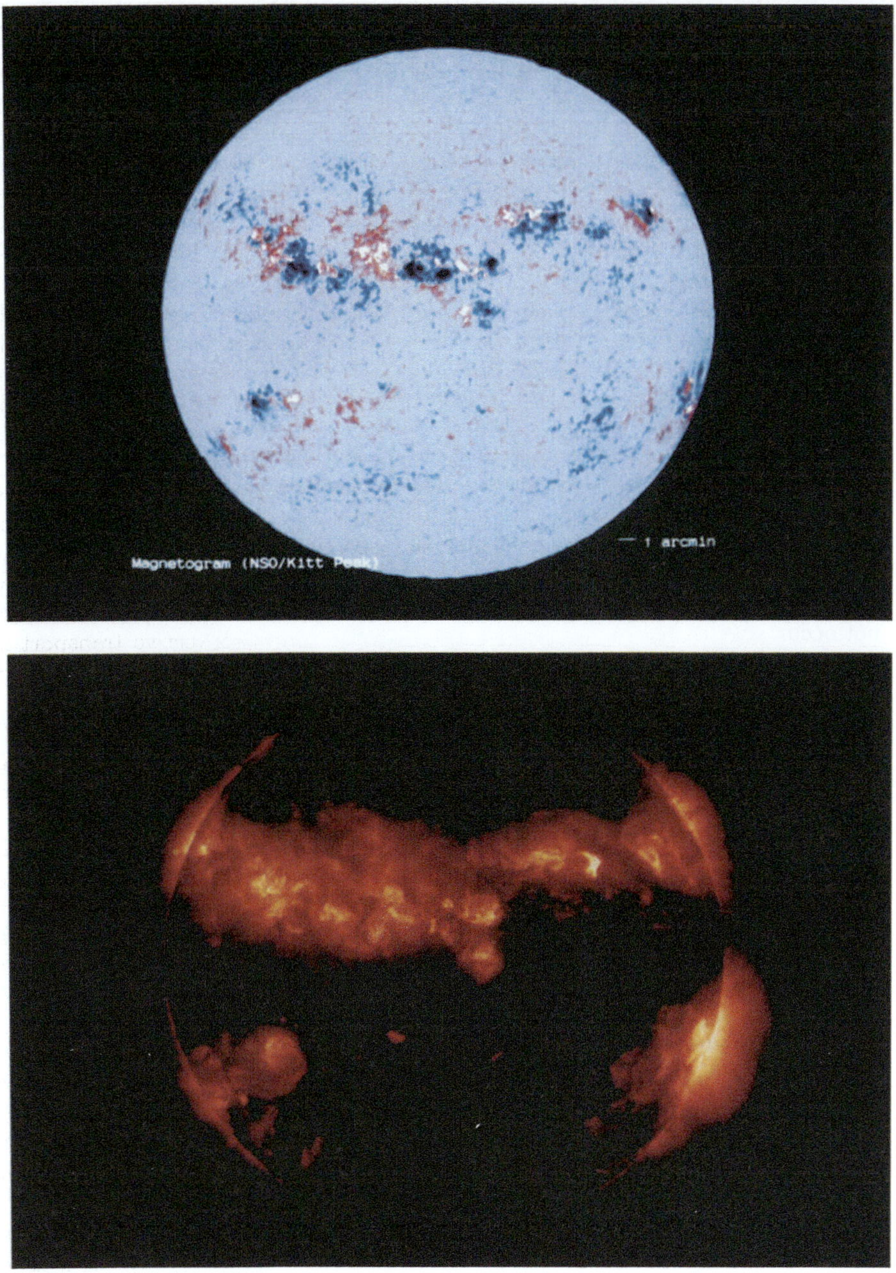

Fig. 1. A magnetogram of the solar disc (courtesy Kitt Peak National Observatory)(*top*) and a simultaneous photograph of the Sun taken in X-rays (courtesy IBM Research and SAO) (*bottom*)

variation in numbers and average latitudes of sunspots that repeats – with reversed field polarities – about every 11 years (Schrijver and Zwaan, 2000).

The prominence of the solar field, and the many ways in which it produces activity, have led to serious efforts both to detect similar fields directly in other, solar-type stars, and to identify analogues of solar activity in such stars. The solar field, covering only about 1% of the solar surface, would not be directly detectable in any other star. Much of the task of looking for stellar analogues of solar magnetism has thus relied on detection of proxy indicators of fields. And indeed, a most gratifying development over the last two or three decades, following on from the pioneering work of Olin Wilson (1978) at Mt. Wilson, has been the positive detection of solar-like activity in many cool stars, summed up in the phrase 'the solar-stellar connection'. For slowly rotating low-mass stars, our Sun appears to be the paradigm: Ca II H and K line emission – a reliable tracer of magnetic activity – is detected and is usually variable. Variations are observed that are produced by stellar rotation, as well as by the intrinsic variations of the emission regions. When observed for several years, the line emission frequently shows clear cyclic behaviour, with variations that are quasi-periodic over a decade or so, but with longer period modulation. In faster rotators, the variations have larger amplitude but are more chaotic (Baliunas et al., 1995; Donahue et al., 1996; Radick et al., 1998). Variability with a rotation period is a clear indicator that magnetic fields are involved; otherwise, the stellar surface would be axisymmetric and rotational modulation would not occur.

Solar X-radiation provides an equally robust tracer of magnetic activity. Satellites pick up thermal X-ray emission, similar to that from the solar corona, and stellar flaring is detected at both X-ray and radio frequencies. A highly significant result is that such activity appears to occur in *all* nearby main sequence stars, and in all yellow (but not in red) giants. The observed emission levels range from roughly the solar level to about four dex stronger, per unit area (Schmitt, 2001). Again, variability (as well as the obvious analogy to the solar corona) strongly hints at the implication of a magnetic field in the X-ray emission.

The Ca II and X-ray results strongly suggest that phenomena very similar to solar magnetic activity are present throughout the lower main sequence and into the giant branch, although with importance varying strongly with rotation rate and evolutionary stage; the situation is in striking contrast to the early-type stars, where the strongly magnetic stars are a minority sub-set (cf. Sect. 6).

Fields are directly detectable in a small minority of low-mass stars, primarily in the most rapidly rotating, active stars. Fields have been detected using polarimetric measurements, averaged over many spectral lines using the method of Least Squares Deconvolution, by Donati et al. (1997). Their list of detected fields includes zero-age main sequence K stars, a K-dwarf flare star, and a number of RS CVn binary systems in which rapid rotation is maintained by tidal interaction. Donati and his collaborators have been observing these stars during multiple rotation cycles (using the MuSiCoS spectropolarimeter at Pic-du-Midi) and then mapping the field structures. The maps to date show rather complex field structures, including toroidal fields near the visible pole (Donati and Collier Cameron, 1997; Donati et al., 1999b). Fields are also detected using Zeeman broadening of Zeeman-sensitive red and IR spectral lines in stars of small $v \sin i$, for example in an X-ray bright member of the Pleiades cluster, and an M Dwarf flare stars. The fields found typically have strengths of a few kG, and are inferred to cover a substantial fraction (of order 50%) of the stellar surface (Valenti and Johns-Krull, 2001). Not much is known yet about the structure of fields detected in this way.

It has long been recognized that among late-type stars, rotation is the crucial parameter fixing the level of activity in a star of given type (Kraft, 1967; Durney, 1972). The rotation period $2\pi/\Omega$ of an active star can be estimated from Doppler line-broadening, or more precisely from rotational modulation of Ca-activity. Noyes et al. (1984) found a striking correlation between time-averaged Ca-activity and an inverse Rossby Number $\sigma = \Omega\tau_c$, where τ_c is the convective turn-over time at the base of the convective envelope, computed from mixing-length theory. Subsequently, Saar and Baliunas (1992) inferred a relation between the cycle frequency Ω_{cycl} and σ.

An eminently plausible picture has emerged. The basic magnetic field is maintained by a contemporary dynamo, probably of the '$\alpha\Omega$'-type, in standard notation (cf. Brandenburg, this volume). The 'α-effect' is a measure of the helicity of the turbulent motions, associated with the star's rotation. Non-uniform rotation in a stellar domain generates a toroidal field $\mathbf{B_t}$ from the poloidal field $\mathbf{B_p}$, while the α-effect completes the $\mathbf{B_t} \rightarrow \mathbf{B_p}$ part of the dynamo cycle. The higher Ω, the stronger the generated magnetic flux and so the more violent the magnetic activity. In this picture, any surviving primeval magnetic flux will have been largely expelled from the convective envelope and concentrated in the radiative core, where however it can still be effective in maintaining near uniform rotation.

The simplest kinematic $\alpha\Omega$-dynamo yields a dynamo number $D \propto \sigma^2$; and although linear theory growth-rates are only a rough guide to the amplitude of the field maintained by a dynamical dynamo, nevertheless the most plausible interpretation is that we are seeing evidence of dynamo generation of a total flux that is systematically higher at higher Ω, for much of the relevant range of Ω. Helioseismology is supplying data on the differential rotation field – crucial for the $\alpha\Omega$ dynamo – within and below the solar convective envelope. The magnetic field couples the sub-photospheric convection zone with the chromosphere and corona; turbulent kinetic energy is converted into excess magnetic energy which is dissipated, so maintaining a hot stellar corona that expands to form a stellar wind. Coupling of the wind with the magnetic field causes magnetic braking, with a consequent decline with age in Ω and in the associated magnetic activity in the optical, radio and X-ray bands.

MHD theory (Mestel, 1967, 1999; Weber and Davis, 1967) predicts that in an axisymmetric, magnetically controlled wind, the Maxwell and Reynolds stresses jointly transport angular momentum J at the rate given by 'effective corotation' – i.e., as if the outflowing gas were to retain the angular velocity of the star as far as the Alfvénic surface S_A, where the accelerating gas reaches the local poloidal Alfvén speed $v_A = B_p/(4\pi\rho)^{1/2}$, and subsequently conserves its angular momentum. An approximate model (Mestel and Spruit, 1987) has the poloidal field with a vacuum dipolar structure out to a radius $\bar{r}$, and radial beyond. The field line passing through the equatorial point $\bar{r}$ separates the wind zone of outflowing gas, linked with the polar caps, from the 'dead zone', containing hot trapped gas. Thus

$$-\frac{dJ}{dt} \simeq \frac{\Phi^2\Omega}{6\pi^2 v_A} , \tag{6}$$

where $\Phi \equiv 2\pi B_A r_A^2 = 2\pi\bar{B}\bar{r}^2$ is the flux of the open field lines that form the wind zone.

With a plausible estimate for $\bar{r}$ and use of the thermo-centrifugal Bernoulli integral, and with a given modelling of the dynamo relation between Φ and Ω, (6) allows one to predict the rotational history of a late-type star from its zero-age main sequence (ZAMS) value Ω_0. The pioneering, observationally inferred relation $\Omega \propto t^{-1/2}$ Skumanich (1972) is easily interpreted as the asymptotic form of the integral of (5), for the linear case $\Phi \propto \Omega$, with $\bar{r}$ constant, and with a purely thermally-driven wind. Something like it appears to be appropriate for stars that have reached the age (6×10^8 yr) of the Hyades cluster, which shows a small spread in Ω for each spectral type. Younger clusters such as the Pleiades (7×10^7 yr) and α Persei (5×10^7 yr) contain both slowly and rapidly rotating dwarf stars. A plausible explanation is that there is a similar scatter in the ZAMS rotation Ω_0, which has not been forgotten by the Pleiades age, but has become irrelevant at the age of the Hyades and the Sun. As noted in Sect. 3, such a scatter is indeed predicted by accretion disc models of T Tauri stars. There is also evidence for saturation of the linear $\Phi - \Omega$ relation at a value $\tilde{\Omega}$ that is a factor 4 or 5 greater than the value at which the chromospheric emission appears to saturate (Collier Cameron and Li, 1994).

Support for the basic premise that rotation is a crucial parameter comes from observations of RS CVn stars – evolved stars in close binary systems – which by contrast remain active as they age (e.g. Baliunas and Vaughan, 1985). They are subject to the same braking process, but tidal or magnetic coupling ensures that spin angular momentum carried off by the magnetically-controlled wind is replenished from the orbital angular momentum, with a consequent modest mutual approach of the two stars, so that magnetic activity is maintained or slightly enhanced. The overall picture does look promising, but one wants to be sure that satellite probes do find that the angular momentum actually carried by the solar wind does accord with theoretical requirements (e.g. Li, 1999).

Over the last decade or so, there have been a number of publications in this area which seem of particularly importance. Support for the theoretical prediction of a multi-component coronal structure comes from combined optical and X-ray studies of the rapidly rotating ($P = 0.5$d) G8-K0 dwarf star AB Doradus (Collier Cameron et al., 1988). A plausible inference from the observations is that the high energy emission comes from a hot dead zone, extending out to (2–3)R, whereas the low energy emission is from a cooler coronal structure, relatively compact and near active regions, and like star-spots, subject to rotational eclipse. Observations in Hα (Collier Cameron and Robinson, 1989) show transient absorption features, consistent with clouds of H I in magnetically enforced corotation with the star, transiting the stellar disc and scattering chromospheric Hα photons. The clouds form at an estimated distance of (3–4)R, outside the Keplerian corotation radius $(GM/\Omega^2)^{1/3}$, where models like that of Mestel and Spruit (1987) suggest that compression of the dead zone gas should cause cooling and recombination.

Doppler imaging of AB Doradus (Unruh et al., 1995), and references cited therein) produced detailed information on the distribution of starspots. *Differential* rotation has been confirmed in AB Doradus and other young, rapidly rotating G and K stars (Donati and Collier Cameron, 1997; Collier Cameron, 2002). In some cases (including AB Doradus), the shear appears to vary over a time-scale of years. Remarkably, the observations yield only a very weak dependence of $\Delta\Omega$ on Ω, over two orders of magnitude.

Donahue and Baliunas (1994) and others concentrate on the Ca H and K lines, which often show variable or multiple rotation periods. The secular changes in rotation are cited as evidence for latitude dependence, showing up through the migration of active regions, as on the Sun. Gray et al. (1996a) and Gray and Baliunas (1997) found similar rotational modulation of H and K in β Comae, a G0 dwarf a little younger than the Sun, inferring an 11–13 day period (rotation about twice the solar). They found also variations in H and K, blue and visual magnitudes, colour index, temperature and granulation, all in a time scale of a few years. Between 1981–94, two rotation periods – 11.6 d and 11.9 d – were seen in H and K and both showed a decline over the 14 years. The interpretation is of two separate centres of magnetic activity, each migrating towards zones of more rapid rotation. However, in contrast to the findings of Collier Cameron and colleagues, the inferred differential rotation is now several times that in the Sun.

Gray and colleagues have studied several other late-type stars, with widely varying levels of activity. The G8 dwarf ξ Bootis A (Gray et al., 1996b), has a 6.43 d rotation, and correspondingly more chaotic behaviour than the Sun. Observed asymmetry in spectral lines is a convenient measure of the granulation (Gray, 1991). At the other extreme, the K0IV subgiant η Cephei, which previously had shown H and K variations, suddenly ceased to be active, suggesting that it may have entered a phase similar to the 17th-century solar Maunder minimum (Gray, 1994).

Massi et al. (1998) have followed earlier work by Mutel and colleagues on the RS CVn binary system UX Arietis, which consists of a spotted K0IV primary and a G5V secondary, with a 6.44 day orbital period, synchronous with the rotation of the spotted star. What is remarkable is that the spotted star has the very short activity cycle of 25.5 days, and this is itself modulated with a period of 158 days. Are we picking up greatly scaled-up analogues of the 22-yr solar cycle and its 90–110 yr modulation? (Shortly after the magnetic Ap stars were discovered, it was at first suggested that those with well-determined periodicity were showing an enormously accelerated solar cycle (from 22 yr to 5–10 d). However, a major difficulty there was the need to generate in a short period a large-scale flux far greater than that found in these late-type stars.)

The radio emission from this system, over the range 21 cm–7 mm, is explained as gyro-synchrotron radiation from mildly relativistic electrons. The observations are well explained by a two-component model for the emitting region: a compact component, generated in the strong field regions in the legs of a magnetic loop, periodically obscured, and responsible for the high ν, high flux (up to 750 mJy) emission; and an extended component, always visible, yielding 50–100 mJy, and associated with the highest parts of loop, or to a larger weakly magnetized volume within the binary system. Such a two-component model had already been inferred from VLA observations of several RS CVn systems (Mutel et al., 1985; Mutel and Morris, 1988). There is a striking similarity with Collier Cameron et al.'s model of AB Doradus, inferred from X-ray and optical observations.

This gratifying plethora of new, high-quality observations must set stringent tests for the theory of magneto-convection in general, and should encourage further the ongoing critical scrutiny of the basics of dynamo theory. One may hazard a guess that something like the phenomenological standard dynamo equations may survive as encapsulating much of what is going on in solar-type stars, especially when combined with a plausible simulation of the back-reaction of the Lorentz forces on the driving motions. For example, by including both quenching of the α-effect and moderation of the differential rotation by the growing field, Weiss et al. (1984) are able to predict a long-term modulation of the basic cycle, including episodes of reduced activity reminiscent of the solar Maunder minimum and the recently discovered analogues in other stars, mentioned above.

Further theoretical predictions are likely to be sensitive to both the horizontal and vertical variations of the rotation field, which in turn will depend on the turbulence model adopted. The cited generalized mixing-length model of Kitchatinov and Rüdiger (1995) brings out the importance of the anisotropy of heat transport. These authors find that even in stars which rotate somewhat more rapidly than the Sun, the Ω-field in the convective zone continues to approximate to the roughly radius-independent form revealed by helioseismology, rather than to the constancy on cylinders (the Taylor-Proudman form) expected in a nearly isentropic domain. The model yields equatorial acceleration, and only a weak variation of $\Delta\Omega$ with Ω. A later paper (Kitchatinov and Rüdiger, 1999) predicts markedly larger differential rotation in late-type giant stars. Besides affecting the Ω-field, it may also be the principal cause of saturation of the dynamo at high Ω.

5 The Main Sequence: Early-type Stars

5.1 The Observed Fields

The first magnetic field discovered beyond that of the Sun (Babcock, 1947) was in a 'peculiar A' (Ap) star, a member of a then obscure classification subset of the main sequence A stars of around 2–3 $M_\odot$. Babcock's initial discovery set off a series of surveys of upper main sequence (UMS) stars which continue to the present. More than 200 UMS stars are now known to have detectable fields (a currently maintained list is found at `http://www.sao.ru/hq/lizm/catalogue/`; an extensive recent bibliography is given by (Bychkov et al., 2003)).

Virtually all the detected fields in UMS stars are found in stars having (suitably extended) Ap classification. Stars of this type have striking and distinctive atmospheric chemical peculiarities that depend in a fairly clear way on effective temperature T_e. The coolest and lowest mass members of this type (Ap SrCrEu stars), with T_e ranging from about 7000 to 9000 K, typically show large overabundances of at least one of Ca, Ti, Cr, and Sr, and usually some rare earths are several dex overabundant (see e.g. Cowley, 1993). Hotter stars (Ap Si and Bp He-weak stars), up to about 18,000 K, are usually found to have strikingly underabundant He and/or overabundant Si. The hottest stars of this type (Bp He-strong) have *over*abundant He but appear otherwise largely normal. These stars collectively are now mostly called Ap or Ap–Bp stars.

In addition to the majority of A and B stars whose atmospheric chemical abundances appear to be roughly solar, several other families of chemical peculiarity, such as Am stars (general mild overabundance of iron peak elements), HgMn star (large overabundances of a few specific elements), and λ Boo stars (pervasive underabundances of iron peak elements) also occur among the A and B main sequence stars. Although occasional reports of fields in such stars appear (e.g. Mathys and Hubrig, 1995), almost none of them have been confirmed.

The fields of the Ap–Bp stars are most often detected through measurements of $\langle B_z \rangle$. Current measurement techniques permit measurement of $\langle B_z \rangle$ fields as small as a few G (Shorlin et al., 2002). In some of the Ap-Bp stars having very little Doppler broadening of spectral lines, measured by a small value of $v \sin i$, it is also possible to measure $\langle B \rangle$. The sensitivity limit of such measures is about 1.5–2 kG in the most favourable cases where $v \sin i$ is below a few km s^{-1} (Mathys et al., 1997).

It is gradually becoming firmly established from observations that essentially all the stars of the Ap–Bp class have detectable magnetic fields. A typical RMS value of $\langle B_z \rangle$ is around 300 G (Bohlender and Landstreet, 1990), but the observed field strengths show an extended distribution running from $\langle B_z \rangle$ values of about 100 G up to about 20 kG, with small $\langle B_z \rangle$ values much more common (in distance- or magnitude-limited samples) than large ones. A significant population of long-period, large-field stars has now been identified, mainly by Mathys and his collaborators (Mathys et al., 1997). The largest value of $\langle B \rangle$ known is 34 kG, in a star discovered many years ago by Babcock (1960).

In contrast, the highest-sensitivity measurements available currently reveal no fields in other classes of A and B stars, either in stars with different peculiarities, such as Am and HgMn stars, or in normal stars, with limits of a few tens of G (Shorlin et al., 2002). The only exceptions to this result are the recent detections of fields with $\langle B_z \rangle$ of the order of $3\,10^2$ G in the pulsating B stars β Cep (Donati et al., 2001) and ζ Cas (Neiner et al., 2003), and in the brightest and most massive O star in the Orion Nebula Cluster, θ^1 Ori C (Donati et al., 2002). It thus appears that there is probably a class of magnetic O and B stars which is *not* announced by the chemical peculiarities of the common magnetic Ap–Bp stars.

The observed fields usually vary periodically, and in the majority of known magnetic Ap stars, the value of $\langle B_z \rangle$ changes sign during the cycle. Frequently the brightness of the star is also variable (by a few times 0.01 mag), and often spectrum lines vary as well. All variations occur with the same period (a recent catalogue of periods is found in (Renson and Catalano, 2001). Since the period, assuming a stellar radius of roughly $3\,R_\odot$, is always found to lead to a value of equatorial velocity $v_{\rm eq}$ which is at least as large as $v \sin i$, it is clear that the period of variation is the rotation period. The observed periods range from 0.5 d to several decades, with most Ap–Bp stars having periods in the range of 1–10 d. These stars thus form a sample of A and B main sequence stars which typically have only 10 or 20% of the specific angular momentum of normal A and B stars, and in a small fraction of cases they have only of order 10^{-3} of the typical specific angular momentum.

The observations thus show that there is a striking contrast in magnetic properties between the upper and the lower main sequence. As noted in Sect. 4, the accumulating evidence on the solar–stellar connection suggests strongly that 'solar activity', and thus a magnetic field, appears in all late-type stars, and there is a broad correlation between the level of activity and the rotation rate Ω of the star. By contrast, only a fraction of the early-type stars show the Ap-phenomenon and its associated magnetism, and they show a broad *anti*-correlation between Ω and surface field $B_{\rm s}$, in the sense that rapidly rotating A stars are mainly non-magnetic, and the magnetic stars are nearly all slow (some, very slow) rotators. The theoretical significance of this will be discussed in Sect. 6.2.

There are now several tens of magnetic Ap–Bp stars for which measurements of both $\langle B_z \rangle$ and $\langle B \rangle$ are available. Typically the ratio of the maximum modulus of $\langle B_z \rangle$ to $\langle B \rangle$ is of the order of 0.2 or 0.3. Modelling experiments clearly show that this fact requires that the overall field topology of the great majority of these stars must be basically dipolar in the sense that magnetic flux mostly emerges from one hemisphere and returns in the other. Thus the field is expected to have at least a roughly defined axis. If the field were basically much more complex than this simple overall structure, emerging line-of-sight flux in one part of the visible hemisphere would largely cancel returning flux in other parts of the same hemisphere, leading to a much smaller ratio of maximum $\langle B_z \rangle$ modulus to $\langle B \rangle$ than is typically observed. Furthermore, the fact that $\langle B_z \rangle$ usually changes sign during the stellar rotation implies that this axis is usually inclined to the rotation axis by an angle β of the order of 60° or more. This simple model is the basic paradigm of magnetic Ap–Bp stars, and is known as the the oblique rotator (Stibbs, 1950; Deutsch, in Lehnert, 1958, p. 209; Deutsch, 1970; Preston, 1970).

In the framework of this model, the variations in spectral line intensity and shape, and in the brightness of a star, reflect non-uniform distribution of one or many elements over the stellar surface, and probably also vertically in the atmosphere. Since among the hotter main sequence stars, such variations – periodic in photospheric spectral lines – are seen essentially only in magnetic stars, it is clear that the horizontal and vertical inhomogeneities are somehow caused by the presence of the magnetic field. Although the way in which the field leads to such inhomogeneities is still mysterious, the fact that such patchiness, once created, is preserved is not: since density falls off exponentially upwards in the atmosphere while field strength probably varies little vertically, much of the atmosphere can be magnetically dominated, with $B^2/8\pi \geq nkT$, leading to the general suppression of horizontal motions that would destroy patches by mixing (cf. Sect. 6.2).

The availability of numerous 'magnetic curves' (measurements of the variation of $\langle B_z \rangle$ and/or $\langle B \rangle$, and possibly other field moments, over the rotation cycle of individual stars) has led to extensive modelling efforts; in fact the oblique rotator model emerged from the pioneering work of (Stibbs, 1950), in which variations of $\langle B_z \rangle$ were modelled with simple oblique dipole fields. Gradually, more precise and extensive observations required more complex field distributions. In recent years data-sets of 'magnetic observables' have been modelled by computing mean values $\langle B_z \rangle$, $\langle B \rangle$, etc. over the visible stellar hemisphere of various low-order multipole expansions, typically using a simple limb darkening and/or line weakening towards the stellar limb, and adjusting model parameters until a good fit to the available data is found. Such models have been constructed using as a basis simple dipoles (Bohlender et al., 1993), collinear dipole, quadrupole, and octupole fields (Landstreet and Mathys, 2000; Strasser et al., 2001), dipole and quadrupole oriented independently (Bagnulo et al., 2002), and even magnetic charges (Glagolevskij and Gerth, 2001). These modelling efforts have usually been reasonably successful, but remaining discrepancies when more than just $\langle B_z \rangle$ data are available make it clear that such low order multipole expansions are at best rather rough models of the actual field structures.

Spectra observed in all four Stokes parameters make possible a very important test of simple multipolar models of field geometry based only on mean observed quantities such as $\langle B_z \rangle$. Model field structures found by the parameter-fitting procedure described above are used to compute the expected intensity and polarization spectral line profiles with the help of the magnetic spectral line synthesis codes currently available (cf. Wade et al., 2001). In general, it appears that the multipole models do very badly in such tests; the actual field structure is clearly far from that assumed in the multipole expansion, even when the phase variation of such observables as $\langle B_z \rangle$ and $\langle B \rangle$ are reproduced faithfully (Bagnulo et al., 2001).

The rapidly improving data situation is stimulating efforts to develop full-fledged Zeeman-Doppler imaging codes to reconstruct abundance and field geometry. These are codes in which the detailed intensity and polarization profiles of a spectral line are synthesized for an assumed finely gridded element distribution and field structure, and both maps are adjusted iteratively until a satisfactory fit to the data is obtained. The new measurements make possible apparently unique reconstruction of the global magnetic field structure over most of the observable surface of a star (currently with spatial resolution of the order of 10° on the surface). The first successful model of this type, for the well-known star 53 Cam (Kochukhov et al., 2004), has revealed a field structure which is considerably more complex in detail than the multipole expansions (see Fig. 2).

However, the basic dipolar topology, of largely constant field sign over regions of the size of a hemisphere, is recovered. On the other hand, the higher moments of the multipole expansions appear to describe statistical attributes of large regions of the stellar surface field (such as average field strength differences between the hemisphere of emerging flux and that of descending flux, or the typical contrast in field strength from near the field axis to near the magnetic equator), but clearly do *not* put the correct flux in specific locations on the stellar surface. This work is certainly going to lead in the near future to a dramatic improvement in our knowledge of field and chemical abundance distributions on magnetic Ap stars.

5.2 The Chemical Peculiarities

Chemical peculiarities that signal the UMS magnetic stars, and that appear in different forms in some other UMS stars, can arise spontaneously as a result of thermal diffusion of trace elements relative to the dominant H medium. In the absence of other effects, the heavier elements would tend to sink. However, as first shown by Michaud (1970), in A and B stars the outward flow of radiation in a star can transfer momentum to some ions through absorption in spectral lines, leading to selective levitation of some species into or even through the stellar atmosphere. This effect is expected in a general way to lead to abundance anomalies qualitatively similar to those observed (Michaud et al., 1976).

However, species separation may have to compete with processes acting to restore homogeneity. For example, any convective region is expected to be well mixed, although a particular species may diffuse up or down through such a region. Any large-scale meridional circulation – laminar or turbulent – will also lead to some mixing effects (cf. Sect. 6.2). If a stellar wind occurs (and such a wind may be either well-mixed or involve selective loss of only certain ions), this will strongly affect which ions reside in the stellar atmosphere (Babel, 1992). From the competition of these various effects, we certainly expect to find vertically inhomogeneous distributions of at least some elements, and this is indeed observed in magnetic stars (Babel, 1994; Bagnulo et al., 2001). From the presence of large-scale and apparently fixed

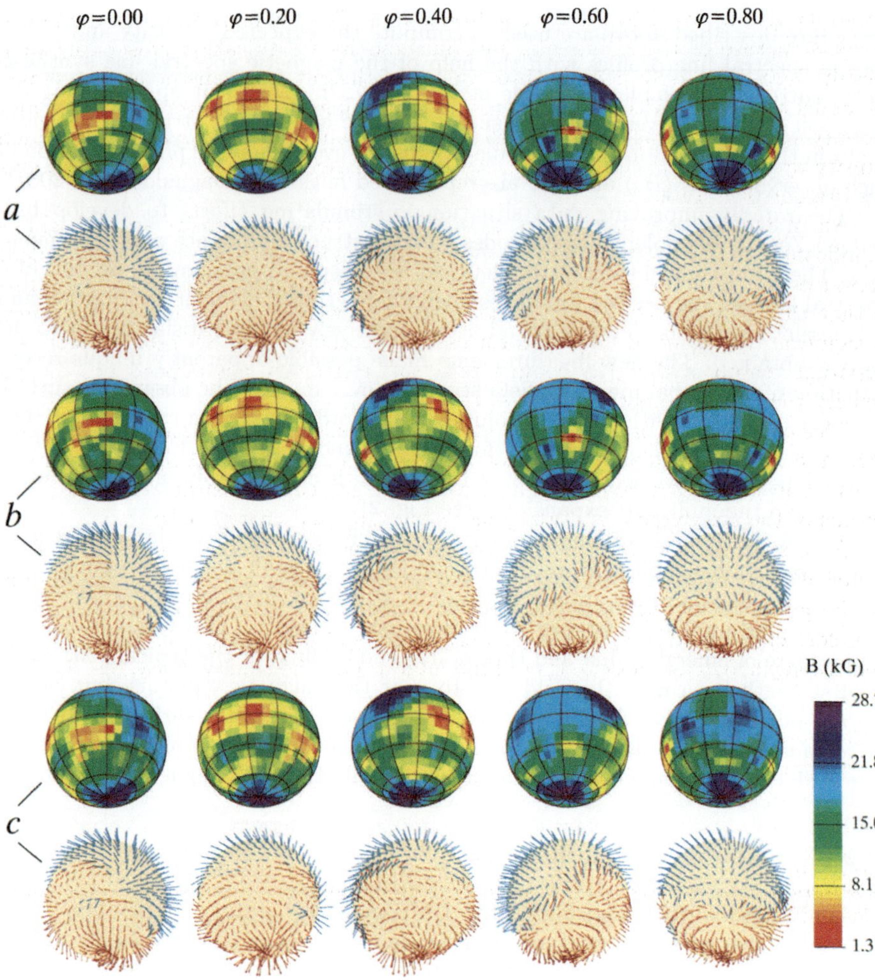

Fig. 2. A map of the magnetic field structure of the magnetic Ap star 53 Cam = HD 65339, reconstructed by Doppler-Zeeman imaging (Kochukhov et al., 2004). The five vertical columns show the field configuration at five evenly spaced rotational phases as seen from Earth. The three double rows of maps labelled a, b, and c show maps as reconstructred from three separate Fe II lines, based on spectropolarimetry in all four Stokes parameters obtained with the Musicos spectropolarimenter (Donati et al., 1999a). In each pair of rows, the upper image (coloured) shows the local field strength $|B|$ (scale at lower right); the lower shows the orientation of field lines (brown lines: emerging flux; blue lines: returning flux). Note the excellent agreement of the three sets of maps

abundance inhomogeneities, sometimes involving abundance contrasts of factors of 10^2 or more (Landstreet, 1988; Strasser et al., 2001), it is clear that the migration of atoms is also strongly influenced by the magnetic field, but the mechanism for this has not been identified – the trapping of ions on field lines only occurs at small optical depths, and does not appear to be a sufficiently important effect to explain the large, and large-scale, abundance contrasts seen.

As a result, it has proven to be quite difficult to explain the observed abundance anomalies and patchiness (e.g. Babel and Michaud, 1991). Even for non-magnetic stars, it appears necessary to invoke important mixing processes in the layers below the atmosphere whose physical origin is not clear (Richer et al., 2000), and efforts to explain abundance anomalies seen on magnetic Ap stars have been at best partially successful (Babel, 1992).

A consequence of this situation that has only begun to be explored is that the atmospheres of magnetic Ap–Bp stars are expected to have structure which is significantly different from that of non-magnetic stars. This will occur in part because of the unusual nature and vertical distribution of opacity sources (which in addition vary from one place to another). It will also occur because of various magnetic effects: Zeeman splitting will change the flux blocking due to spectral lines, and there may also be some Lorentz forces in the atmosphere, changing the pressure structure (Landstreet, 1987). Recent observations indicating that the Balmer lines of H have anomalous profiles (Cowley et al., 2001; Bagnulo et al., 2003), an effect which may be due to an abnormal $T(\tau)$ relationship (Kochukhov et al., 2002), confirms that this is probably an important consideration. Very little work has been done on this topic to date.

It is clear that the task of explaining the details of the atmospheric compositions of the different classes of Ap stars and of the Hg–Mn and the Am stars is formidable. The growing importance of this area is brought out well by this quotation from (Bagnulo et al., 2001): 'Any realistic study of the photosphere of magnetic CP stars must consider the atmosphere as a three-dimensional structure permeated by a complex magnetic field, taking into account not only the horizontal non-uniformities of chemical abundances, but also their important vertical variations as well. This implies that we need more accurate model atmospheres, accounting for element stratification and magnetic forces, and more sophisticated modelling techniques for stellar magnetic fields.'

5.3 Correlations

As already noted, there is a striking contrast in magnetic properties between the upper and the lower main sequence. The accumulating evidence on the solar–stellar connection suggests strongly that 'solar activity', and thus a magnetic field, appears in all late-type stars, and there is a broad correlation between the level of activity and the rotation rate Ω of the star. By contrast, only a fraction of the early-type stars show the Ap-phenomenon and its associated magnetism, and they show a broad *anti*-correlation between Ω and surface field B_s, in the sense that rapidly rotating A stars are mainly non-magnetic, and the magnetic stars are nearly all slow (some, very slow) rotators. This shows up in particular among the A stars in binary systems, where it is those of intermediate separation that show abuncance peculiarities (Abt, 1965). A plausible explanation is that tidal friction has led to spin-orbit synchronization in both close and moderately close systems. The close systems will behave like normal, rapidly rotating A stars, whereas the moderately close will rotate fairly slowly, and so may be expected to show the Ap phenomenon.

In wide binaries, tidal synchronization takes too long, and an initially rapid rotation will persist, unless the star is subject to strong magnetic braking.

The accumulating observational data yield strong support to the oblique rotator as the paradigm model for the strongly magnetic stars. Over the decades, there have been suggested further correlations, restricting the allowed domains of the parameters, which if validated could be helpful guides to theorists; but often they had a habit of disappearing with improved observation and critical scrutiny. A few decades ago there was a claim that the distribution of obliquities was non-random, in fact bimodal, with a marked preference for β to be near either zero or $\pi/2$. Subsequent work led most observers to the much more cautious statement that any non-randomness was at most marginal. Again, an early report of a marked correlation between surface field strength and period (an extension of the broad anti-correlation already noted between rapid rotation and surface field) was not confirmed.

However, major surveys carried out over the last decade in particular using Hipparcos data, have greatly extended the number of Ap stars with high-quality observational data, and have yielded new correlations. A very interesting result that has emerged from simple modelling of the observations is that in most of the magnetic Ap stars, the axis of the dipolar topology is inclined to the rotation axis by a rather large angle β (Preston, 1967; Landstreet, 1970), not inconsistent with a near random distribution; but in the small fraction with rotation periods of more than about one month, β is typically rather small, of the order of 10 or 20° (Landstreet and Mathys, 2000). Clearly, a convincing discussion of the angular momentum of the Ap–Bp stars must also account for the preferred distribution of β-values.

A topic on which observations should be able to shed some light is the evolution of field strength and structure with age in main sequence stars, but the results to date are rather contradictory. One fairly secure method of exploring this question is to use Ap stars in clusters. It was found by (Hartoog, 1976, 1977) and Abt (1979) that Ap and Bp stars appear in such clusters, but they did not observe any low-mass Ap stars in the younger clusters, and Abt even suggested that such stars do not become chemically peculiar until about 10^8 yr have passed. This failure to find low-mass Ap stars in clusters may be a selection effect discriminating against these relatively faint stars in favour of brighter cluster member, and in fact isolated reports of low-mass cluster members do occur (summarized in Bagnulo et al., 2004). The occurrence of Ap stars of all masses with ages from the ZAMS to the end of the main sequence phase has recently been supported by the results of the Hipparcos mission; using distances from this experiment, Gomez et al. (1998) have shown that the nearby Ap stars, including the coolest ones, appear to be spread approximately uniformly between these two limits.

An early effort to study the evolution of fields by comparing Ap members of the young Sco–Cen association to field stars (Thompson et al., 1987) suggested that evolutionary field changes are not large, but may have been adversely affected by uncertain membership of some of the stars in the association. Since then, few data on magnetic fields in cluster stars have been published, although a major survey is currently in progress (see Bagnulo et al., 2003) which will soon shed new light on this question.

Recently, Hubrig et al. (2000) have used Hipparcos data to study the evolutionary state of the particular sample of magnetic Ap stars with observed values of $\langle B \rangle$. They report a marginal trend for $\langle B \rangle$ to be lower in the slowest of the slow rotators: the strongest fields – 6.5 kG or more – are found only in the stars with rotation periods less than 1000 days. (However, the fields in this sample of stars are in general considerably larger than the typical fields of the much more common magnetic Ap stars with rotation periods of a few days.) Stars with higher temperature and higher mass also tend to have stronger fields. In addition, they find that the stars in this sample seem to be concentrated towards the middle of the main sequence band, and suggest on the basis of this result that magnetic fields may not appear in Ap stars of $M \leq 3\,M_\odot$ until these stars have completed at least 30% of their main sequence lifetimes. Also, no correlation is observed between P and the fraction of main sequence lifetime completed, suggesting that slow rotation is achieved before they have become observably magnetic. However, their suggestions remain controversial. Bagnulo et al. (2003) report finding a counter-example – a large magnetic field in a cluster star with a mass of about $2.1\,M_\odot$ and which has completed only $16 \pm 5\%$ of its main sequence life, and their general claim is challenged in a new study by Pöhnl et al. (2003). Rather than being a general result, the Hubrig et al. conclusion may be a consequence of the sample studied, which appears to be strongly skewed towards large fields and long rotation periods compared to those of typical magnetic Ap stars. Thus to date, observations have not yet shed as much light on evolutionary issues as one would desire.

6 The Early-type Magnetic Stars: Basic Theoretical Questions

6.1 Origin of Field

There are two broad possibilities.

(a) The field is a 'fossil' – a slowly decaying, dynamically stable relic from an earlier epoch.

(b) The star has effectively forgotten its previous magnetic history, but is being maintained by contemporary dynamo action.

The contemporary dynamo theory, which assumes the flux to be continuously generated by d-c dynamo action in the convective core must show how the flux can rise quickly enough through the subadiabatic, radiative envelope, ultimately to manifest itself at the star's surface as a large-scale ordered field with a well-defined obliquity angle (cf. Moss, 1989). Equally, the fossil theory must show convincingly that flux of the observed order will remain effectively trapped over an Ap star lifetime (cf. Meyer, in Lynden-Bell, 1994a, p. 67), and must take cognizance of the relation between the total internal magnetic flux and the fraction crossing the star's photosphere (cf. Sect. 6.2).

In our view, the non-universality of $\langle B \rangle$ in stars of a given spectral type suggests strongly that there is a least one extra parameter required in the theory, with no simple, quickly established connection to the instantaneous stellar rotation, as is believed to be the case for the late-type stars. As discussed in Sect. 3, there are plausible arguments for the presence of observable relic magnetic fields inside stars, especially those massive enough to have retained a radiative core during the Hayashi phase and correspondingly become early-type stars with radiative envelopes on the main sequence. The simple fossil theory has the virtue of flexibility, as the relic flux in a proto-star will depend on the conditions determining the loss of primeval flux, which may very well vary from one cloud domain to another (cf. Nakano et al., 2002 and references therein), yielding sometimes observably magnetic Ap stars, and sometimes normal A and B stars. Likewise, the rotational evolution of a star will depend not only on the magnetic flux but on the moment of inertia of the external gas magnetically coupled with the star. Thus, for example, the existence of stars with similar Zeeman patterns but differing in P by a factor 10 or more certainly requires elucidation, but is not paradoxical.

Some forms of the contemporary dynamo explanation can be thought of as fossil variants. One model (Rüdiger et al., 2001) appeals to a postulated non-uniform rotation $\Omega(\mathbf{r})$ that acts on a seed poloidal field – a 'dynamo instability' in a radiative envelope. There is no clear justification for the assumed 'fossil' Ω-field. The steadily growing magnetic field will exert torques that in general will act to suppress the imposed shear: the dynamo action will ultimately be suppressed, leaving the system with a slowly decaying fossil field that is however self-built rather than primeval.

A definitive answer to the basic question – the origin of the Ap star fields – must await much more research on dynamos and buoyancy in stars with convective cores and radiative envelopes, and on possible dissipation-dependent hydromagnetic instabilities. Input from observation is thought-provoking but as yet unable to yield unambiguous interpretation. Thus Hubrig et al. remark that their reported (but controversial) finding of an inverse correlation among very slow rotators between B_s and P – no stars with periods exceeding 1000 days have fields above 6.5 kG – does fit in with a simple dynamo interpretation. Our feeling is that any clearly unambiguous results of this sort can certainly be helpful, but must be considered along with the rather bewildering corpus of data on the majority of Ap stars with much shorter periods, where it is difficult to discern any such convincing correlation between observable flux and period.

It could even be that sometimes we are observe a confusing combination of primeval and dynamo-generated flux. Prima facie, any braking model, e.g. that of pre-main sequence magnetic braking outlined in Sect. 3, should predict a stronger fossil field to yield a *lower* Ω than a weaker field acting for the same time. But equally, one expects some dynamo action to occur in the convective core (probably of the α^2-type – cf. Sect. 7), and continue after the pre-main sequence braking has ceased. Could some flux from the fossil field in the surrounding radiative envelope penetrate into a boundary layer and and act as a 'seed', accelerating slightly the spontaneous core dynamo action? The tentative suggestion is that a lower fossil field could yield a higher Ω, but that subsequent Ω-dependent dynamo action increases B, so that finally there might be at least a marginal $B - \Omega$ correlation.

6.2 Stellar MHD and the Oblique Rotator

Theoretical problems emerging from the required modifications to basic stellar structure, outlined in Sect. 1 are of interest in themselves, are to a great extent independent of the origin of the field, and may throw some light on the often bewildering plethora of data. As noted in Sect. 1.1, any non-spherical perturbing force – centrifugal, magnetic – in general upsets the local radiative equilibrium normally holding in a non-turbulent domain. The consequent buoyancy forces lead to a thermally-driven circulation – the generalized Eddington-Sweet circulation currents (Sweet, 1950; Mestel, 1999). In the low-density surface regions of a rapidly rotating but non-magnetic A star, the predicted circulation speeds are fast enough to develop shear turbulence, which can keep the atmosphere mixed (Kippenhahn, 1959), so nullifying the spontaneous diffusion discussed in Sect. 5.2; whereas if a magnetic field is present, maintaining near uniform rotation, any surviving circulation should be slow enough for element diffusion to persist (Mestel and Moss, 1977; Moss, 1984). In a slowly rotating A star, e.g. a synchronized member of a moderately close binary, whether magnetic or not, again the circulation should be slow enough for peculiarities to develop.

Magnetic braking is presumably the reason for the slower rotations of the magnetic Ap stars. If some angular momentum loss occurs during the main sequence phase, then accretion braking is a possible process for stars with moderate rotation periods. Infalling gas, halted at the Alfvénic radius r_A, where the magnetic and kinetic energy densities are comparable, and spun up approximately to corotate with the star, is then shot back centrifugally into the interstellar medium. This has the advantage of cutting off spontaneously when Ω has fallen to the Keplerian value $(GM/r_A^3)^{1/2}$, yielding a typical rotation period of a few days. The estimated time of braking $\propto B_s^{-8/7}$ is below a stellar lifetime even for B_s as low as 100 G. There are however observational objections against the hypothesis that all CP stars arrive on the main sequence with the same angular momentum as a normal A star of the same mass, and are subsequently braked (Stępień 2000 and references therein). Magnetic coupling of a strong primeval field with a pre-main sequence wind leads to an exponential braking law, without a cut-off, and could explain the small minority of 100–1000 d periods (Mestel, 1968, 1984; Schüssler, 1980). Stępień's preferred pre-main sequence model is similar to that for T Tauri stars, summarized above, involving accretion of mass and angular momentum from a dense rotating disc, and magnetic interaction with the disc as well as with a wind. The predicted main sequence rotations depend on the mass of the star and the lifetime of the disc.

There are several physical processes that can in principle alter the obliquity angle β in an oblique rotator. If a star is being braked, the magnetic torques responsible will in general have an associated precessional component. A qualitative argument suggests that in general the torque will cause the instantaneous axis of rotation to approach the region on the stellar surface where the magnetic field is strongest, so tending to align the rotation and magnetic axes. For a specially simple case (Mestel and Selley, 1970), with the field supposed drawn out by a wind into a nearly radial structure, detailed treatment confirms this conclusion, but finds the effect is small. However, the same analysis shows that a star that is being magnetically *spun up* will have β increased, yielding a correlation between high obliquity and short period. This could be relevant to stars that have suffered strong braking via a thermally-driven wind in the pre-main sequence phase, but are subsequently *spun up* by magnetically-controlled accretion (Mestel and Moss, in preparation).

Another process depends rather on the gross structure of the star. An obliquely rotating magnetic star has some properties in common with a top (Spitzer, in Lehnert, 1958, p. 169). The Lorentz forces exerted by the field of total flux F cause small but finite density perturbations that are at least approximately symmetric about the magnetic axis. To keep the angular momentum vector invariant in space, as required by Newtonian mechanics, there must be superposed on the basic rotation Ω the Eulerian nutation – a rotation ω about the magnetic axis, analogous to the geophysicist's Chandler wobble, of order $\omega \simeq (F^2/\pi^2 GM^2)\Omega$. However, the density-pressure field contains also the usually much larger perturbations due to the centrifugal forces, which are symmetric about the rotation axis. To maintain hydrostatic equilibrium in a radiative domain, the changes in the (ρ, p) field due to the Eulerian nutation must be offset by nearly divergence-free internal motions (Mestel et al., 1981; Mestel, 1999). These dynamically-forced, oscillatory 'ξ-motions' have the period of the Eulerian nutation which is much longer than the free oscillation periods of the star, but may be shorter than the Kelvin–Helmholtz or the nuclear time-scales. In a rapid rotator, the amplitudes of the ξ-motions are large and could cause significant mixing of nuclear-processed material from a convective core through a radiative envelope. As against this, the slow but persisting Ohmic dissipation of the associated distortions to the magnetic field would make the star approach the state of rotation about its maximum moment of inertia. A dynamically stable magnetic field must have mutually linking poloidal ($F_{\rm p}$) and toroidal ($F_{\rm t}$) flux. The density field could then be either globally oblate or prolate about the magnetic axis, depending on the ratio of the fluxes, yielding $\beta \to 0, \pi/2$ respectively. However, tentative estimates for the time-scale suggest that if the field strength increases inward moderately, e.g. as in the principal Cowling decay mode, then only in a rapid rotator will β be changed significantly in a stellar lifetime.

Another possibility appeals to studies of the MHD of the outermost stellar regions, of which those by (Moss, 1984, 1986, 1990) seem particularly important. In a rapid rotator of small or moderate obliquity, it is easy to picture the slow but inexorable Eddington–Sweet circulation, symmetric about the rotation axis, tending to bury flux deep in the star. If this occurs in a time short compared with the magnetic braking time, it would tend to yield a correlation between observable field strength and rotation period P. However, in a rotator with $\beta > 55°$, Moss shows that the flux is not buried but tends to be concentrated towards the magnetic equator. The simultaneous distortion of the field by the same circulation, horizontal near the surface, will increase the apparent obliquity. After braking of the rotation by the emerging field lines the circulation would become negligibly slow, allowing the field in the surface regions to begin diffusion back into its undistorted structure. The generation in a rapid rotator of an apparent large obliquity from an initial small β does fit in with the Landstreet–Mathys results. The equatorial concentration of flux would presumably also affect the interpretation of an observed field strength. However, Moss's calculations to date suggest that the effects are important only for P below about 5 days.

There are other interesting categories of phenomena requiring further theoretical study. Low-amplitude pulsations occur in some of the coolest magnetic Ap stars (Kurtz, 2000; Balmforth et al., 2001). These are apparently non-radial pulsations of high radial order but of low harmonic degree and azimuthal order. Frequently several frequencies are observed, typically in the range of 1 to 4 mHz (periods of 4 to 15 min). The pulsations are usually modulated by the rotation period of the star, and it appears that they are probably symmetric about the magnetic axis rather than the rotation axis (the 'oblique pulsator' model – Kurtz, 2000). It is not yet clear how the frequencies observed are excited, why other possible frequencies are not excited, or indeed why some cool Ap stars pulsate when other similar ones do not. Once these matters become clearer, the pulsations offer an extremely important possibility for studying the interior structure of some of the magnetic Ap stars. In a few magnetic stars there occur magnetically controlled stellar winds or trapped circumstellar material. This effect was first observed in σ Ori E (Landstreet and Borra, 1978), and has since been observed in other stars (Shore et al., 1990). The occurrence of periodic variations in wind phenomena led to the discovery of fields in non-peculiar O and B stars (Donati et al., 2001, 2002; Neiner et al., 2003).

To sum up: the steadily growing quantity of high quality observational data will set more and more stringent tests for the theorist, who however has available continually escalating computing facilities.

7 Giant Stars

One expects dynamo action and associated magnetic activity in rotating late-type giant stars, even if they have evolved from early-type main sequence stars. X-ray emission is again an excellent diagnostic of a hot stellar corona, presumably magnetically heated. The observations of Haisch et al. (1992) pointed to an X-ray dividing line (XDL) in the H–R Diagram, nearly vertical and near spectral type K3, with X-ray emission occurring only on the left-hand side. Hünsch et al. (1996) confirmed that all G and early K giants of luminosity class III are X-ray emitters. More recently, Hünsch et al. (1998) found that only 11 out of 482 M giants observed had detectable X-ray emission, and of these, only 4 could not be eliminated through their definitely having accretion discs or X-ray emitting G type companions, or through a suspiciously large mutual off-set of the optical and X-ray sources. The possibility exists that these 4 differ from most M giants in having a supply of orbital angular momentum able to maintain dynamo action, as in RS CVn binaries, although the known RS CVn systems do not contain M stars. But for the majority of giants, the evidence points to a cut-off in coronal X-ray emission, and so the presumed associated dynamo action, occurring at late K-type.

Interpretation of the observations is clearly harder than for dwarf stars, for which the magnetic spin-down time-scale is shorter than that of nuclear-driven stellar evolution. Thus whereas for the dwarf stars, the sequence in spectral type is a sequence in mass, for the giants it is more nearly a constant-mass sequence in time, with the spectral type normally increasing monotonically. Recent work (Schröder et al., 1998) combines ROSAT observations of stellar activity with the most recent stellar evolution calculations, and uses Hipparcos parallaxes for more reliable positioning in the H–R diagram. Of particular interest is a clump of fairly long-lived post-helium flash (i.e. core helium-burning) K-giants which show activity at a solar level. Their masses are between about $1.3\,M_\odot$ and $2.3\,M_\odot$. Lower masses have had a much longer main sequence active period with a correspondingly greater early loss of angular momentum, and so should indeed show weaker activity in the giant domain.

There is a possible conflict here with standard $\alpha\Omega$ dynamo theory. Gray (1989, 1991) finds a precipitous drop in surface rotation between giants of type GO and G3, yet there is strong activity observed in G-stars that are approaching the first giant branch, as well as this moderately strong activity persisting among stars that have evolved into the K-giant clump. Gray argues for a source of angular momentum below the convective envelope, which is tapped as the lower boundary of the envelope deepens with time. We find more attractive the suggestion that in these late phases, the dynamo process is of α^2-type, requiring enough rotation to give the turbulence helicity, but depending on local cyclonic motions rather than large-scale shear to generate a toroidal field component.

8 Degenerate Stars

Burnt-out stars of moderate mass – either from birth, or through sufficient mass loss during their subsequent evolution – end their lives either as white dwarfs or neutron stars, while the more massive dying stars are presumed plausibly to become black holes. The existence of the observed class of magnetic white dwarfs, with surface field strengths up to a few times 10^8 G, is not unexpected if some of their main sequence precursors have a significant fossil flux, and indeed can be regarded as observational evidence that appropriate magnetic structures (e.g. a mixed poloidal-toroidal configuration) in gaseous stars can be stable on long time-scales (Spruit, 1999). Likewise, a neutron star, a factor 10^6 denser but with a similar flux, will have a surface field of $\simeq 10^{12}$ G. This is in fact the canonical figure for most radio pulsars, inferred originally by equating the observed spin-down energy loss rate to that carried off by an obliquely rotating dipole in vacuo (Gold, 1969). For all but the minority of milli-second pulsars, this estimate has persisted through the many subsequent attempts, beginning with Goldreich and Julian (1969), to model the pulsar magnetosphere more realistically as a charged domain, in which special (or even general) relativistic effects are paramount (e.g. Mestel, 1999). A possible alternative to the fossil field picture is the spontaneous build-up of flux by a 'battery'-type process (e.g. Blandford et al., 1983). In their 1995 review, Bhattacharya and Srinivasan concluded that 'at present there is no compelling observational evidence to suggest that the magnetic fields of neutron stars are generated after their birth'.

The first white dwarf magnetic field was detected in 1970 (Kemp et al., 1970), when a field (now thought to be about $3\,10^8$ G) was found in the star Grw+70° 8247 by detection of continuum circular polarization of more than 1%. Searches for other white dwarf fields have continued since. Currently known fields are observed to range from a few tens of kG (detected by circular polarization measurements across Balmer lines), to some MG (detected by visible Zeeman splitting of spectral lines of H or He), to fields of hundreds of MG (detected via continuum circular and sometimes linear polarization, or fits to strange and unique intensity spectra). The frequency of occurrence of detectable fields in white dwarfs is of the order 1% per decade of field strength between 10^5 and 10^9 G (Schmidt, 2001; Fabrika et al., 1997). More than 60 white dwarfs are now known to have detectable magnetic fields.

White dwarf fields are either constant, or vary periodically with periods of hours or days. These fields thus appear to be fossil fields, like those of magnetic Ap–Bp stars. From the fact that the longitudinal fields inferred from circular polarization in lines is typically of the order of 25% of the field strength deduced from Zeeman splitting in the intensity spectrum, we deduce that the fields have a simple dipolar topology like that found on Ap stars. This result is confirmed by satisfactory fits to data with simple dipole-like models (Wickramasinghe, 2001; Euchner et al., 2002).

For fields closer to 10^8 G, modelling requires detailed knowledge of atomic level splitting in fields whose effects on the atomic energy levels is comparable to the effect of the Coulomb field. Such calculations are notoriously difficult, and have only been carried through for H and He. The available atomic data now make it possible to obtain plausible fits to the intensity spectra of some of the very high-field white dwarfs (e.g. Wickramasinghe and Ferrario, 1988), but the detailed field geometry is not easily inferred from such modelling.

One very interesting connection between the presence of a magnetic field and basic stellar parameters is found. From the sample of roughly 20 magnetic white dwarfs for which accurate masses are available, it seems quite clear that the typical mass of a magnetic white dwarf is substantially larger than that of a non-magnetic white dwarf, about $0.9\,M_\odot$ compared to $0.6\,M_\odot$ (Liebert et al., 2003). This suggests either that magnetic white dwarfs descend from a different population than normal white dwarfs, or that such massive stars are produced by binary mergers which somehow also lead to magnetic fields.

Studies of the evolution of white dwarf fields are possible because the history of an isolated white dwarf since its formation is simply one of cooling, so enabling estimation of the star's age (Mestel, 1952; Mestel and Ruderman, 1967) and e.g. (Chabrier et al., 2000), and references therein). One can therefore obtain information about field evolution by comparing samples of young and old magnetic white dwarfs. It appears that fields in white dwarfs are more common in older white dwarfs than in younger ones (Valyavin and Fabrika, 1998; Liebert et al., 2003). This suggests that surface field strength may actually increase with time, perhaps as a result of flux leakage to the surface of the star.

References

Abt, H. A.: 1965, ApJS 11, 429
Abt, H. A.: 1979, ApJ 230, 485
Armitage, P. J., Clarke, C. J.: 1996, MNRAS 280, 458
Babcock, H. W.: 1947, ApJ 105, 105
Babcock, H. W.: 1960, ApJ 132, 521
Babcock, H. W., Babcock, H. D.: 1952, PASP 64 282
Babel, J.: 1992, A&A 258, 449
Babel, J.: 1994, A&A 283, 189
Babel, J., Michaud, G.: 1991, ApJ 366, 560
Bagnulo, S., Wade, G. A., Donati, J.-F. et al.: 2001, A&A 369, 889
Bagnulo, S., Landi Degl'Innocenti, M., Landolfi, M., Mathys, G.: 2002, A&A 394, 1023
Bagnulo, S., Landstreet, J. D., Lo Curto, G., Szeifert, T., Wade, G. A.: 2003, A&A 403, 645
Bagnulo, S., Hensberge, H., Landstreet, J. D., Szeifert, T., Wade, G. A.: 2004, submitted to A&A
Balbus, S., Hawley, J. F.: 1991, ApJ 376, 214
Baliunas, S. L., Vaughan, A. H.: 1985, ARA&A 23, 379
Baliunas, S. L., Donahue, R. A., Soon, W. H. et al.: 1995, ApJ 438, 269

Balmforth, N. J., Cunha, M. S., Dolez, N., Gough, D. O., Vauclair, S.: 2001, MNRAS 323, 362
Barker, D. M., Mestel, L.: 1996, MNRAS 282, 317
Bhattacharya, D., Srinivasan, G.: 1995, in *X-ray Binaries*, ed. W. H. G. Lewin et al. (Cambridge: Cambridge University Press), p. 502
Biermann, L.: 1935, Astron. Nachrichten 257, 269
Blandford, R. D.: 1989, in *Theory of Accretion Discs*, ed. F. Meyer et al. (Dordrecht: Kluwer), p. 35
Blandford, R. D., Applegate, J., Hernquist, L.: 1983, MNRAS 204, 1025
Bohlender, D. A., Landstreet, J. D.: 1990, MNRAS 247, 606
Bohlender, D. A., Landstreet, J. D., Thompson, H. I. B.: 1993, A&A 269, 355
Braithwaite, J., Spruit, H. C.: 2004, Nature 431, 819
Brandenburg, A., Nordlund, Aa., Stein, R. F., Torkelsson, U.: 1995, ApJ 446, 741
Brandenburg, A., Nordlund, Aa., Stein, R. F., Torkelsson, U.: 1996, ApJ 458, L45
Bychkov, V. D., Bychkova, L. V., Madej, J.: 2003, A&A 407, 631
Campbell, C. G.: 1997, *Magnetohydrodynamics in Binary Stars* (Dordrecht: Kluwer)
Chabrier, G., Brassard, P., Fontaine, G., Saumon, D.: 2000, ApJ 543, 216
Christensen-Dalsgaard, J., Gough, D. O., Thompson, M. J.: 1991, ApJ 378, 413
Collier Cameron, A.: 2002, Astron. Nachrichten 323, 336
Collier Cameron, A., Li, J.: 1994, MNRAS 269, 1099
Collier Cameron, A., Robinson, R. D.: 1989, MNRAS 236, 57 and 238, 657
Collier Cameron, A., Bedford, D. K., Rucinski, S. M., Vilhu, O., White, N. E.: 1988, MNRAS 231, 131
Collier Cameron, A., Campbell, C. G., Quaintrell, H.: 1995, A&A 298, 133
Cowley, C. R.: 1993, in *Peculiar Versus Normal Phenomena in A-type and Related Stars*, IAU Coll. 138, ed. M. M. Dworetsky, F. Castelli, R. Farragiana (San Francisco: Astronomical Society of the Pacific), p. 18
Cowley, C. R., Hubrig, S., Ryabchikova, T. A. et al.: 2001, A&A 367, 939
Crutcher, R. M.: 1999, ApJ 520, 706
Deutsch, A. J.: 1970, ApJ 159, 985
Donahue, R. A., Baliunas, S. L.: 1994, in *Eighth Cambridge Workshop on Cool Stars, Stellar Systems and the Sun*, ed. J. P. Caillault, ASP Conf. Ser. 64, p. 396
Donahue, R. A., Saar, S. H., Baliunas, S. L.: 1996, ApJ 466, 384
Donati, J. F., Collier Cameron, A.: 1997, MNRAS 291, 1
Donati, J.-F., Semel, M., Carter, B., Rees, D. E., Collier Cameron, A.: 1997, MNRAS 291, 658
Donati, J.-F., Collier Cameron, A., Hussain, G. A. J., Semel, M.: 1999a, MNRAS 302, 437
Donati, J.-F., Catala, C., Wade, G. A. et al.: 1999b, A&AS 134, 149
Donati, J.-F., Wade, G. A., Babel, J. et al.: 2001, MNRAS 326, 1265
Donati, J.-F., Babel, J., Harries, T. J. et al.: 2002, MNRAS 333, 55
Durney, B. R.: 1972, in *The Solar Wind*, ed. C. P. Sonnett et al. (Washington: NASA), p. 282
Euchner, F., Jordan, S., Beuermann, K., Gänsicke, B. T., Hessman, F. V.: 2002, A&A 390, 633
Fabrika, S. N., Shtol, V. G., Valyavin, G. G., Bychkov, V. D.: 1997, Astronomy Letters 23, 43
Ferraro, V. C. A.: 1937, MNRAS 97, 458

Glagolevskij, Yu. V., Gerth, E.: 2001, in *Magnetic Fields Across the H-R Diagram*, ed. G. Mathys, S. T. Solanki, D. T. Wickramasinghe, ASP Conf. Ser. 248, 337
Gold, T.: 1969, Nature 221, 25
Goldreich, P., Julian, W. H.: 1969, ApJ 157, 869
Gomez, A. E., Luri, X., Grenier, S. et al.: 1998, A&A 336, 953
Gray, D. F.: 1989, ApJ 347, 1021
Gray, D. F.: 1991, in *The Sun and Cool Stars*, ed. I. Tuominen et al. (Berlin: Springer), p. 336
Gray, D. F.: 1994, ApJ 428, 765
Gray, D. F., Baliunas, S. L.: 1997, ApJ 475, 303
Gray, D. F., Baliunas, S. L., Lockwood, G. W., Skiff, B. A.: 1996a, ApJ 456, 365
Gray, D. F., Baliunas, S. L., Lockwood, G. W., Skiff, B. A.: 1996b, ApJ 465, 945
Haisch, B. M., Schmitt, J. H. M. M., Fabian, A. C.: 1992, Nature 360, 269
Hale, G. E.: 1908, PASP 20, 220
Hartoog, M. R.: 1976, ApJ 205, 897
Hartoog, M. R.: 1977, ApJ 212, 723
Heiles, C.: 1997, ApJS 111, 245
Heitsch, F., Zweibel, E. G., Slyz, A. D., Devriendt, J. E. G.: 2004, ApJ 603, 165
Hubrig, S., North, P., Mathys, G.: 2000, ApJ 539, 352
Hünsch, M., Schmitt, J. H. M. M., Schröder, K.-P., Reimers, D.: 1996, A&A 310, 801
Hünsch, M., Schmitt, J. H. M. M., Schröder, K.-P., Zickgraf, F.-J.: 1998, A&A 330, 225
Johns-Krull, C. M., Valenti, J. A., Koresko, C.: 1999, ApJ 516, 900
Johns-Krull, C. M., Valenti, J. A., Piskunov, N. E., Saar, S. H., Hatzes, A. P.: 2001, in *Magnetic Fields Across the Hertzsprung-Russell Diagram*, ed. G. Mathys, S. K. Solanki, D. T. Wickramasinghe, ASP Conf. Ser. 248, p. 527
Kemp, J. C.: 1970, ApJ 162, 169
Kemp, J. C, Swedlund, J. B., Landstreet, J. D., Angel, J. R. P.: 1970, ApJ 161, L77
Kippenhahn, R.: 1959, Zeitschrift f. Astrophysik 48, 203
Kitchatinov, L. L., Rüdiger, G.: 1995, A&A 299, 446
Kitchatinov, L. L., Rüdiger, G.: 1999, A&A 344, 911
Kochukhov, O., Bagnulo, S., Barklem, P. S.: 2002, Astrophys. J. Lett. 578, L75
Kochukhov, O., Bagnulo, S., Wade, G. A. et al.: 2004, A&A 414, 613
Kraft, R. P.: 1967, ApJ 150, 551
Kurtz, D. W.: 2000, in *Variable Stars as Essential Astrophysical Tools*, ed. C. Ibanoglu (Dordrecht: Kluwer), p. 313
Landstreet, J. D.: 1967, PhD Thesis, Columbia University
Landstreet, J. D.: 1970, ApJ 159, 1001
Landstreet, J. D.: 1982, ApJ 258, 639
Landstreet, J. D.: 1987, MNRAS 225, 437
Landstreet, J. D.: 1988, ApJ 326, 967
Landstreet, J. D., Borra, E. F.: 1978, ApJL 224, L5
Landstreet, J. D., Mathys, G.: 2000, A&A 359, 213
Lehnert, B. (ed.): 1958, *Electromagnetic Processes in Cosmical Physics* (Cambridge: Cambridge University Press)
Li, J.: 1999, MNRAS 302, 203

Liebert, J., Bergeron, P., Holberg, J. B.: 2003, AJ 125, 348
Lynden-Bell, D. (ed.): 1994a, *Cosmical Magnetism* (Dordrecht: Kluwer)
Lynden-Bell, D. (ed.): 1994b, *Cosmical Magnetism*, contributed papers (Cambridge: IoA)
Lynden-Bell, D., Pringle, J. E.: 1974, MNRAS 168, 603
Markey, P., Tayler, R. J.: 1973, MNRAS 163, 77
Markey, P., Tayler, R. J.: 1974, MNRAS 168, 505
Massi, M., Neidhöfer, J., Torricelli-Ciamponi, G., Chiuderi-Drago, F.: 1998, A&A 332, 149
Mathys, G.: 1989, Fund. Cos. Phys. 13, 143
Mathys, G., Hubrig, S.: 1995, A&A 293, 810
Mathys, G., Hubrig, S., Landstreet, J. D., Lanz, T., Manfroid, J.: 1997, A&AS 123, 353
Mestel, L.: 1952, MNRAS 113, 583
Mestel, L.: 1967, in *Plasma Astrophysics*, ed. P. A. Sturrock (London: Academic Press), p. 185
Mestel, L.: 1968, MNRAS 138, 359
Mestel, L.: 1984, Astron. Nachrichten 305, 301
Mestel, L.: 1999, *Stellar Magnetism* (Oxford: Clarendon Press) Paperback edition 2003
Mestel, L., Moss, D. L.: 1977, MNRAS 178, 27
Mestel, L., Ruderman, M. A.: 1967, MNRAS 136, 27
Mestel, L., Selley, C. S.: 1970, MNRAS 149, 197
Mestel, L., Spitzer Jr., L.: 1956, MNRAS 116, 583
Mestel, L., Spruit, H. C.: 1987, MNRAS 226, 57
Mestel, L., Weiss, N. O.: 1987, MNRAS 226, 123
Mestel, L., Nittman, J., Wood, W. P., Wright, G. A. E.: 1981, MNRAS 195, 979
Michaud, G.: 1970, ApJ 160, 641
Michaud, G., Charland, Y., Vauclair, S., Vauclair, G.: 1976, ApJ 210, 447
Moss, D.: 1984, MNRAS 209, 607
Moss, D.: 1986, Phys. Reports 140, 1
Moss, D.: 1989, MNRAS 236, 629
Moss, D.: 1990, MNRAS 244, 272
Moss, D.: 1992, MNRAS 257, 593
Moss, D.: 2001, in *Magnetic Fields Across the H-R Diagram*, ed. G. Mathys, S. T. Solanki, D.T. Wickramasinghe, ASP Conf. Ser. 248, p. 305
Moss, D.: 2003, A&A 403, 693
Moss, D., Mestel, L., Tayler, R. J.: 1990, MNRAS 245, 550
Mouschovias, T. Ch.: 1987, in *Physical Processes in Interstellar Clouds*, ed. G. E. Morfill, M. Scholer (Dordrecht: Reidel), p. 453
Mouschovias, T. Ch.: 2001, in *Magnetic Fields Across the H-R Diagram*, ed. G. Mathys, S. T. Solanki, D.T. Wickramasinghe, ASP Conf. Ser. 248, p. 515
Mutel, R. L., Morris, D. H.: 1988, in *Activity in Cool Star Envelopes*, ed. O. Havnes et al. (Dordrecht: Kluwer), p. 283
Mutel, R. L., Lestrade, J. F., Preston, R. A. et al.: 1985, ApJ 289, 262
Myers, P. C., Goodman, A. A., Güsten, R., Heiles, C.: 1995, ApJ 220, 864
Nakano, T.: 1976, PASJ 28, 355
Nakano, T., Nishi, R., Umebayashi, T.: 2002, ApJ 573, 199

Neiner, C., Geers, V. C., Henrichs, H. F. et al.: 2003, A&A 406, 1019
Nishi, R., Nakano, T., Umebayashi, T.: 1991, ApJ 368, 181
Nordlund, Aa., Brandenburg, A., Jennings, R. L. et al.: 1992, ApJ 392, 647
Noyes, R. W., Hartmann, L. W., Baliunas, S. L., Duncan, D. K., Vaughan, A. H.: 1984, ApJ 279, 763
Okamoto, I.: 1999, MNRAS 307, 253
Okamoto, I.: 2003, ApJ 589, 671
Ostriker, E., Shu, F. H.: 1995, ApJ 447, 813
Palla, F., Stahler, S. W.: 1993, ApJ 418, 414
Pöhnl, H., Maitzen, H. M., Paunzen, E.: 2003, A&A 402, 247
Preston, G. W.: 1967, ApJ 150, 547
Preston, G. W.: 1970, in *Stellar Rotation*, ed. A. Slettebak (Dordrecht: Reidel), p. 254
Proctor, M. R. E., Weiss, N. 0.: 1982, Reports Progress Physics 45, 1317
Radick, R. R., Lockwood, G. W., Skiff, B. A., Baliunas, S. L.: 1998, ApJS 118, 239
Rädler, K.-H.: 1986, in *Plasma Astrophysics*, ed. T. D. Guyenne (ESA SP-251), p. 569
Renson, P., Catalano, F. A.: 2001, A&A 378, 113
Richer, J., Michaud, G., Turcotte, S.: 2000, ApJ 529, 338
Rüdiger, G., Arlt, R., Hollerbach, R.: 2001, in *Magnetic Fields Across the H-R Diagram*, ed. G. Mathys, S. T. Solanki, D.T. Wickramasinghe, ASP Conf. Ser. 248, p. 315
Saar, S. H., Baliunas, S. L.: 1992, in *The Solar Cycle*, ed. K. L. Harvey, ASP Conf. Ser. 27, p. 150
Schmidt, G. D.: 2001, in *Magnetic Fields Across the Hertzsprung-Russell Diagram*, ed. G. Mathys, S. K. Solanki, D. T. Wickramasinghe, ASP Conf. Ser. 248, p. 443
Schmitt, J. H. M. M.: 2001, in *Magnetic Fields Across the Hertzsprung-Russell Diagram*, ed. G. Mathys, S. K. Solanki, D. T. Wickramasinghe, ASP Conf. Ser. 248, p. 199
Schrijver, C. J., Zwaan, C.: 2000, *Solar and Stellar Magnetic Activity* (Cambridge: Cambridge University Press)
Schrijver, C. J., Title, A. M., van Ballegooijen, A. A., Hagenaar, H. J., Shine, R. A.: 1997, ApJ 487, 424
Schröder, K.-P., Hünsch, M., Schmitt, J. H. M. M.: 1998, A&A 335, 591
Schüssler, M.: 1980, Nucleonika 25, 1425
Shore, S. N., Brown, D. N., Sonneborn, G., Landstreet, J. D., Bohlender, D. A.: 1990, ApJ 348, 242
Shorlin, S. L. S., Wade, G. A., Donati, J.-F. et al.: 2002, A&A 392, 637
Shu, F. H., Adams, F. C., Lizano, S.: 1987, ARA&A 25, 23
Skumanich, A.: 1972, ApJ 171, 565
Solanki, S. K.: 2001, in *Magnetic Fields Across the Hertzsprung-Russell Diagram*, ed. G. Mathys, S. K. Solanki, D. T. Wickramasinghe, ASP Conf. Ser. 248, p. 45
Spitzer Jr., L.: 1957, ApJ 125, 525
Spruit, H. C.: 1999, A&A 349, 189
Spruit, H. C., Foglizzo, T., Stehle, R.: 1997, MNRAS 288, 333
Stępień, K.: 2000, A&A 353, 227
Stibbs, D. W. N.: 1950, MNRAS 110, 395
Strasser, S., Landstreet, J. D., Mathys, G.: 2001, A&A 378, 153

Sweet, P. A.: 1950, MNRAS 110, 548
Tayler, R. J.: 1980, MNRAS 191, 151
Tayler, R. J.: 1987, MNRAS 227, 553
Thompson, I. B., Brown, D. N., Landstreet, J. D.: 1987, ApJS 64, 219
Tout, C. A., Pringle, J. E.: 1992, MNRAS 259, 604
Unruh, Y. C., Collier Cameron, A., Cutispoto, G.: 1995, MNRAS 277, 1145
Valenti, J. A., Johns-Krull, C.: 2001, in *Magnetic Fields Across the Hertzsprung-Russell Diagram*, ed. G. Mathys, S. K. Solanki, D. T. Wickramasinghe, ASP Conf. Ser. 248, p. 179
Valyavin, G. G., Fabrika, S. N.: 1998, Bull. Spec. Ap. Obs. 45, 69
Wade, G. A., Donati, J.-F., Landstreet, J. D., Shorlin, S. L. S.: 2000, MNRAS 313, 823
Wade, G. A., Bagnulo, S., Kochukhov, O. et al.: 2001, A&A 374, 265
Weber, E. J., Davis Jr., L.: 1967, ApJ 148, 217
Weiss, N. O., Cattaneo, F., Jones, C. A.: 1984, Geophys. Astrophys. Fluid Dynamics 30, 305
Wickramasinghe, D. T.: 2001, in *Magnetic Fields Across the Hertzsprung-Russell Diagram*, ed. G. Mathys, S. K. Solanki, D. T. Wickramasinghe, ASP Conf. Ser. 248, p. 453
Wickramasinghe, D. T., Ferrario, L.: 1988, ApJ 327, 222
Wilson, O. C.: 1978, ApJ 226, 379
Wright, G. A. E.: 1973, MNRAS 162, 329

Importance of Magnetic Helicity in Dynamos

Axel Brandenburg

Nordita, Blegdamsvej 17, 2100 Copenhagen Ø, Denmark
brandenb@nordita.dk

Magnetic helicity is nearly conserved and its evolution equation provides a dynamical feedback on the alpha effect that is distinct from the conventional algebraic alpha quenching. The seriousness of this dynamical alpha quenching is particularly evident in the case of closed or periodic boxes. The explicit connection with catastrophic alpha quenching is reviewed and the alleviating effects of magnetic and current helicity fluxes are discussed.

1 Introduction

Let us begin by defining dynamos and helicity. Dynamos are a class of velocity fields that allow a weak seed magnetic field to be amplified until some saturation process sets in. Mathematically, this is described by exponentially growing solutions of the induction equation. Simulations have shown that any sufficiently complex flow field can act as a dynamo if the resistivity is below a certain threshold. It is in principle not even necessary that the flow is three-dimensional, only the magnetic field must be three-dimensional because otherwise one of several antidynamo theorems apply (Cowling, 1934; Zeldovich, 1957).

Helicity, on the other hand, quantifies the swirl in a vector field. There is kinetic helicity, which describes the degree to which vortex lines follow a screw-like pattern, and it is positive for right-handed screws. Examples of helical flows are the highs and lows on the weather map. For both highs and lows the kinetic helicity has the same sign and is negative (positive) in the northern (southern) hemisphere. For example, in an atmospheric low, air flows inward, i.e. toward the core of the vortex, and down to the bottom of the atmosphere, but the Coriolis force makes it spin anti-clockwise, causing left-handed spiraling motions and hence negative helicity.

A connection between helicity and dynamos has been established already quite some time ago when Steenbeck et al. (1966) calculated the now famous α effect in mean field dynamo theory and explained its connection with kinetic helicity. In this paper we are not so much concerned with kinetic helicity, but mostly with the magnetic and current helicities. Quantifying the swirl of magnetic field lines has diagnostic significance, because magnetic helicity is a topological invariant of the ideal (non-resistive) equations. Especially in the solar community the *diagnostic* properties of magnetic helicity have been exploited extensively over the past decade. However, the use of magnetic helicity as a *prognostic* quantity for understanding the governing nonlinearity of α effect dynamos has only recently been noted in connection with the magnetic helicity constraint (Brandenburg, 2001, hereafter referred to as B01).

We should emphasize from the beginning that dynamos do not have to have helicity. The small-scale dynamo of Kazantsev (1968) is an example of a dynamo that works even without helicity. Nonhelical dynamos are generally harder to excite than helical dynamos, but both can generate fields of appreciable strength if the magnetic Reynolds number is large. The stretch-twist-fold dynamo also operates with twist (as the name suggests!), but the orientation of twist can be random, so the net helicity can be zero. Simulations have shown that even with zero helicity density everywhere, dynamos can work (Hughes et al., 1996).

It is also possible to generate magnetic fields of large scale once there is strong shear, even if there is no helicity (Vishniac and Brandenburg, 1997). This case is very much a topic of current research. One of the possibilities is is the so-called shear-current effect (Rogachevskii and Kleeorin, 2003, 2004), but such dynamos still produce helical large-scale magnetic fields. There is also the possibility of an intrinsically nonlinear dynamo operating with magnetic helicity flux alone (Vishniac and Cho, 2001). Thus, it is not necessarily clear that large-scale dynamos have to work with kinetic helicity and the corresponding α effect. However, there is as yet no convincing example of a dynamo without the involvement of kinetic helicity that generates large-scale magnetic fields with a degree of coherence that is similar to that observed in stars and in galaxies, e.g. cyclically migrating magnetic fields in the sun and grand magnetic spirals in some nearby galaxies. Such fields can potentially be generated by dynamos with an α effect, as has been shown in many papers over the past 40 years; see Chaps. 2, 4, and 6.

There is however a major problem with α effect dynamos; see Brandenburg (2003); Brandenburg and Subramanian (2005) for recent reviews on the issue. The degree of severity depends on the nature of the problem. It is most severe in the case of a homogeneous α effect in a periodic box, which is also when the problem shows up most pronouncedly. Cattaneo and Hughes (1996) fo und that the α effect is quenched to resistively small values once the mean field becomes a fraction of the equipartition field strength. In response to such difficulties three different approaches have been pursued. The most practical one is to simply ignore the problem and the proceed as if we can still use the α effect with a quenching that only sets in at equipartition field strengths. This can partially be justified by the apparent success in applying this theory; see the recent reviews by Beck et al. (1996); Kulsrud (1999), and Widrow (2002). The second approach is to resort to direct three dimensional simulations of the turbulence in such astrophysical bodies. In the solar community this approach has been pioneered by Gilman (1983) and Glatzmaier (1985), and more recently by Brun et al. (2004). The third approach is a combination of the first two, i.e. to use direct simulations of problems where mean-field theory should give a definitive answer. This is also the approach taken in the present work. The hope is ultimately to find guidance toward a revised mean-field theory and to test it quantitatively. A lot of progress has already been made which led to the suggestion that only a dynamical (i.e. explicitly time dependent) theory of α quenching is compatible with the simulation results. In the present paper we review some of the simulations that have led to this revised understanding of mean-field theory.

The dynamical quenching theory is now quite successful in reproducing the results from simulations in a closed or periodic box with and without shear. In these cases super-equipartition fields are possible, but only after a resistively long time scale. In the case of an open box without shear the dynamical quenching theory is also successful in reproducing the results of simulations, but here the root mean-field strength decreases with increasing magnetic Reynolds number, suggesting that such a dynamo is unimportant for astrophysical applications. Open boxes with shear appear now quite promising, but the theory is still incomplete and, not surprisingly, there are discrepancies between simulations and theory. In fact, it is quite possible that it is not even the α effect that is important for large-scale field regeneration. Alternatives include the shear-current effect of Rogachevskii and Kleeorin (2003, 2004) and the Vishniac and Cho (2001) magnetic and current helicity fluxes. In both cases strong helicity fluxes are predicted by the theory and such fluxes are certainly also confirmed observationally for the sun (Berger and Ruzmaikin, 2000; DeVore, 2000; Chae, 2000; Low, 2001). For the galaxy the issue of magnetic helicity is still very much in its infancy, but some first attempts in this direction are already being discussed (Shukurov, 2005).

2 Dynamos in a Periodic Box

To avoid the impression that all dynamos have to have helicity, we begin by commenting explicitly on dynamos that do not have net kinetic helicity, i.e. $|\langle\boldsymbol{\omega}\cdot\boldsymbol{u}\rangle|/(k_\mathrm{f}\langle\boldsymbol{u}^2\rangle) \ll 1$, where k_f is the wavenumber of the forcing (corresponding to the energy carrying scale). Unless the flow also possesses some large-scale shear flow (discussed separately in Sect. 4.5 below), such dynamos are referred to as small-scale dynamos. The statement made in the introduction that *any* sufficiently complex flow field can act as a dynamo is really only based on experience, and the statement may need to be qualified for small-scale dynamos. Indeed, whether or not turbulent small-scale dynamos work in stars where the magnetic Prandtl numbers are small ($\mathrm{Pr_M} \approx 10^{-4}$) is unclear (Schekochihin et al., 2004; Boldyrev and Cattaneo, 2004). Simulations suggest that the critical magnetic Reynolds numbers increase with decreasing magnetic Prandtl number like $R_\mathrm{m,crit} \approx 35\mathrm{Pr_M}$ (Haugen et al., 2004).

Throughout the rest of this review, we want to focus attention on large scale dynamos. This is where magnetic helicity plays an important role. Before we explain why in a periodic box nonlinear dynamos operate only on a resistively slow time scale, it may be useful to illustrate the problem with some numerical facts.

In the simulations of B01 the flow was forced at an intermediate wavenumber, $k \approx k_\mathrm{f} = 5$, while the smallest wavenumber in the computational domain corresponds to $k = k_1 = 1$. The kinetic energy spectrum peaks at $k \approx k_\mathrm{f}$, which is therefore also the wavenumber of the energy carrying scale. The turbulence is nearly fully helical with $\langle\boldsymbol{\omega}\cdot\boldsymbol{u}\rangle/(k_\mathrm{f}\langle\boldsymbol{u}^2\rangle) \approx 0.7...0.9$. The initial field is random, but as time goes on it develops a large-scale component at wavenumber $k \approx k_1 = 1$; see Fig. 1. In Fig. 2 we show the evolution of the magnetic energy of the mean field from the same simulation.[1] Here the mean field is defined as two-dimensional averages over planes perpendicular to the direction in which the mean field varies. There are of course three such directions, but there is usually only one direction for which there is a significant mean field.

[1] Here the time unit is $[t] = (c_\mathrm{s}k_1)^{-1}$, where c_s is the isothermal speed of sound, and the magnetic field is measured in units of $[B] = \sqrt{\mu_0\rho_0}c_\mathrm{s}$.

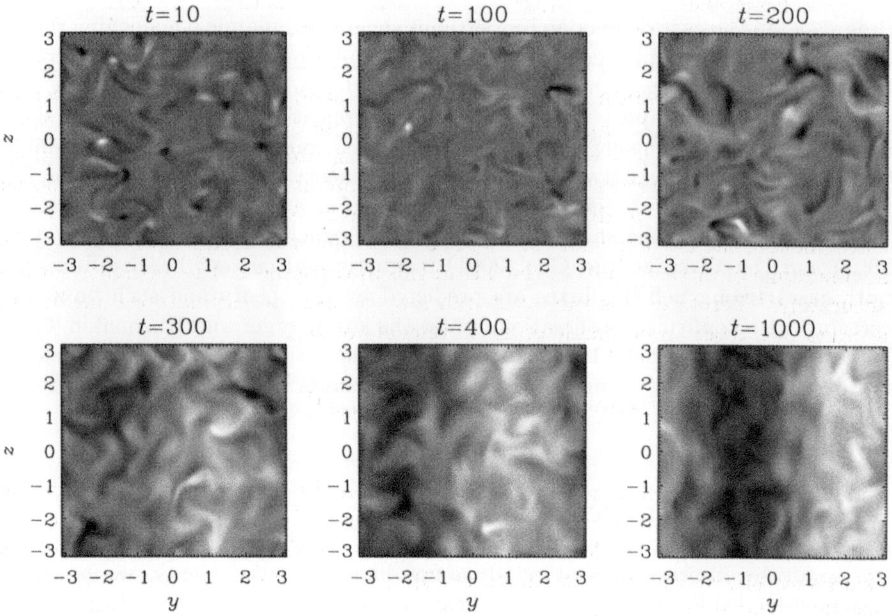

Fig. 1. Cross-sections of $B_x(0,y,z)$ for Run 3 of B01 at different times showing the gradual build-up of the large-scale magnetic field after $t=300$. The diffusive time scale for this run is $(\eta k_1^2)^{-1}=500$. Dark (light) corresponds to negative (positive) values. Each image is scaled with respect to its min and max values. The final state corresponds to the second eigenfunction given in (33), but with some smaller scale turbulence superimposed [Adapted from Brandenburg and Subramanian (2005)]

While the saturation field strength increases with increasing magnetic Reynolds number, the time scale on which this nonlinear dynamo saturates increases too. To avoid misunderstandings, it is important to emphasize that this result applies only when we are in the *nonlinear regime* and when the flows are *helical*.

In turbulence one is used to situations where the microscopic values of viscosity ν and magnetic diffusivity η do not matter in the sense that, for almost all practical purposes, they are superseded by turbulent effective values, $\nu_{\rm t}$ and $\eta_{\rm t}$, respectively. This is because in turbulence there is spectral energy all the way down to the viscous/resistive length scale, $(\eta\tau)^{1/2}$, where τ is the turnover time.[2] Thus, even when ν is very small, the rate of viscous dissipation, $\langle 2\nu\rho\mathbf{S}^2\rangle$, is in general finite ($\mathbf{S}$ is the trace-less rate of strain tensor). Likewise, even when η is very small, the rate of Joule dissipation, $\eta\mu_0\langle \boldsymbol{J}^2\rangle$, is in general finite ($\mu_0$ is the magnetic permeability). This is because the current density diverges with decreasing η like $|\boldsymbol{J}|\sim\eta^{-1/2}$, so the energy dissipation stays finite and asymptoticly independent of how small η is. The trouble is that the value of magnetic helicity dissipation is proportional

[2] The turnover time at the wavenumber k is $(u_k k)^{-1}$. Using Kolmogorov scaling, $u_k\sim k^{-1/3}$, one finds the familiar formula $k_\eta=k_{\rm f}R_{\rm m}^{3/4}$, where $k_{\rm f}$ is the wavenumber of the energy carrying eddies.

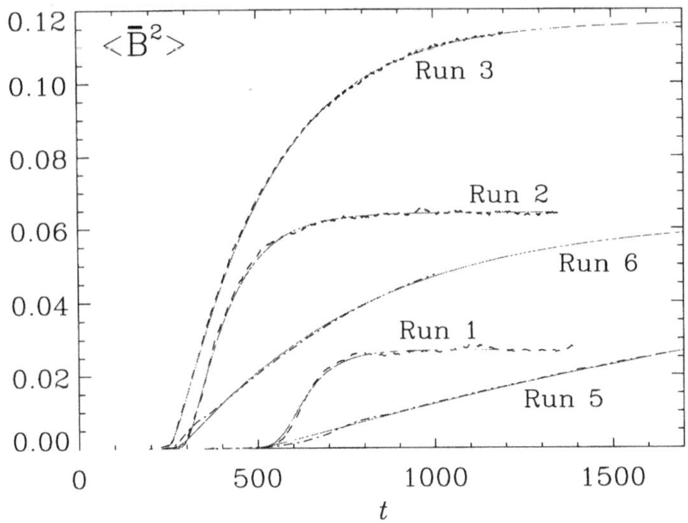

Fig. 2. Evolution of $\langle \overline{\boldsymbol{B}}^2 \rangle$ for Runs 1–3 and 5, 6 (*dashed lines*). The magnetic Reynolds numbers are $R_\mathrm{m} \equiv u_\mathrm{rms}/\eta k_\mathrm{f} = 2.4$, 6, 18, 100, and 16, respectively; see B01. The *solid lines* denote the solution of the associated mean-field dynamo problem where *both* α and and turbulent diffusivity η_t are quenched in a magnetic Reynolds number dependent fashion [Adapted from B01]

to $\eta\langle \boldsymbol{J} \cdot \boldsymbol{B} \rangle$ (see below), and in the limit $\eta \to 0$ we have $\eta\langle \boldsymbol{J} \cdot \boldsymbol{B} \rangle \to \eta^{1/2} \to 0$, so resistive magnetic helicity dissipation becomes impossible in the limit of large R_m. In the following section we derive and discuss the evolution equation for magnetic helicity.

3 Magnetic Helicity Evolution

3.1 The Two-scale Property of Helical Turbulence

Usually in mean-field dynamo theory one talks about the two-scale *assumption* made in order to derive the mean-field equations (e.g. Moffatt, 1978; Krause and Rädler, 1980). This has to do with the fact that higher order derivatives in the mean field equation can only be neglected when the mean field is sufficiently smooth. Here, instead, we use the two-scale *properties* of helical turbulence as demonstrated in the previous section. These properties emerge automatically when the size of the whole body is at least several times larger than the scale of the turbulent eddies. As Fig. 1 shows explicitly, a large-scale field (wavenumber k_1) emerges in addition to the forcing scale (wavenumber $k_\mathrm{f} \gg k_1$).

In this section we discuss the magnetic helicity equation and use it together with the two-scale property of helical turbulence to derive the so-called magnetic helicity constraint that allows the result of Fig. 2 to be understood quantitatively.

3.2 Definition of Helicity

The helicity of any solenoidal vector field $\boldsymbol{f}$, i.e. with $\boldsymbol{\nabla}\cdot\boldsymbol{f}=0$, is defined as the volume integral of $\boldsymbol{f}$ dotted with its inverse curl, i.e. $\mathrm{curl}^{-1}\boldsymbol{f}\equiv\boldsymbol{g}$. As pointed out by Moffatt (1969), the helicity quantifies the topological linkage between tubes in which $\boldsymbol{f}$ is non-vanishing. In the following the linkage aspect of helicity will not be utilized, but rather the mathematical evolution equation that the helicity obeys (see the next section). However, the calculation of $\boldsymbol{g}$ is problematic because it involves a gauge ambiguity in that the curl of $\boldsymbol{g}'=\boldsymbol{g}+\boldsymbol{\nabla}\phi$ also gives the same $\boldsymbol{f}=\mathrm{curl}\,\boldsymbol{g}'$.

In the special case of periodic boundary conditions or for $\hat{\boldsymbol{n}}\cdot\boldsymbol{f}=0$ on the boundaries, where $\hat{\boldsymbol{n}}$ is the normal vector on the boundary, the helicity is actually gauge-invariant, because

$$\begin{aligned}\int\boldsymbol{f}\cdot\boldsymbol{g}'\,\mathrm{d}V&=\int\boldsymbol{f}\cdot\boldsymbol{g}\,\mathrm{d}V+\int\boldsymbol{f}\cdot\boldsymbol{\nabla}\phi\,\mathrm{d}V\\&=\int\boldsymbol{f}\cdot\boldsymbol{g}\,\mathrm{d}V-\int\phi\boldsymbol{\nabla}\cdot\boldsymbol{f}\,\mathrm{d}V=\int\boldsymbol{f}\cdot\boldsymbol{g}\,\mathrm{d}V\,,\end{aligned}\tag{1}$$

where we have used $\boldsymbol{\nabla}\cdot\boldsymbol{f}$. Since the magnetic field is divergence free, the magnetic helicity, $\int\boldsymbol{B}\cdot\mathrm{curl}^{-1}\boldsymbol{B}\,\mathrm{d}V$ is gauge invariant. For other boundary conditions this is unfortunately not the case.

For vector fields whose inverse curl is a physically meaningful quantity, such as the vorticity $\boldsymbol{\omega}$, whose inverse curl is the velocity $\boldsymbol{u}$, the gauge question never arises. In this and similar cases the helicity density, $\boldsymbol{\omega}\cdot\boldsymbol{u}$ in this case, is physically meaningful. Other examples are the cross helicity, $\int\boldsymbol{B}\cdot\mathrm{curl}^{-1}\boldsymbol{\omega}\,\mathrm{d}V$, which describes the linkage between magnetic flux tubes and vortex tubes, and the current helicity, $\int\boldsymbol{J}\cdot\mathrm{curl}^{-1}\boldsymbol{J}\,\mathrm{d}V$, which quantifies the linkage of current tubes. In these two cases it is natural to use $\mathrm{curl}^{-1}=\boldsymbol{B}$ and $\mathrm{curl}^{-1}\boldsymbol{\omega}=\boldsymbol{u}$. For the magnetic field one can define the magnetic vector potential, $\mathrm{curl}^{-1}\boldsymbol{B}=\boldsymbol{A}$, but $\boldsymbol{A}$ is not a physically meaningful quantity and hence the magnetic helicity,

$$H=\int_V\boldsymbol{A}\cdot\boldsymbol{B}\,\mathrm{d}V\equiv\langle\boldsymbol{A}\cdot\boldsymbol{B}\rangle\,V\tag{2}$$

is gauge-dependent, unless the boundaries of the volume V are periodic or perfectly conducting. Here and below, angular brackets denote volume averages. Occasionally, however, we simply refer to $\langle\boldsymbol{A}\cdot\boldsymbol{B}\rangle$ as the magnetic helicity, but this is strictly speaking only the magnetic helicity per unit volume.

In the following section we derive the evolution equation for $\langle\boldsymbol{A}\cdot\boldsymbol{B}\rangle$ and focus first on the case where the boundary conditions are indeed periodic, so $\langle\boldsymbol{A}\cdot\boldsymbol{B}\rangle$ is gauge-invariant.

3.3 Derivation of the Magnetic Helicity Equation

The homogeneous Maxwell equations are

$$\frac{\partial\boldsymbol{B}}{\partial t}=-\boldsymbol{\nabla}\times\boldsymbol{E},\quad\boldsymbol{\nabla}\cdot\boldsymbol{B}=0\,.\tag{3}$$

Expressing this in terms of the magnetic vector potential, $\boldsymbol{A}$, where $\boldsymbol{B}=\boldsymbol{\nabla}\times\boldsymbol{A}$, we have

$$\frac{\partial \boldsymbol{A}}{\partial t} = -\boldsymbol{E} - \boldsymbol{\nabla}\phi \,, \tag{4}$$

where ϕ is the scalar potential. Dotting (3) and (4) with $\boldsymbol{A}$ and $\boldsymbol{B}$, respectively, and adding them, we have

$$\frac{\partial}{\partial t}(\boldsymbol{A}\cdot\boldsymbol{B}) = -2\boldsymbol{E}\cdot\boldsymbol{B} - \boldsymbol{\nabla}\cdot(\boldsymbol{E}\times\boldsymbol{A} + \phi\boldsymbol{B}) \,. \tag{5}$$

Here, $\boldsymbol{A}\cdot\boldsymbol{B}$ is the magnetic helicity density, but since it is not gauge invariant (see below) it is not a physically meaningful quantity. After integrating (5) over a periodic volume, the divergence term does not contribute. Furthermore, using Ohm's law, $\boldsymbol{E} = -\boldsymbol{U}\times\boldsymbol{B} + \eta\mu_0\boldsymbol{J}$, where $\boldsymbol{J} = \boldsymbol{\nabla}\times\boldsymbol{B}/\mu_0$ is the current density, we have

$$\frac{\mathrm{d}}{\mathrm{d}t}\langle\boldsymbol{A}\cdot\boldsymbol{B}\rangle = -2\eta\mu_0\langle\boldsymbol{J}\cdot\boldsymbol{B}\rangle \,, \tag{6}$$

i.e. the magnetic helicity, $\langle\boldsymbol{A}\cdot\boldsymbol{B}\rangle$, changes only at a rate that is proportional to $\eta\langle\boldsymbol{J}\cdot\boldsymbol{B}\rangle$. (Here and elsewhere, angular brackets denote volume averaging.) As discussed in the previous section, this rate converges to zero in the large R_{m} limit. Here, angular brackets denote volume averages, i.e. $\langle\boldsymbol{A}\cdot\boldsymbol{B}\rangle = \frac{1}{V}\int_V \boldsymbol{A}\cdot\boldsymbol{B}\,\mathrm{d}V$.

We recall that for periodic boundary conditions, $\langle\boldsymbol{A}\cdot\boldsymbol{B}\rangle$ is invariant under the transformation $\boldsymbol{A}\to\boldsymbol{A}' = \boldsymbol{A} + \boldsymbol{\nabla}\Lambda$, which does not change the value of $\boldsymbol{B}' = \boldsymbol{\nabla}\times\boldsymbol{A}' = \boldsymbol{\nabla}\times\boldsymbol{A} = \boldsymbol{B}$. Here, Λ is a gauge potential. Thus, for periodic boundary conditions, $\langle\boldsymbol{A}\cdot\boldsymbol{B}\rangle$ is a physically meaningful quantity. The same is also true for perfectly conducting boundaries (see Brandenburg and Dobler, 2002, for corresponding simulations). For open boundaries, however, $\int_V \boldsymbol{A}\cdot\boldsymbol{B}\,\mathrm{d}V$ is not gauge invariant, but one can derive a gauge-invariant relative magnetic helicity (Berger and Field, 1984).

3.4 The Magnetic Helicity Constraint

A very simple argument can be made to explain the saturation level and the resistively slow saturation behavior observed in Fig. 2. The only assumption is that the turbulence is helical, i.e. $\langle\boldsymbol{\omega}\cdot\boldsymbol{u}\rangle \neq 0$, where $\boldsymbol{\omega}$ is the vorticity, and that this introduces current helicity $\langle\boldsymbol{j}\cdot\boldsymbol{b}\rangle$, at the same scale and of the same sign as the kinetic helicity. Here we have split the field into large and small-scale fields, i.e. $\boldsymbol{B} = \overline{\boldsymbol{B}} + \boldsymbol{b}$ and hence also $\boldsymbol{J} = \overline{\boldsymbol{J}} + \boldsymbol{j}$ and $\boldsymbol{A} = \overline{\boldsymbol{A}} + \boldsymbol{a}$.

The first remarkable thing to note is that, even though we are dealing with *helical* dynamos, there is no net current helicity in the steady state, i.e.

$$\langle\boldsymbol{J}\cdot\boldsymbol{B}\rangle = 0 \,; \tag{7}$$

see (6). However, using the decomposition into large and small-scale fields, we can write

$$\langle\boldsymbol{J}\cdot\boldsymbol{B}\rangle = \langle\overline{\boldsymbol{J}}\cdot\overline{\boldsymbol{B}}\rangle + \langle\boldsymbol{j}\cdot\boldsymbol{b}\rangle = 0 \,, \tag{8}$$

so we have

$$\langle\overline{\boldsymbol{J}}\cdot\overline{\boldsymbol{B}}\rangle = -\langle\boldsymbol{j}\cdot\boldsymbol{b}\rangle \tag{9}$$

in the steady state. We now introduce the approximations[3]

[3] Here and elsewhere we use units where $\mu_0 = 1$ or, following R. Blandford (private communication), we use units in which pi is one quarter.

$$\langle \overline{\boldsymbol{J}} \cdot \overline{\boldsymbol{B}} \rangle \approx k_{\rm m} \langle \overline{\boldsymbol{B}}^2 \rangle \quad \text{and} \quad \langle \boldsymbol{j} \cdot \boldsymbol{b} \rangle \approx -k_{\rm f} \langle \boldsymbol{b}^2 \rangle \tag{10}$$

where $k_{\rm m}$ and $k_{\rm f}$ are the typical wavenumbers of the mean and fluctuating fields, respectively. These approximations are only valid for fully helical turbulence, but can easily be generalized to the case of fractional helicity; see Sect. 4.1 and Blackman and Brandenburg (2002, hereafter BB02). We have furthermore assumed that the sign of the kinetic helicity is negative, as is the case in the northern hemisphere of the sun, for example. (The case of positive kinetic helicity is straightforward; see below.) The wavenumber $k_{\rm f}$ of the fluctuating field is for all practical purposes equal to the wavenumber of the forcing function. (In more general situations, such as convection or shear flow turbulence, $k_{\rm f}$ would be the wavenumber of the energy carrying eddies.) We also note that for large values of the magnetic Reynolds number, $R_{\rm m}$, the $k_{\rm f}$ factor in (10) gets attenuated by an $R_{\rm m}^{1/4}$ factor (BB02). On the other hand, the wavenumber of the mean field is in practice the wavenumber of the box, i.e. $k_{\rm m} = k_1$. Inserting now (10) into (9) yields

$$\langle \overline{\boldsymbol{B}}^2 \rangle = \frac{k_{\rm f}}{k_{\rm m}} \langle \boldsymbol{b}^2 \rangle \ , \tag{11}$$

i.e. the energy of the mean field can *exceed* the energy of the fluctuating field – in contrast to earlier expectations (e.g. (Vainshtein and Cattaneo, 1992; Kulsrud and Anderson, 1992; Gruzinov and Diamond, 1994). Indeed, in the two-dimensional case there is an exact result due to Zeldovich (1957),

$$\langle \overline{\boldsymbol{B}}^2 \rangle = R_{\rm m}^{-1} \langle \boldsymbol{b}^2 \rangle \quad \text{(2-dimensional case)} \ . \tag{12}$$

This result has also be derived in three dimensions using first order smoothing (Krause and Rädler, 1980), but it is important to realize that this result can break down in the nonlinear case in three dimensions, where (11) is in good agreement with the simulations results. However, the assumption of periodic or closed boundaries is an essential one. We return to the more general case in Sects 4.4–4.5.

The time dependence near the saturated state can be approximated by using

$$\langle \overline{\boldsymbol{J}} \cdot \overline{\boldsymbol{B}} \rangle \approx k_{\rm m}^2 \langle \overline{\boldsymbol{A}} \cdot \overline{\boldsymbol{B}} \rangle \quad \text{and} \quad \langle \boldsymbol{j} \cdot \boldsymbol{b} \rangle \approx k_{\rm f}^2 \langle \boldsymbol{a} \cdot \boldsymbol{b} \rangle \ . \tag{13}$$

These equations are still valid in the case of fractional helicity (BB02). Only the two-scale assumption is required. Near saturation,

$$|\langle \overline{\boldsymbol{A}} \cdot \overline{\boldsymbol{B}} \rangle| = \left(\frac{k_{\rm f}}{k_{\rm m}} \right)^2 \langle \boldsymbol{a} \cdot \boldsymbol{b} \rangle \ , \tag{14}$$

i.e. $|\langle \overline{\boldsymbol{A}} \cdot \overline{\boldsymbol{B}} \rangle| \gg |\langle \boldsymbol{a} \cdot \boldsymbol{b} \rangle|$ and so we can neglect $\langle \boldsymbol{a} \cdot \boldsymbol{b} \rangle$, and the magnetic helicity equation (6) becomes therefore an approximate evolution equation for the magnetic helicity of the mean field,

$$\frac{\partial}{\partial t} \langle \overline{\boldsymbol{A}} \cdot \overline{\boldsymbol{B}} \rangle = -2\eta\mu_0 \langle \overline{\boldsymbol{J}} \cdot \overline{\boldsymbol{B}} \rangle - 2\eta\mu_0 \langle \boldsymbol{j} \cdot \boldsymbol{b} \rangle \ , \tag{15}$$

or, by using (10),

$$k_{\rm m}^{-1} \frac{\partial}{\partial t} \langle \overline{\boldsymbol{B}}^2 \rangle = -2\eta k_{\rm m} \langle \overline{\boldsymbol{B}}^2 \rangle + 2\eta k_{\rm f} \langle \boldsymbol{b}^2 \rangle \ . \tag{16}$$

Note the plus sign in front of the $\langle \boldsymbol{b}^2 \rangle$ term resulting from (10). The plus sign leads to growth while the minus sign in front of the $\langle \overline{\boldsymbol{B}}^2 \rangle$ term leads to saturation (but both terms are proportional to the microscopic value of η; see below). Once the small-scale field has saturated, which will happen after a few dynamical time scales such that $\langle \boldsymbol{b}^2 \rangle \approx B_{\rm eq}^2 \equiv \mu_0 \langle \rho \boldsymbol{u}^2 \rangle$, the large-scale field will continue to evolve slowly according to

$$\langle \overline{\boldsymbol{B}}^2 \rangle = \frac{k_{\rm f}}{k_{\rm m}} \langle \boldsymbol{b}^2 \rangle \left[1 - e^{-2\eta k_{\rm m}^2 (t - t_{\rm sat})} \right] , \qquad (17)$$

where $t_{\rm sat}$ is the time at which $\langle \boldsymbol{b}^2 \rangle$ has reached approximate saturation. In practice, $t_{\rm sat}$ can be determined such that (17) describes the simulation data best. We refer to (17) as the *magnetic helicity constraint*. The agreement between this and the actual simulations (Fig. 3) is quite striking.

The significance of this remarkable and simple equation and the almost perfect agreement with simulations is that the constraint can be extrapolated to large values of $R_{\rm m}$ where it would provide a benchmark, against which all analytic dynamo theories, when subjected to the same periodic boundary conditions, should be compared to. In particular the late saturation behavior should be equally slow. We return to this in Sect. 5.

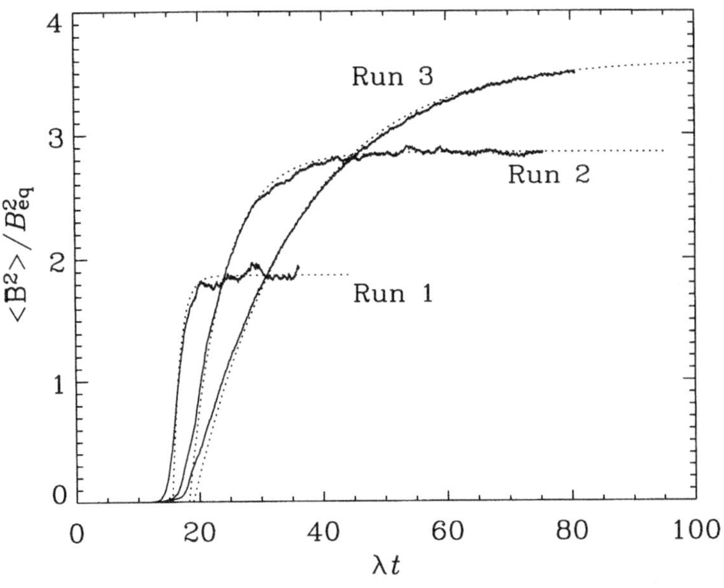

Fig. 3. Late saturation phase of fully helical turbulent dynamos for three different values of the magnetic Reynolds number: $R_{\rm m} \equiv u_{\rm rms}/\eta k_{\rm f} = 2.4$, 6, and 18 for Runs 1, 2, and 3 respectively; see B01. The mean magnetic field, $\overline{\boldsymbol{B}}$, is normalized with respect to the equipartition value, $B_{\rm eq} = \sqrt{\mu_0 \rho_0} u_{\rm rms}$, and time is normalized with respect to the kinematic growth rate, λ. The dotted lines represent the fit formula (17) which tracks the simulation results rather well [Adapted from Brandenburg et al. (2003)]

An important question is whether anything can be learned about stars and galaxies. Before this can be addressed, we need to understand the differences between dynamos in real astrophysical bodies and dynamos in periodic domains.

4 What Do Stars and Galaxies Do Differently?

We begin with a discussion of fractional helicity, shear and other effects that cause the magnetic helicity to be reduced. We then address the possibility of helicity fluxes through boundaries, which can alleviate the helicity constraint (Blackman and Field, 2000).

4.1 Fractional Helicity

When the turbulence is no longer fully helical, (10) is no longer valid and needs to be generalized to

$$\langle \overline{\boldsymbol{J}} \cdot \overline{\boldsymbol{B}} \rangle = \epsilon_{\mathrm{m}} k_{\mathrm{m}} \langle \overline{\boldsymbol{B}}^2 \rangle \quad \text{and} \quad \langle \boldsymbol{j} \cdot \boldsymbol{b} \rangle = -\epsilon_{\mathrm{f}} k_{\mathrm{f}} \langle \boldsymbol{b}^2 \rangle \,, \tag{18}$$

where $\epsilon_{\mathrm{m}} < 1$ and $\epsilon_{\mathrm{f}} < 1$ are coefficients denoting the degree to which the mean and fluctuating fields are helical. Equation (13) is still approximately valid in the fractionally helical case.

Maron and Blackman (2002) found that there is a certain threshold of ϵ_{f} below which the large-scale dynamo effect stops working. Qualitatively, this could be understood by noting that the large scale magnetic field comes from the helical part of the flow, so the velocity field can be though of as having a helical and a nonhelical component, i.e.

$$\boldsymbol{U} = \boldsymbol{U}_{\mathrm{hel}} + \boldsymbol{U}_{\mathrm{nohel}} \,. \tag{19}$$

However, the dynamo effect has to compete with turbulent diffusion which comes from both the helical and the nonhelical parts of the flow. Thus, when $|\boldsymbol{U}_{\mathrm{nohel}}|$ becomes too large compared with $|\boldsymbol{U}_{\mathrm{hel}}|$ the large-scale dynamo effect will stop working.

Although we have not yet discussed mean-field theory we may note that the value of the threshold can be understood quantitatively (Brandenburg et al., 2002), hereafter BDS02) and one finds that large-scale dynamo action is only possible when

$$\epsilon_{\mathrm{f}} > \frac{k_{\mathrm{m}}}{k_{\mathrm{f}}} \quad \text{(for large-scale dynamos)} \,. \tag{20}$$

In many three-dimensional turbulence simulations or in astrophysical bodies, this threshold criterion may not be satisfied, and so mean-field dynamo of the type described above (α^2 dynamo) may not be excited. If there is shear, however, this criterion will be modified to

$$\epsilon_{\mathrm{f}} > \epsilon_{\mathrm{m}} \frac{k_{\mathrm{m}}}{k_{\mathrm{f}}} \,, \tag{21}$$

where ϵ_{f} is the degree to which the large-scale field is helical. In dynamos with strong shear, $|\epsilon_{\mathrm{f}}|$ may be very small, making mean-field dynamo action in fractionally helical flows more likely. This will be discussed in the next section.

4.2 Dynamos with Shear

In the presence of shear, the streamwise component of the field can be amplified by winding up the poloidal (cross-stream) field. Again, the resulting saturation field strength can be estimated based on magnetic helicity conservation arguments.

Note first of all that for closed or periodic domains, (8) is still valid and therefore $\langle \overline{\boldsymbol{J}} \cdot \overline{\boldsymbol{B}} \rangle = -\langle \boldsymbol{j} \cdot \boldsymbol{b} \rangle$ in the steady state.[4] However, while $\langle \boldsymbol{j} \cdot \boldsymbol{b} \rangle$ still depends on the helicity of the small-scale field, the corresponding value of $\langle \overline{\boldsymbol{J}} \cdot \overline{\boldsymbol{B}} \rangle$ no longer provides such a stringent bound on $\langle \overline{\boldsymbol{B}}^2 \rangle$ as before. This is because shear can amplify the toroidal field independently of any magnetic helicity considerations. The component of $\overline{\boldsymbol{B}}^2$ that is amplified by shear is nonhelical and so we have

$$\epsilon_\mathrm{m} = |\langle \overline{\boldsymbol{J}} \cdot \overline{\boldsymbol{B}} \rangle| \Big/ \left(k_\mathrm{m} \langle \overline{\boldsymbol{B}}^2 \rangle \right) \ll 1 \tag{22}$$

(or at least $|\epsilon_\mathrm{m}| \ll 1$ when the helicity of the forcing is negative and ϵ_m therefore negative). The value of ϵ_m is proportional to the ratio of poloidal to toroidal field

$$\epsilon_\mathrm{m} \approx \pm 2 \langle B_\mathrm{pol}^2 \rangle^{1/2} / \langle B_\mathrm{tor}^2 \rangle^{1/2} , \tag{23}$$

where the numerical pre-factor can be different for different examples.[5] With these preparations the magnetic helicity constraint can be generalized to

$$2 B_\mathrm{pol}^\mathrm{rms} B_\mathrm{tor}^\mathrm{rms} \approx \frac{k_\mathrm{f}}{k_\mathrm{m}} \langle \boldsymbol{b}^2 \rangle \left[1 - e^{-2\eta k_\mathrm{m}^2 (t - t_\mathrm{sat})} \right] . \tag{24}$$

This form of the constraint was proposed and confirmed using three-dimensional simulations of forced helical turbulence with large-scale shear (Brandenburg et al., 2001, hereafter BBS01); see also Fig. 4.

The main conclusion to be drawn from this is that the magnetic helicity constraint is still valid in the presence of shear, i.e. the timescale of saturation is still controlled by the microscopic magnetic diffusivity. The only difference is that stronger field strengths are now possible.

Another interesting aspect is that dynamos with shear allow for oscillatory solutions of the magnetic field. This is expected from mean-field dynamo theory (Steenbeck and Krause, 1969a,b), but it is also borne out by simulations (BBS01). The main result is that the resulting cycle frequency seems to scale with the microscopic magnetic diffusivity, not the turbulent magnetic diffusivity. This confirms again that in a closed domain the magnetic helicity constraint plays a crucial role in controlling the timescale of nonlinear dynamos.

4.3 Hall Effect Dynamos

In recent years the importance of the Hall effect has been emphasized by a number of groups, especially in applications to protostellar accretion discs (Balbus and

---

[4] This is because in the $\boldsymbol{E} \cdot \boldsymbol{B}$ term in the magnetic helicity equation the induction term, $\boldsymbol{U} \times \boldsymbol{B}$, drops out after dotting with $\boldsymbol{B}$. (For this reason, also ambipolar diffusion and the Hall effect do not change magnetic helicity conservation.)

[5] Take as an example $\overline{\boldsymbol{B}}(z) = (\overline{B}_\mathrm{pol}, \overline{B}_\mathrm{tor}, 0)^T = (\epsilon \cos k_1 z, \sin k_1 z, 0)^T$ for $\epsilon \ll 1$, so $\langle \overline{\boldsymbol{B}}^2 \rangle \approx 1/2$ and $\langle \overline{\boldsymbol{J}} \cdot \overline{\boldsymbol{B}} \rangle \approx \epsilon k_1$ and therefore $\epsilon_\mathrm{m} = 2\epsilon$.

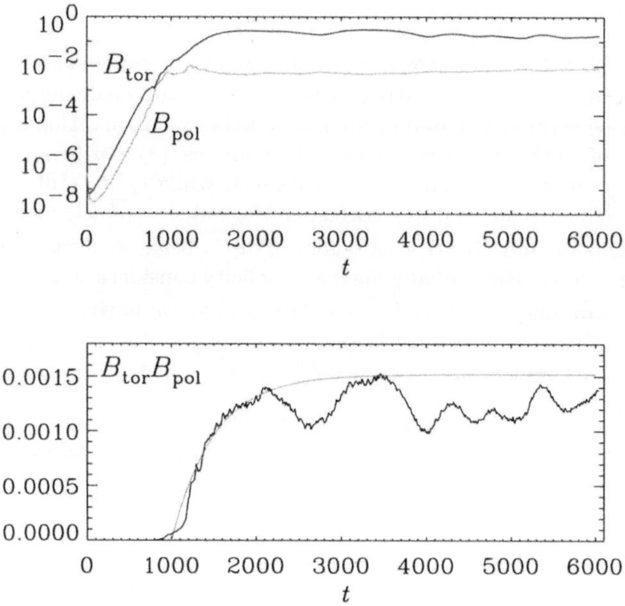

Fig. 4. Growth of poloidal and toroidal magnetic fields on a logarithmic scale (*upper panel*), and product of poloidal and toroidal magnetic fields on a linear scale. For the fit we have used $k_1^2 = 2$, $B_{eq} = 0.035$, and $\epsilon_0 = 1.3$ [Adapted from BBS01]

Terquem, 2001). The hall effect can lead to strong nonlinear steepening of field gradients (Vainshtein et al., 2000), and is therefore important for fast reconnection (e.g. Rogers et al., 2001), which in turn is relevant for neutron stars (Hollerbach and Rüdiger, 2004). Nevertheless, since magnetic helicity generation (and removal) is proportional to the dot product of electric and magnetic fields, and since the Hall current is proportional to $\boldsymbol{J} \times \boldsymbol{B}$, the Hall term does not affect magnetic helicity conservation. Therefore the resistively limited saturation behavior should not be affected by the Hall term. Nevertheless, some degree of extra field amplification of the large-scale field has been reported (Mininni et al., 2003), and it will be interesting to identify exactly the processes that led to this amplification.

4.4 Magnetic Helicity Exchange Across the Equator or with Depth

The presence of an equator provides a source of magnetic helicity exchange between domains of negative helicity in the northern hemisphere (upper disc plane in an accretion disc) and positive helicity in the southern hemisphere (lower disc plane). A similar situation can also arise in convection zones where the helicity is expected to change with depth (Yoshimura, 1975).

So far, simulations have not yet shown that the losses of small-scale magnetic fields are actually stronger than those of large-scale fields. In Fig. 5 we show the saturation behavior of a system that is periodic, but the helicity of the forcing is

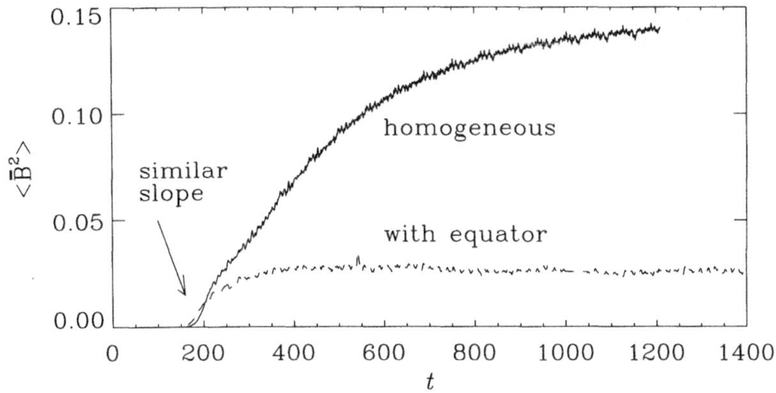

Fig. 5. Evolution of the magnetic energy for a run with homogeneous forcing function (*solid line*) and a forcing function whose helicity varies sinusoidally throughout the domain (*dotted line*) simulating the effects of equators at the two nodes of the sinusoidal helicity profile [Adapted from Brandenburg et al. (2001)]

modulated in the z-direction such that the sign of the kinetic helicity changes in the middle. One can therefore view this system as two subsystems with a boundary in between. This boundary would correspond to the equator in a star or the midplane in a disc. It can also model the change of sign of helicity at some depth in a convection zone.

As far as the magnetic helicity constraint is concerned, the divergence term of current helicity flux is likely to be important when there is a boundary between two domains with different helicities. Naively, one might expect there to be current helicity fluxes that are proportional to the current helicity gradient, analogous to Fick's diffusion law. These current helicity fluxes should be treated separately for large and small-scale components of the field, so we introduce approximations to the current helicity fluxes from the mean and fluctuating fields as

$$\overline{\boldsymbol{\mathcal{F}}}_{\rm m} \approx -\eta_{\rm m} \boldsymbol{\nabla} C_{\rm m}, \quad \overline{\boldsymbol{\mathcal{F}}}_{\rm f} \approx -\eta_{\rm f} \boldsymbol{\nabla} C_{\rm f} \; . \tag{25}$$

The rate of magnetic helicity loss is here proportional to some turbulent diffusivity coefficient, $\eta_{\rm m}$ or $\eta_{\rm f}$ for the losses from mean or fluctuating parts, respectively. We assume that the small and large-scale fields are maximally helical (or have known helicity fractions $\epsilon_{\rm m}$ and $\epsilon_{\rm f}$) and have opposite signs of magnetic helicity at small and large scales. The details can be found in BDS02 and Blackman and Brandenburg (2003). The strength of this approach is that it is quite independent of mean-field theory.

We proceed analogously to the derivation of (17) where we used the magnetic helicity equation (6) for a closed domain to estimate the time derivative of the magnetic helicity of the mean field by neglecting the time derivative of the fluctuating field. This is a good approximation after the fluctuating field has reached saturation, i.e. $t > t_{\rm sat}$. Thus, we have

$$k_{\rm m}^{-1} \frac{\partial}{\partial t} \overline{\boldsymbol{B}}^2 = -2\eta_{\rm m} k_{\rm m} \overline{\boldsymbol{B}}^2 + 2\eta_{\rm f} k_{\rm f} \overline{\boldsymbol{b}^2} \; , \tag{26}$$

where $\eta_{\rm m}=\eta_{\rm f}=\eta$ corresponds to the case of a closed domain. Note also that we have here ignored the volume integration, so we are dealing with horizontal averages that depend still on height and on time.

After the time when the small-scale magnetic field saturates, i.e. when $t>t_{\rm sat}$, we have $\langle \boldsymbol{b}^2\rangle \approx$ constant. After that time, (26) can be solved to give

$$\langle \overline{\boldsymbol{B}}^2\rangle = \langle \boldsymbol{b}^2\rangle \frac{\eta_{\rm f} k_{\rm f}}{\eta_{\rm m} k_{\rm m}} \left[1-{\rm e}^{-2\eta_{\rm m} k_{\rm m}^2 (t-t_{\rm sat})}\right], \quad \text{for } t>t_{\rm sat} \; . \tag{27}$$

This equation demonstrates three remarkable properties (Brandenburg et al., 2003; Brandenburg and Subramanian, 2005):

- Large-scale helicity losses are needed ($\eta_{\rm m}>\eta$) to shorten the typical time scale. This is required to prevent resistively long cycle periods.
- However, the saturation amplitude is proportional to $\eta_{\rm f}/\eta_{\rm m}$, so the large-scale field becomes weaker as $\eta_{\rm m}$ is increased. Thus,
- also small-scale losses are needed to prevent the saturation amplitude from becoming too small.

Future work can hopefully verify that these conditions are indeed obeyed by a working large-scale dynamo. Simulations without shear have been unsuccessful to demonstrate that small-scale losses are important (Brandenburg and Dobler, 2001), but new simulations with shear now begin to show significant small-scale losses of current helicity, an enhanced α effect (Brandenburg and Sandin, 2004), and strong large-scale dynamo action (see below).

4.5 Open Surfaces and Shear

The presence of an outer surface is in many respects similar to the presence of an equator. In both cases one expects magnetic and current helicity fluxes via the divergence term. A particularly instructive system is helical turbulence in an infinitely extended horizontal slab with stress-free boundary conditions and a vertical field condition, i.e.

$$u_{x,z}=u_{y,z}=u_z=B_x=B_y=0 \; . \tag{28}$$

Such simulations have been performed by Brandenburg and Dobler (2001) who found that a mean magnetic field is generated, similar to the case with periodic boundary conditions, but that the energy of the mean magnetic field, $\langle \overline{\boldsymbol{B}}^2\rangle$, decreases with magnetic Reynolds number. Nevertheless, the energy of the total magnetic field, $\langle \boldsymbol{B}^2\rangle$, does not decrease with increasing magnetic Reynolds number. Although they found that $\langle \overline{\boldsymbol{B}}^2\rangle$ decreases only like $R_{\rm m}^{-1/2}$, new simulations confirm that a proper scaling regime has not yet been reached and that the current data may well be compatible with an $R_{\rm m}^{-1}$ dependence; see Fig. 6.

Clearly, an asymptotic decrease of the mean magnetic field must mean that the small-scale dynamo does not work with such boundary conditions. Thus, the anticipated advantages of open boundary conditions are not borne out by this type of simulations.

At this point we can mention some new simulations in a cartesian domain where differential rotation has been modeled as a region of the convection zone without explicitly allowing for convection; see Fig. 7. Instead, an external forcing term has

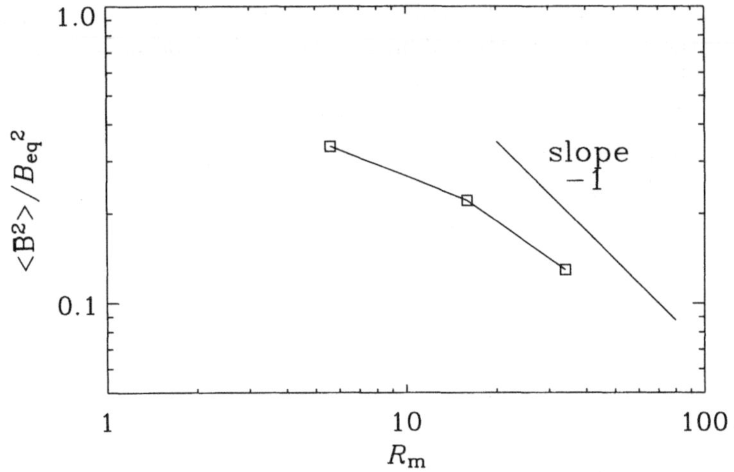

Fig. 6. Dependence of the energy of the mean magnetic field on the magnetic Reynolds number for a run with open boundary conditions and no shear

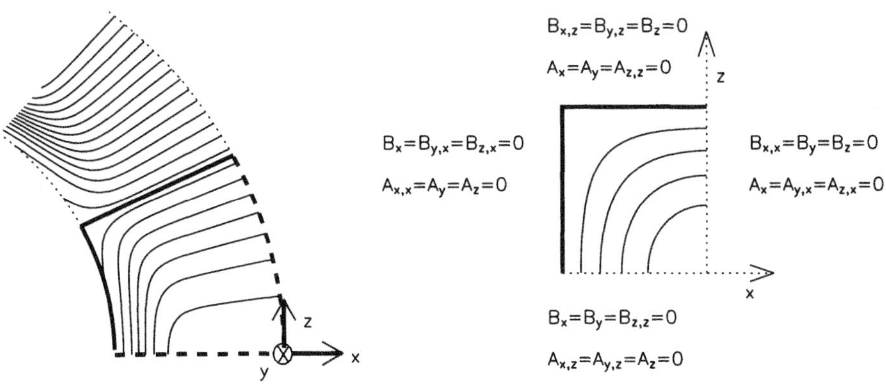

Fig. 7. *Left*: A sketch of the solar angular velocity at low latitudes with spoke-like contours in the bulk of the convection zone merging gradually into uniform rotation in the radiative interior. The low latitude region, modeled in this paper, is indicated by thick lines. *Right*: Differential rotation in our cartesian model, with the equator being at the bottom, the surface to the right, the bottom of the convection zone to the left and mid-latitudes at the top [Adapted from Brandenburg and Sandin (2004)]

been applied that also drives the differential rotation. (Studies of the α effect have already been published; see Sect. 5.6 for details of the simulations and Sect. 5 for a discussion of the direct correspondence between the helicity constraint and the so-called catastrophic α quenching.) Here we briefly report on recent explicit dynamo simulations that have been carried out in this geometry.

The size of the computational domain is $\frac{1}{2}\pi \times 2\pi \times \frac{1}{2}\pi$ and the numerical resolution is $128 \times 512 \times 128$ meshpoints. The magnetic Reynolds number based on

Fig. 8. Visualization of the toroidal magnetic field during three different times during the growth and saturation for the run without kinetic helicity

the forcing wavenumber and the turbulent flow is around 80 and shear flow velocity exceeds the rms turbulent velocity by a factor of about 5. We have carried out experiments with no helicity in the forcing (labeled by $\alpha = 0$), as well as positive and negative helicity in the forcing (labeled by $\alpha < 0$ and $\alpha > 0$, respectively); see Fig. 8 for a visualization of the run without kinetic helicity. We emphasize that no explicit α effect has been invoked. The labeling just reflects the fact that, in isotropic turbulence, negative kinetic helicity (as in the northern hemisphere of a star or the upper disc plane in galaxies) leads to a positive α effect, and vice versa.

We characterize the relative strength of the mean field by the ratio $q = \langle\overline{\boldsymbol{B}}^2\rangle/\langle\boldsymbol{B}^2\rangle$, where overbars denote an average in the toroidal (y) direction; see Fig. 9. There are two surprising results emerging from this work. First, in the presence of shear rather strong mean fields can be generated, where up to 70% of the energy can be in the mean field; see Fig. 9. Second, even without any kinetic helicity in the flow there is strong large-scale field generation. Obviously, this cannot be an

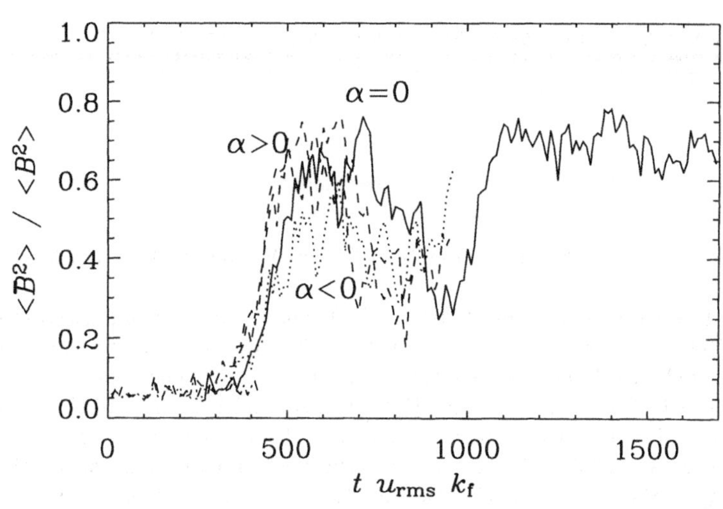

Fig. 9. Saturation behavior of the ratio $q = \langle\overline{\boldsymbol{B}}^2\rangle/\langle\boldsymbol{B}^2\rangle$ for runs with different kinetic helicity of the flow. *Solid line*: zero helicity, *dotted line*: positive helicity (opposite to the sun) *dashed line*: negative helicity (as in the sun)

$\alpha\Omega$ dynamo in the usual sense. One possibility is the $\boldsymbol{\delta} \times \boldsymbol{J}$ effect, which emerged originally in the presence of the Coriolis force; see Rädler (1969) and Krause and Rädler (1980). In the present case with no Coriolis force, however, a $\boldsymbol{\delta} \times \boldsymbol{J}$ effect is possible even in the presence of shear alone, because the vorticity associated with the shear contributes directly to $\delta \propto \overline{\boldsymbol{W}} = \boldsymbol{\nabla} \times \overline{\boldsymbol{U}}$ (Rogachevskii and Kleeorin, 2003).

There is evidence that the strong dynamo action seen in the simulations is only possible due to the combined presence of open boundaries and shear. This however has so far only been checked explicitly for the α effect that is present when the forcing is helical; see Sect. 5.6. In the case of the solar surface such losses are actually observed to occur in the form of coronal mass ejections and in active regions. In the sun, coronal mass ejections are quite vigorous events that are known to shed large amounts of helical magnetic fields (Berger and Ruzmaikin, 2000; DeVore, 2000; Chae, 2000; Low, 2001). This kind of physics is not at all represented by adopting vacuum or pseudo-vacuum (vertical field) boundary conditions, as was done in Brandenburg and Subramanian (2005).

5 Connection with the α Effect

5.1 Preliminary Considerations

The α effect formalism provides so far the only workable mathematical framework for describing the large-scale dynamo action seen in simulations of helically forced turbulence. (In this section we retain the μ_0 factor.) The governing equation for the mean magnetic field is

$$\frac{\partial \overline{\boldsymbol{B}}}{\partial t} = \boldsymbol{\nabla} \times \left(\overline{\boldsymbol{U}} \times \overline{\boldsymbol{B}} + \overline{\boldsymbol{\mathcal{E}}} - \eta \mu_0 \overline{\boldsymbol{J}} \right) , \tag{29}$$

where $\overline{\boldsymbol{\mathcal{E}}} = \overline{\boldsymbol{u} \times \boldsymbol{b}}$ is the electromotive force resulting from the $\boldsymbol{u} \times \boldsymbol{b}$ nonlinearity in the averaged Ohm's law. Without mean flow, $\overline{\boldsymbol{U}} = 0$, and an electromotive force given by a homogeneous isotropic α effect and turbulent diffusion $\eta_{\rm t}$, i.e.

$$\overline{\boldsymbol{\mathcal{E}}} = \alpha \overline{\boldsymbol{B}} - \eta_{\rm t} \mu_0 \overline{\boldsymbol{J}} , \tag{30}$$

we have

$$\frac{\partial \overline{\boldsymbol{B}}}{\partial t} = \alpha \boldsymbol{\nabla} \times \overline{\boldsymbol{B}} + (\eta + \eta_{\rm t}) \nabla^2 \overline{\boldsymbol{B}} , \tag{31}$$

which has solutions of the form $\overline{\boldsymbol{B}} = \hat{\overline{\boldsymbol{B}}} \mathrm{e}^{\mathrm{i}\boldsymbol{k}\cdot\boldsymbol{x} + \lambda t}$ with the dispersion relation

$$\lambda_{\pm} = -\eta_{\rm T} k^2 \pm |\alpha k| , \tag{32}$$

and three possible eigenfunctions (appropriate for the periodic box)

$$\overline{\boldsymbol{B}}(\boldsymbol{x}) = \begin{pmatrix} \cos k_{\rm m} z \\ \sin k_{\rm m} z \\ 0 \end{pmatrix} , \quad \begin{pmatrix} 0 \\ \cos k_{\rm m} x \\ \sin k_{\rm m} x \end{pmatrix} , \quad \text{or} \quad \begin{pmatrix} \sin k_{\rm m} y \\ 0 \\ \cos k_{\rm m} y \end{pmatrix} , \tag{33}$$

where $k_{\rm m} = k_1 = 1$. Obviously, when the coefficients α and $\eta_{\rm T} \equiv \eta + \eta_{\rm t}$ remain constant, and there is an exponentially growing solution (for $|\alpha| > \eta_{\rm T} k_1$), the solution must eventually grow beyond any bound. At the latest when the magnetic field reaches equipartition with the kinetic energy, α and η_t must begin to depend on the magnetic field itself. However, the present case is sufficiently simple so that we can continue to assume that $\overline{\boldsymbol{B}}^2$, as well as α and η_t, are uniform in space and depend only on time.

Comparison with simulations has enabled us to eliminate a large number of various quenching models where $\alpha = \alpha(\overline{\boldsymbol{B}})$. The only quenching model that seems reasonably well compatible with simulations of α^2-like dynamo action in a periodic box *without* shear is

$$\alpha = \frac{\alpha_0}{1 + R_{\mathrm{m}} \overline{\boldsymbol{B}}^2 / B_{\mathrm{eq}}^2}, \quad \eta_{\mathrm{t}} = \frac{\eta_{\mathrm{t}0}}{1 + R_{\mathrm{m}} \overline{\boldsymbol{B}}^2 / B_{\mathrm{eq}}^2} \quad \text{(empirical)}\ , \tag{34}$$

see Fig. 3. However, this type of quenching is not fully compatible with magnetic helicity conservation, as has been shown by Field and Blackman (2002). This will be discussed in the next section.

5.2 Dynamical α Quenching

The basic idea is that magnetic helicity conservation must be obeyed, but the presence of an α effect leads to magnetic helicity of the mean field which has to be balanced by magnetic helicity of the fluctuating field. This magnetic helicity of the fluctuating (small-scale) field must be of opposite sign to that of the mean (large-scale) field.

We begin with the uncurled mean-field induction equation, written in the form

$$\frac{\partial \overline{\boldsymbol{A}}}{\partial t} = \overline{\boldsymbol{\mathcal{E}}} - \eta \mu_0 \overline{\boldsymbol{J}}\ , \tag{35}$$

dot it with $\overline{\boldsymbol{B}}$, add the result to $\overline{\boldsymbol{A}} \cdot \partial \overline{\boldsymbol{B}} / \partial t$, average over the periodic box, and obtain

$$\frac{\partial}{\partial t} \langle \overline{\boldsymbol{A}} \cdot \overline{\boldsymbol{B}} \rangle = 2 \langle \overline{\boldsymbol{\mathcal{E}}} \cdot \overline{\boldsymbol{B}} \rangle - 2 \eta \mu_0 \langle \overline{\boldsymbol{J}} \cdot \overline{\boldsymbol{B}} \rangle\ . \tag{36}$$

To satisfy the helicity equation for the full field, $\langle \boldsymbol{A} \cdot \boldsymbol{B} \rangle = \langle \overline{\boldsymbol{A}} \cdot \overline{\boldsymbol{B}} \rangle + \langle \boldsymbol{a} \cdot \boldsymbol{b} \rangle$, we must have

$$\frac{\partial}{\partial t} \langle \boldsymbol{a} \cdot \boldsymbol{b} \rangle = -2 \langle \overline{\boldsymbol{\mathcal{E}}} \cdot \overline{\boldsymbol{B}} \rangle - 2 \eta \mu_0 \langle \boldsymbol{j} \cdot \boldsymbol{b} \rangle\ . \tag{37}$$

Note the minus sign in front of the $2 \langle \overline{\boldsymbol{\mathcal{E}}} \cdot \overline{\boldsymbol{B}} \rangle$ term, indicating once again that the α effect produces magnetic helicity of opposite sign at the mean and fluctuating fields. The sum of the two equations yields (6).

The significance of (37) is that it contains the $\langle \boldsymbol{j} \cdot \boldsymbol{b} \rangle$ term which contributes to the α effect, as was first shown by Pouquet et al. (1976). Specifically, they found (see also Blackman and Field, 2002)

$$\alpha = \alpha_{\mathrm{K}} + \alpha_{\mathrm{M}}, \quad \text{with} \quad \alpha_{\mathrm{K}} = -\tfrac{1}{3} \tau \langle \boldsymbol{\omega} \cdot \boldsymbol{u} \rangle, \quad \alpha_{\mathrm{M}} = +\tfrac{1}{3} \tau \langle \boldsymbol{j} \cdot \boldsymbol{b} \rangle\ , \tag{38}$$

where τ is the correlation time of the turbulence, $\boldsymbol{\omega} = \boldsymbol{\nabla} \times \boldsymbol{u}$ is the vorticity, and $\langle \boldsymbol{\omega} \cdot \boldsymbol{u} \rangle$ is the kinematic helicity.

Using $\langle \boldsymbol{j} \cdot \boldsymbol{b} \rangle \approx k_{\mathrm{f}}^2 \langle \boldsymbol{a} \cdot \boldsymbol{b} \rangle$, see (13), we can rewrite (37) in a form that can directly be used in mean-field calculations:

$$\frac{\mathrm{d}\alpha}{\mathrm{d}t} = -2\eta_t k_f^2 \left(\frac{\alpha\langle \overline{\boldsymbol{B}}^2\rangle - \eta_t\langle \overline{\boldsymbol{J}}\cdot\overline{\boldsymbol{B}}\rangle}{B_{eq}^2} + \frac{\alpha - \alpha_K}{R_m} \right) , \tag{39}$$

Here we have used $\eta_t = \frac{1}{3}\tau u_{rms}^2$ to eliminate τ in favor of η_t and $B_{eq}^2 = \mu_0\rho_0 u_{rms}^2$ to eliminate u_{rms}^2 in favor of B_{eq}^2.

So, α is no longer just an algebraic function of $\overline{\boldsymbol{B}}$, but it is related to $\overline{\boldsymbol{B}}$ via a *dynamical*, explicitly time-dependent equation. In the context of dynamos in periodic domains, where magnetic helicity conservation is particularly important, the time dependence of α can hardly be ignored, unless one wants to describe the final stationary state, which can be at the end of a very slow saturation phase. However, in order to make contact with earlier work, it is useful to consider the stationary limit of (39), i.e. set $\partial\alpha/\partial t$.

5.3 Steady Limit and its Limitations

In the steady limit the term in brackets in (39) can be set to zero, so this equation reduces to

$$R_m \frac{\alpha\langle \overline{\boldsymbol{B}}^2\rangle - \eta_t\langle \overline{\boldsymbol{J}}\cdot\overline{\boldsymbol{B}}\rangle}{B_{eq}^2} + \alpha = \alpha_K \quad (\text{for } \mathrm{d}\alpha/\mathrm{d}t = 0) . \tag{40}$$

Solving this equation for α yields (Kleeorin et al., 1982; Gruzinov and Diamond, 1994)

$$\alpha = \frac{\alpha_K + \eta_t R_m \langle \overline{\boldsymbol{J}}\cdot\overline{\boldsymbol{B}}\rangle/B_{eq}^2}{1 + R_m\langle \overline{\boldsymbol{B}}^2\rangle/B_{eq}^2} \quad (\text{for } \mathrm{d}\alpha/\mathrm{d}t = 0) . \tag{41}$$

And, sure enough, for the numerical experiments with an imposed large scale field over the scale of the box (Cattaneo and Hughes, 1996), where $\overline{\boldsymbol{B}}$ is spatially uniform and therefore $\overline{\boldsymbol{J}} = 0$, one recovers the 'catastrophic' quenching formula,

$$\alpha = \frac{\alpha_K}{1 + R_m\langle \overline{\boldsymbol{B}}^2\rangle/B_{eq}^2} \quad (\text{for } \overline{\boldsymbol{J}} = 0) , \tag{42}$$

which implies that α becomes quenched when $\langle \overline{\boldsymbol{B}}^2\rangle/B_{eq}^2 = R_m^{-1} \approx 10^{-8}$ for the sun, and for even smaller fields in the case of galaxies.

On the other hand, if the mean field is not imposed, but maintained by dynamo action, $\overline{\boldsymbol{B}}$ cannot be spatially uniform and then $\overline{\boldsymbol{J}}$ is finite. In the case of a Beltrami field (33), $\langle \overline{\boldsymbol{J}}\cdot\overline{\boldsymbol{B}}\rangle/\langle \overline{\boldsymbol{B}}^2\rangle \equiv \tilde{k}_m$ is some effective wavenumber of the large-scale field [$\tilde{k}_m = \epsilon_m k_m$; see (22)]. Since R_m enters both the numerator and the denominator, α tends to $\eta_t k_m$, i.e.

$$\alpha \longrightarrow \eta_t\tilde{k}_m \quad (\text{for } \overline{\boldsymbol{J}} \neq 0 \text{ and } \overline{\boldsymbol{J}} \parallel \overline{\boldsymbol{B}}) . \tag{43}$$

Compared with the kinematic estimate, $\alpha_K \approx \eta_t k_f$, α is only quenched by the modified scale separation ratio. More importantly, α is quenched to a value that is just slightly above the critical value for the onset of dynamo action, $\alpha_{crit} = \eta_T\tilde{k}_m$. How is it then possible that the fit formula (34) for α *and* η_t produced reasonable agreement with the simulations? The reason is that in the simple case of an α^2 dynamo the solutions are degenerate in the sense that $\overline{\boldsymbol{J}}$ and $\overline{\boldsymbol{B}}$ are parallel to each other. Therefore, the term $\langle \overline{\boldsymbol{J}}\cdot\overline{\boldsymbol{B}}\rangle\overline{\boldsymbol{B}}$ is the same as $\langle \overline{\boldsymbol{B}}^2\rangle\overline{\boldsymbol{J}}$, which means that in the mean EMF the term $\alpha\overline{\boldsymbol{B}}$, where α is given by (41), has a component that can be expressed as being parallel to $\overline{\boldsymbol{J}}$. In other words, the roles of turbulent diffusion (proportional to $\overline{\boldsymbol{J}}$) and α effect (proportional to $\overline{\boldsymbol{B}}$) cannot be disentangled. This is the *force-free degeneracy* of α^2 dynamos in a periodic box (BB02). This degeneracy is also the reason why for α^2 dynamos the late saturation behavior can also be described by an algebraic (non-dynamical, but catastrophic) quenching formula proportional to $1/(1+R_\mathrm{m}\langle\overline{\boldsymbol{B}}^2\rangle)$ for *both* α and η_t, as was done in B01. To see this, substitute the steady state quenching expression for α, from (41), into the expression for $\overline{\boldsymbol{\mathcal{E}}}$. We find

$$\overline{\boldsymbol{\mathcal{E}}} = \alpha\overline{\boldsymbol{B}} - (\eta+\eta_\mathrm{t})\overline{\boldsymbol{J}} = \frac{\alpha_\mathrm{K} + R_\mathrm{m}\eta_\mathrm{t}\langle\overline{\boldsymbol{J}}\cdot\overline{\boldsymbol{B}}\rangle/B_\mathrm{eq}^2}{1+R_\mathrm{m}\langle\overline{\boldsymbol{B}}^2\rangle/B_\mathrm{eq}^2}\overline{\boldsymbol{B}} - \eta_\mathrm{t}\overline{\boldsymbol{J}}$$

$$= \frac{\alpha_\mathrm{K}\overline{\boldsymbol{B}}}{1+R_\mathrm{m}\langle\overline{\boldsymbol{B}}^2\rangle/B_\mathrm{eq}^2} - \frac{\eta_\mathrm{t}\overline{\boldsymbol{J}}}{1+R_\mathrm{m}\langle\overline{\boldsymbol{B}}^2\rangle/B_\mathrm{eq}^2}, \tag{44}$$

which shows that in the force-free case the adiabatic approximation, together with constant (unquenched) turbulent magnetic diffusivity, becomes equal to the pair of expressions where both α and η_t are catastrophically quenched. This force-free degeneracy is lifted in cases with shear or when the large-scale field is no longer fully helical (e.g. in a nonperiodic domain, and in particular in the presence of open boundaries).

5.4 The Keinigs Relation and its Relevance

Applying (37) to the steady state using $\overline{\boldsymbol{\mathcal{E}}} = \alpha\overline{\boldsymbol{B}} - \eta_\mathrm{t}\mu_0\overline{\boldsymbol{J}}$ (and retaining μ_0 factor), we get

$$\alpha = -\eta\mu_0\frac{\langle\boldsymbol{j}\cdot\boldsymbol{b}\rangle}{\langle\overline{\boldsymbol{B}}^2\rangle} + \eta_\mathrm{t}k_\mathrm{m} \quad \text{(for periodic domain)}\,, \tag{45}$$

where we have defined an effective wavenumber of the large-scale field, $k_\mathrm{m} = \mu_0\langle\overline{\boldsymbol{J}}\cdot\overline{\boldsymbol{B}}\rangle/\langle\overline{\boldsymbol{B}}^2\rangle$; see (9). This relation applies only to a closed or periodic box, because otherwise there would be boundary terms. Moreover, if the mean field is defined as a volume average, i.e. $\overline{\boldsymbol{B}} = \langle\boldsymbol{B}\rangle \equiv \boldsymbol{B}_0$, then $\mu_0\overline{\boldsymbol{J}} = \boldsymbol{\nabla}\times\boldsymbol{B}_0 = 0$, so $k_\mathrm{m}=0$ and one has simply

$$\alpha = -\eta\mu_0\frac{\langle\boldsymbol{j}\cdot\boldsymbol{b}\rangle}{\langle\overline{\boldsymbol{B}}^2\rangle} \quad \text{(for imposed field)}\,. \tag{46}$$

This equation is due to Keinigs (1983). For the more general case with $k_\mathrm{m}\neq 0$ this equation has been discussed in more detail by Brandenburg and Subramanian (2005) and Brandenburg and Matthaeus (2004).

Let us now discuss the significance of this relation relative to (41). Both equations apply only in the strictly steady state, of course. Since we have assumed stationarity, we can replace $\langle\boldsymbol{j}\cdot\boldsymbol{b}\rangle$ by $-\langle\overline{\boldsymbol{J}}\cdot\overline{\boldsymbol{B}}\rangle$; see (9). Thus, (45) reduces to

$$\alpha = \eta_\mathrm{T}k_\mathrm{m} \tag{47}$$

where $\eta_\mathrm{T} = \eta + \eta_\mathrm{t}$ is the total (turbulent and microscopic) magnetic diffusivity. This relation is just the condition for a marginally excited dynamo; see (32), so it does *not* produce any independent estimate for the value of α. In particular, it does not provide a means of independently testing (41). The two can however be used to calculate the mean-field energy in the saturated state and we find (BB02)

$$\frac{\langle \overline{\boldsymbol{B}}^2 \rangle}{B_{\rm eq}^2} = \frac{\alpha - \eta_{\rm T} k_{\rm m}}{\eta_{\rm t} k_{\rm m}} \quad \text{(periodic domain)} . \tag{48}$$

By replacing $k_{\rm m}$ by an effective value $\tilde{k}_{\rm m}$, this equation can be generalized to apply also to the case with shear (for details see BB02).

5.5 Blackman's Multi-scale Model: Application to Helical Turbulence with Imposed Field

The restriction to a two scale model may in some cases turn out to be insufficient to capture the variety of scales involved in astrophysical bodies. This is already important in the kinematic stage when the small-scale dynamo obeys the Kazantsev (1968) scaling with a $k^{3/2}$ spectrum that peaks at the resistive scale. As the dynamo saturates, the peak moves to the forcing scale. This lead Blackman (2003) to develop a four scale model where he includes, in addition to the wavenumbers of the mean field $k_{\rm m}$ ($\equiv k_1$) and the wavenumber of the energy carrying scale of the velocity fluctuations $k_{\rm f}$ ($\equiv k_2$), also the viscous wavenumber k_ν ($\equiv k_3$) and the resistive wavenumber k_η ($\equiv k_4$). The set of helicity equations for the four different scales is

$$\left(\partial_t + 2\eta k_1^2\right) H_1 = 2\langle \overline{\boldsymbol{\mathcal{E}}}_1 \cdot \overline{\boldsymbol{B}}_1 \rangle + 2\langle \overline{\boldsymbol{\mathcal{E}}}_2 \cdot \overline{\boldsymbol{B}}_1 \rangle , \tag{49}$$

$$\left(\partial_t + 2\eta k_2^2\right) H_2 = -2\langle \overline{\boldsymbol{\mathcal{E}}}_1 \cdot \overline{\boldsymbol{B}}_1 \rangle + 2\langle \overline{\boldsymbol{\mathcal{E}}}_2 \cdot \overline{\boldsymbol{B}}_2 \rangle , \tag{50}$$

$$\left(\partial_t + 2\eta k_3^2\right) H_3 = -2\langle \overline{\boldsymbol{\mathcal{E}}}_2 \cdot \overline{\boldsymbol{B}}_1 \rangle - 2\langle \overline{\boldsymbol{\mathcal{E}}}_2 \cdot \overline{\boldsymbol{B}}_2 \rangle , \tag{51}$$

$$\left(\partial_t + 2\eta k_4^2\right) H_4 = 0 , \tag{52}$$

where $\overline{\boldsymbol{\mathcal{E}}}_1$ is the usual electromotive force based on kinetic helicity at the forcing scale, k_2, with feedback proportional to H_2, and $\overline{\boldsymbol{\mathcal{E}}}_2$ has no kinetic helicity input but only reacts to the automatically generated magnetic helicity H_3 produced at the viscous scale k_3. These equations are constructed such that

$$\frac{\partial}{\partial t} \sum_{i=1}^{4} H_i = -2\eta \sum_{i=1}^{4} k_i^2 H_i , \tag{53}$$

which is consistent with the magnetic helicity equation (6) for the total field. An important outcome of this model is that in the limit of large $R_{\rm m}$ the magnetic peak travels from k_3 to k_2 on a dynamical timescale, i.e. a timescale that is independent of $R_{\rm m}$.

Brandenburg and Matthaeus (2004) have applied the general idea to the case of a model with an applied field. Here the new scale is the scale of the applied field, but since this scale is infinite, this field is fixed and not itself subject to an evolution equation. Nevertheless, the electromotive force from this field acts as a sink on the next smaller scale with wavenumber k_1, which is the largest wavenumber in the domain of the simulation. They thus arrive at the following set of evolution equations,

$$\left[\left(\partial_t + 2\eta k_0^2\right) H_0 = ... + 2\langle \overline{\boldsymbol{\mathcal{E}}}_0 \cdot \overline{\boldsymbol{B}}_0 \rangle, \right] \tag{54}$$

$$\left(\partial_t + 2\eta k_1^2\right) H_1 = -2\langle \overline{\boldsymbol{\mathcal{E}}}_0 \cdot \overline{\boldsymbol{B}}_0 \rangle + 2\langle \overline{\boldsymbol{\mathcal{E}}}_1 \cdot \overline{\boldsymbol{B}}_1 \rangle , \tag{55}$$

$$\left(\partial_t + 2\eta k_2^2\right) H_2 = -2\langle \overline{\boldsymbol{\mathcal{E}}}_1 \cdot \overline{\boldsymbol{B}}_1 \rangle . \tag{56}$$

The square brackets around the first equation indicate that this equation is not explicitly included. From the second equation (55) one can see that there is a competition between two opposing effects: the α effect operating on the imposed field $\boldsymbol{B}_0$ and the α effect operating on the $\boldsymbol{B}_1$ field on the scale of the box. When the imposed field exceeds a certain field strength, $B_0 > B_*$, the former will dominate, reversing the sign of the magnetic helicity at wavenumber k_1. This is actually seen in the simulations of helically forced turbulence with an imposed field $\boldsymbol{B}_0$; see Fig. 10. We return to this at the end of this section.

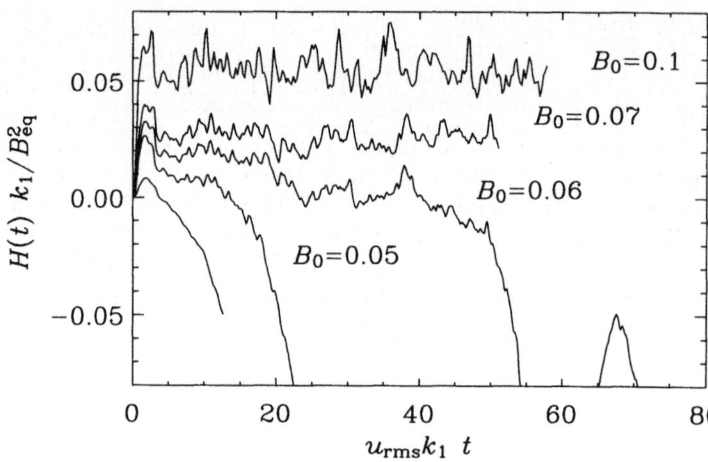

Fig. 10. Evolution of the total magnetic helicity, $H = H_1 + H_{\rm f}$, as a function of t for different values of B_0, as obtained from the three-dimensional simulation. Note the change of sign at $B_0 \approx B_* \approx 0.07$ [Adapted from Brandenburg and Matthaeus (2004)]

The work of Brandenburg and Matthaeus (2004) was motivated by earlier work of Montgomery et al. (2002) and Milano et al. (2003) who showed that, if the imposed magnetic field is weak or absent, there is a strong nonlocal transfer of magnetic helicity and magnetic energy from the forcing scale to larger scales. This leads eventually to the accumulation of magnetic energy at the scale of the box (Meneguzzi et al., 1981; Balsara and Pouquet, 1999, B01). As the strength of the imposed field (wavenumber $k = 0$) is increased, the accumulation of magnetic energy at the scale of the box ($k = 1$) becomes more and more suppressed (Montgomery et al., 2002).

In order to solve the model equations, we have to make some assumptions about the electromotive force operating at k_0 and k_1. The large-scale magnetic helicity production from the α effect operating on the imposed field is $\boldsymbol{\mathcal{E}}_0 \cdot \boldsymbol{B}_0 = \alpha_1 \boldsymbol{B}_0^2$. On the other hand, $\boldsymbol{\mathcal{E}}_1$ at wavenumber k_1 is given by

$$\boldsymbol{\mathcal{E}}_1 = \alpha_{\rm f} \boldsymbol{B}_1 - \eta_{\rm t} \mu_0 \boldsymbol{J}_1 \,. \tag{57}$$

To calculate $\langle \boldsymbol{\mathcal{E}}_1 \cdot \boldsymbol{B}_1 \rangle$ in (55) and (56) we dot (57) with $\boldsymbol{B}_1$, volume average, and note that $\mu_0 \langle \boldsymbol{J}_1 \cdot \boldsymbol{B}_1 \rangle = k_1^2 H_1$ and $\langle \boldsymbol{B}_1^2 \rangle = k_1 |H_1|$. The latter relation assumes that the field at wavenumber k_1 is fully helical, but that it can have either sign. Thus, we have

$$\langle \boldsymbol{\mathcal{E}}_1 \cdot \boldsymbol{B}_1 \rangle = \alpha_{\rm f} k_1 |H_1| - \eta_{\rm t} k_1^2 H_1 \ . \tag{58}$$

The α effects on the two scales are proportional to the residual magnetic helicity of Pouquet et al. (1976); see (38). In terms of H_1 and $H_2 \equiv H_{\rm f}$ we write

$$\alpha_1 = \alpha_{\rm K} + \tfrac{1}{3}\tau k_1^2 H_1 \ , \tag{59}$$

$$\alpha_2 = \alpha_{\rm K} + \tfrac{1}{3}\tau k_2^2 H_2 \ , \tag{60}$$

for the α effect with feedback from H_1 and H_2, respectively.

For finite values of B_0, the final value of H_1 is particularly sensitive to the value of $\alpha_{\rm K}$ and turns out to be too large compared with the simulations. This disagreement with simulations is readily removed by taking into account that $\alpha_{\rm K} = -\frac{1}{3}\tau\langle \boldsymbol{\omega} \cdot \boldsymbol{u} \rangle$ should itself be quenched when B_0 becomes comparable to $B_{\rm eq}$. Thus, we write

$$\alpha_{\rm K} = \alpha_{\rm K0}/(1 + B_0^2/B_{\rm eq}^2) \ , \tag{61}$$

which is a good approximation to more elaborate expressions (Rüdiger and Kitchatinov, 1993). We emphasize that this equation only applies to $\alpha_{\rm K}$ and is therefore distinct from (34), (39), or (41).

Under the assumption that the turbulence is fully helical, the critical value B_* of the imposed field can be estimated by balancing the two terms on the right hand side of (56) and by approximating, $\alpha \approx \eta_{\rm t} k_{\rm f}$ and $\langle \boldsymbol{j} \cdot \boldsymbol{b} \rangle \approx k_{\rm f} B_{\rm eq}^2$. This yields

$$B_*^2/B_{\rm eq}^2 \approx \eta/\eta_{\rm t} \equiv R_{\rm m}^{-1} \ , \tag{62}$$

where the last equality is to be understood as a definition of the magnetic Reynolds number, see also BB02. For $B_0 > B_*$ the sign of the magnetic helicity is the same both at $k = 1$ and at $k = k_{\rm f}$, while for $B_0 < B_*$ the signs are opposite.

In Fig. 11 we show the result of a numerical integration of (55) and (56). Both the three-dimensional simulation and the two-scale model show a similar value of $B_0 \approx 0.06...0.07$, above which H_1 changes sign. This confirms the validity of our estimate of the critical value B_* obtained from (62). Secondly, the time evolution is slow when $B_0 < B_*$ and faster when $B_0 > B_*$. In the simulation, however, the field attains its final level for $B_0 > B_*$ almost instantaneously, which is not the case in the model. The significance of this discrepancy remains unclear. Nevertheless, the level of agreement between the simulations and 3-scale model is surprising, suggesting that the approach can indeed be quite useful.

5.6 Alpha Effect with Open Boundaries and Shear

In a recent paper, Brandenburg and Sandin (2004) have carried out a range of simulations for different values of the magnetic Reynolds number, $R_{\rm m} = u_{\rm rms}/(\eta k_{\rm f})$, for both open and closed boundary conditions using the geometry depicted on the right hand panel of Fig. 7. In order to measure α, a uniform magnetic field, $\boldsymbol{B}_0 =$ const, is imposed, and the magnetic field is now written as $\boldsymbol{B} = \boldsymbol{B}_0 + \boldsymbol{\nabla} \times \boldsymbol{A}$. They have determined α by measuring the turbulent electromotive force, and hence $\alpha = \langle \boldsymbol{\mathcal{E}} \rangle \cdot \boldsymbol{B}_0 / B_0^2$. Similar investigations have been done before both for forced turbulence (Cattaneo and Hughes, 1996, B01) and for convective turbulence (Brandenburg et al., 1990; Ossendrijver et al., 2001).

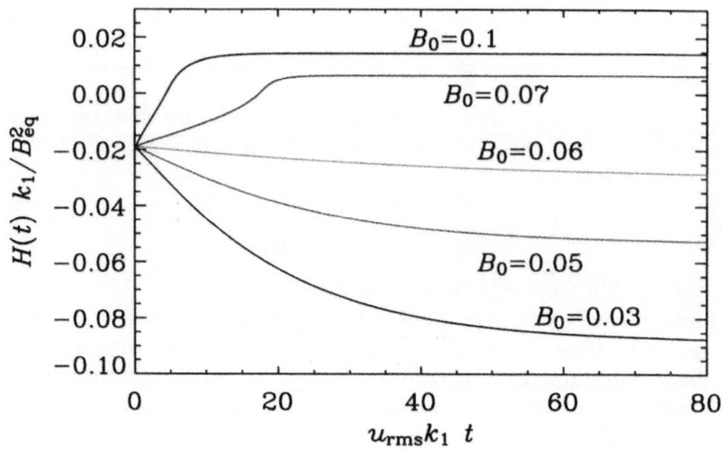

Fig. 11. Evolution of magnetic helicity as a function of t for different values of B_0, as obtained from the two-scale model [Adapted from Brandenburg and Matthaeus (2004)]

As expected, α is negative when the helicity of the forcing is positive, and α changes sign when the helicity of the forcing changes sign. The magnitudes of α are however different in the two cases: $|\alpha|$ is larger when the helicity of the forcing is negative. In the sun, this corresponds to the sign of helicity in the northern hemisphere in the upper parts of the convection zone. This is here the relevant case, because the differential rotation pattern of the present model also corresponds to the northern hemisphere.

There is a striking difference between the cases with open and closed boundaries which becomes particularly clear when comparing the averaged values of α for different magnetic Reynolds numbers; see Fig. 12. With closed boundaries α tends to zero like R_m^{-1}, while with open boundaries α shows no such decline. There is also a clear difference between the cases with and without shear together with open boundaries in both cases. In the absence of shear (dashed line in Fig. 12) α declines with increasing R_m, even though for small values of R_m it is larger than with shear. The difference between open and closed boundaries will now be discussed in terms of a current helicity flux through the two open open boundaries of the domain.

5.7 Current Helicity Flux

It is suggestive to interpret the above results in terms of the dynamical α quenching model. However, (39) has to be generalized to take the divergence of the flux into account. In order to avoid problems with the gauge, it is advantageous to work directly with $\overline{\boldsymbol{j}\cdot\boldsymbol{b}}$ instead of $\overline{\boldsymbol{a}\cdot\boldsymbol{b}}$. Using the evolution equation, $\partial\boldsymbol{b}/\partial t = -\boldsymbol{\nabla}\times\boldsymbol{e}$, for the fluctuating magnetic field, where $\boldsymbol{e} = \boldsymbol{E} - \overline{\boldsymbol{E}}$ is the small-scale electric field and $\overline{\boldsymbol{E}} = \eta\mu_0\overline{\boldsymbol{J}} - \overline{\boldsymbol{\mathcal{E}}}$ the mean electric field, one can derive the equation

$$\frac{\partial}{\partial t}\overline{\boldsymbol{j}\cdot\boldsymbol{b}} = -2\,\overline{\boldsymbol{e}\cdot\boldsymbol{c}} - \boldsymbol{\nabla}\cdot\overline{\boldsymbol{\mathcal{F}}}_C^{\mathrm{SS}}\,, \tag{63}$$

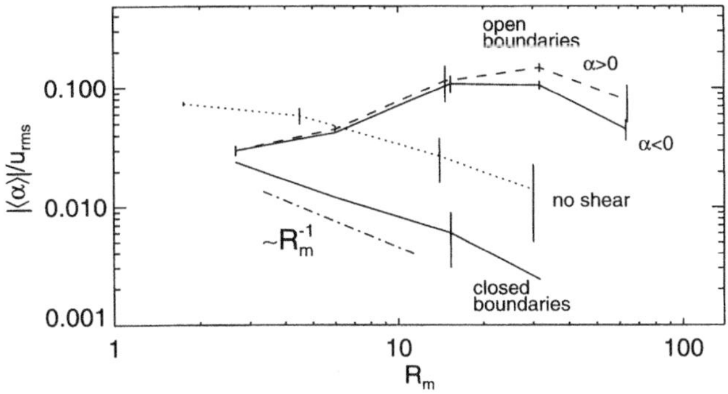

Fig. 12. Dependence of $|\langle\alpha\rangle|/u_{\rm rms}$ on $R_{\rm m}$ for open and closed boundaries. The case with open boundaries and negative helicity is shown as a dashed line. Note that for $R_{\rm m} \approx 30$ the α effect is about 30 times smaller when the boundaries are closed. The *dotted line* gives the result with open boundaries but no shear. The *vertical lines* indicate the range obtained by calculating α using only the first and second half of the time interval [Adapted from Brandenburg and Subramanian (2005)]

where

$$\overline{\mathcal{F}}_C^{\rm SS} = \overline{2\boldsymbol{e}\times\boldsymbol{j}} + \overline{(\boldsymbol{\nabla}\times\boldsymbol{e})\times\boldsymbol{b}}/\mu_0\,, \tag{64}$$

is the current helicity flux from the small-scale field, and $\boldsymbol{c} = \boldsymbol{\nabla}\times\boldsymbol{j}$ the curl of the small-scale current density, $\boldsymbol{j} = \boldsymbol{J} - \overline{\boldsymbol{J}}$. In the isotropic case, $\overline{\boldsymbol{e}\cdot\boldsymbol{c}} \approx k_{\rm f}^2\overline{\boldsymbol{e}\cdot\boldsymbol{b}}$, where $k_{\rm f}$ is the typical wavenumber of the fluctuations, here assumed to be the forcing wavenumber. Ignoring the effect of the mean flow on $\overline{\mathcal{E}}$ [as is usually done; but see Krause and Rädler (1980) and the recent on the shear current effect by Rogachevskii and Kleeorin (2003, 2004); see Sect. 4.5], we obtain

$$\overline{\boldsymbol{e}\cdot\boldsymbol{b}} \approx -\overline{(\boldsymbol{u}\times\boldsymbol{B}_0)\cdot\boldsymbol{b}} + \eta\mu_0\overline{\boldsymbol{j}\cdot\boldsymbol{b}} = \overline{\mathcal{E}}\cdot\overline{\boldsymbol{B}} + \eta\mu_0\overline{\boldsymbol{j}\cdot\boldsymbol{b}}\,, \tag{65}$$

where we have used $\overline{\boldsymbol{u}\times\boldsymbol{b}} = \overline{\mathcal{E}}$ and $\boldsymbol{B}_0 = \overline{\boldsymbol{B}}$. Using standard expressions for the turbulent magnetic diffusivity, $\eta_{\rm t} = \frac{1}{3}\tau u_{\rm rms}^2$, and the equipartition field strength, $B_{\rm eq} = \sqrt{\mu_0\rho}\,u_{\rm rms}$, we eliminate τ via

$$\tfrac{1}{3}\tau = \mu_0\rho_0\eta_{\rm t}/B_{\rm eq}^2\,. \tag{66}$$

This leads to an explicitly time dependent formula for α,

$$\frac{\partial\alpha}{\partial t} = -2\eta_{\rm t}k_{\rm f}^2\left(\frac{\overline{\mathcal{E}}\cdot\overline{\boldsymbol{B}} + \frac{1}{2}k_{\rm f}^{-2}\boldsymbol{\nabla}\cdot\mu_0\overline{\mathcal{F}}_C^{\rm SS}}{B_{\rm eq}^2} + \frac{\alpha-\alpha_{\rm K}}{R_{\rm m}}\right)\,. \tag{67}$$

This equation is similar to that of Kleeorin et al. (2000, 2002, 2003) who considered the flux of magnetic helicity instead of current helicity.

Making use of the adiabatic approximation, i.e. putting the rhs of (67) to zero, one arrives at the algebraic steady state quenching formula ($\partial\alpha/\partial t = 0$)

$$\alpha = \frac{\alpha_{\rm K} + R_{\rm m}\left(\eta_{\rm t}\mu_0\overline{\boldsymbol{J}}\cdot\overline{\boldsymbol{B}} - \frac{1}{2}k_{\rm f}^{-2}\boldsymbol{\nabla}\cdot\mu_0\overline{\mathcal{F}}_C^{\rm SS}\right)/B_{\rm eq}^2}{1 + R_{\rm m}\overline{\boldsymbol{B}}^2/B_{\rm eq}^2}\,. \tag{68}$$

In the absence of a mean current, e.g. if the mean field is defined as an average over the whole box, then $\overline{\boldsymbol{B}} \equiv \boldsymbol{B}_0 = \text{const}$, and $\overline{\boldsymbol{J}} = 0$, so (68) reduces to

$$\alpha = \frac{\alpha_{\rm K} - \frac{1}{2}k_{\rm f}^{-2} R_{\rm m} \boldsymbol{\nabla} \cdot \mu_0 \overline{\boldsymbol{\mathcal{F}}}_C^{\rm SS} / B_{\rm eq}^2}{1 + R_{\rm m} \boldsymbol{B}_0^2 / B_{\rm eq}^2} \,. \tag{69}$$

This expression applies to the present case, because we consider only the statistically steady state and we also define the mean field as a volume average.

For closed boundaries, $\langle \boldsymbol{\nabla} \cdot \overline{\boldsymbol{\mathcal{F}}}_C^{\rm SS} \rangle = 0$, and so (69) clearly reduces to a catastrophic quenching formula, i.e. α vanishes in the limit of large magnetic Reynolds numbers as

$$\alpha^{(\rm closed)} = \frac{\alpha_{\rm K}}{1 + R_{\rm m} \boldsymbol{B}_0^2 / B_{\rm eq}^2} \to R_{\rm m}^{-1} \quad (\text{for } R_{\rm m} \to \infty) \,. \tag{70}$$

The $R_{\rm m}^{-1}$ dependence is confirmed by the simulations (compare with the dash-dotted line in Fig. 12). On the other hand, for open boundaries the limit $R_{\rm m} \to \infty$ gives

$$\alpha^{(\rm open)} \to -(\boldsymbol{\nabla} \cdot \mu_0 \overline{\boldsymbol{\mathcal{F}}}_C^{\rm SS}) / (2 k_{\rm f}^2 \boldsymbol{B}_0^2) \quad (\text{for } R_{\rm m} \to \infty) \,, \tag{71}$$

which shows that losses of negative helicity, as observed in the northern hemisphere of the sun, would enhance a positive α effect (Kleeorin et al., 2000). In the simulations, the current helicity flux is found to be independent of the magnetic Reynolds number. This explains why the α effect no longer shows the catastrophic $R_{\rm m}^{-1}$ dependence (see Fig. 12). In principle it is even conceivable that with $\alpha_{\rm K} = 0$ a current helicity flux can be generated, for example by shear, and that this flux divergence could drive a dynamo, as was suggested by Vishniac and Cho (2001). It is clear, however, that for finite values of $R_{\rm m}$ this would be a non-kinematic effect requiring the presence of an already finite field (at least of the order of $B_{\rm eq}/R_{\rm m}^{1/2}$). This is because of the $1 + R_{\rm m} \boldsymbol{B}_0^2 / B_{\rm eq}^2$ term in the denominator of (69). At the moment we cannot say whether this is perhaps the effect leading to the nonhelically forced turbulent dynamo discussed in Sect. 4.5, or whether it is perhaps the $\boldsymbol{\delta} \times \overline{\boldsymbol{J}}$ or shear-current effect that was also mentioned in that section.

6 What about η Quenching?

As we have seen above, in a closed domain the value of α in the saturated state cannot conclusively be determined without also determining at the same time the turbulent magnetic diffusivity. There are different ways of determining $\eta_{\rm t}$. The values are not necessarily all in agreement with each other, because one does not know whether the mean-field equation, where $\eta_{\rm t}$ enters, is correct and applicable. We report here a few different examples where $\eta_{\rm t}$ has been determined.

6.1 Direct Measurements in a Working Dynamo

We first consider the case of a helical turbulent dynamo without shear (B01) and compare it with a simple mean-field α^2 dynamo. Assuming that α is uniform, we can use (31) and, assuming furthermore that $\alpha < 0$ (which is the case when the helicity of the forcing is positive, as in B01), the solution is

$$\overline{\boldsymbol{B}} = (b_x \cos k_1 z, b_y \sin k_1 z, 0)^T \; . \tag{72}$$

The time-dependent equations can then be written as

$$\frac{\mathrm{d}b_x}{\mathrm{d}t} = |\alpha| b_y - \eta_\mathrm{T} k_1^2 b_x \; , \tag{73}$$

$$\frac{\mathrm{d}b_y}{\mathrm{d}t} = |\alpha| b_x - \eta_\mathrm{T} k_1^2 b_y \; . \tag{74}$$

In an isotropic, homogeneous α^2 dynamo, the eigenfunction obeys $b_x = b_y$.

We now assume that, at some particular time, we put $b_x = 0$, for example. This means that $b_x(t)$ will first grow linearly in time at a rate that is proportional to α like $b_x \approx |\alpha| k_1 b_y$. At the same time as b_x grows, b_y will first decrease at a rate that is proportional to η_T. This allows an independent estimate of b_x and b_y by solving the matrix equation

$$\begin{pmatrix} k_1 b_y & -k_1^2 b_x \\ k_1 b_x & -k_1^2 b_y \end{pmatrix} \begin{pmatrix} |\alpha| \\ \eta_\mathrm{T} \end{pmatrix} = \begin{pmatrix} \mathrm{d}b_x/\mathrm{d}t \\ \mathrm{d}b_y/\mathrm{d}t \end{pmatrix} . \tag{75}$$

The result for α is found to be roughly consistent with that of Cattaneo and Hughes (1996), and the result for η_T is reproduced in Fig. 13, and can be described by the fit formula

$$\eta_\mathrm{t} = \frac{\eta_{\mathrm{t}0}}{1 + \tilde{g} |\overline{\boldsymbol{B}}| / B_\mathrm{eq}} \tag{76}$$

with $\tilde{g} \approx 16$. This expression needs to be compared with that obtained from other approaches.

The fact that the results obtained for α by using this approach are consistent with that for a uniform field is quite surprising and unexpected. This agreement probably indicates that in this type of simulation α is independent of scale – at least in the scale range corresponding to wavenumbers $k = k_1$ (= 1) and $k = 0$. In general, this may not be true. Indeed, in the case of accretion discs some numerical evidence for scale dependence of α and η_t has been found (Brandenburg and Sokoloff, 2002).

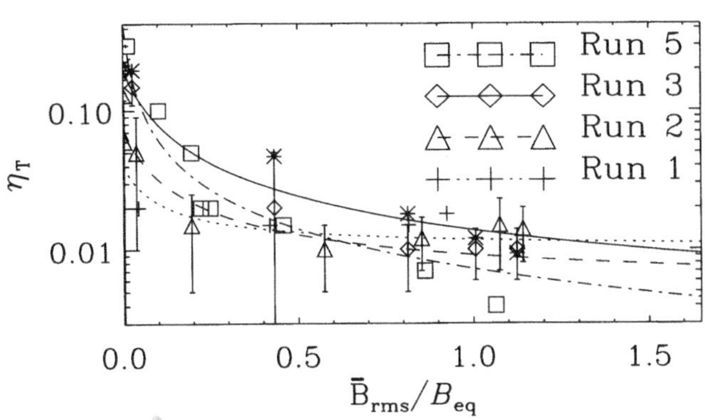

Fig. 13. Result for η_T for different values of R_m. The lines represent the fits described in the text. In the plot of η_T the asterisks denote $|\alpha| - \lambda$ for the $R_{\mathrm{m,forc}} = 120$ run, which agrees reasonably well with η_T [Adapted from B01]

Table 1. Summary of the main properties of the three-dimensional simulations with shear. Here, $\eta/(c_s k_1)$ is the magnetic diffusivity in units of the sounds speed and the wavenumber of the domain, and $\omega_{\rm cyc} = 2\pi/T_{\rm cyc}$ is the cycle frequency. In Run (iii) there is no clear cycle visible [Adapted from BDS02]

Run	(i)	(ii)	(iii)
$\eta/(c_s k_1)$	10^{-3}	5×10^{-4}	2×10^{-4}
ν/η	5	10	25
$R_{\rm m}=\langle \boldsymbol{u}^2\rangle^{1/2}/(\eta k_1)$	30	80	200
$C_\Omega=\langle \overline{\boldsymbol{U}}^2\rangle^{1/2}/(\eta k_1)$	1000	2000	4000
$\langle \boldsymbol{b}^2\rangle/B_{\rm eq}^2$	4	6	20
$\langle \overline{\boldsymbol{B}}^2\rangle/B_{\rm eq}^2$	20	30	60
$\epsilon_{\rm m}=\mu_0\langle \overline{\boldsymbol{J}}\cdot\overline{\boldsymbol{B}}\rangle/\langle \overline{\boldsymbol{B}}^2\rangle$	0.11	0.06	0.014
$\omega_{\rm cyc}/(\eta k_1^2)$	8...9	6...12	≥ 10?

6.2 Measurements in an $\alpha\Omega$ Dynamo

In the case of an $\alpha\Omega$ dynamo the cycle frequency $\omega_{\rm cyc}$ depends directly on the nonlinearly suppressed value of $\eta_{\rm T}$

$$\omega_{\rm cyc} = \eta_{\rm T} k_1^2 \quad (\eta_{\rm T} \text{ is quenched})\ , \tag{77}$$

see BB02 (their Sect. 4.2). The estimates of BBS01 indicated that the dynamo numbers based on shear, $C_\Omega = S/(\eta_{\rm T} k_1^2)$, is between 40 and 80, whilst the total dynamo number ($\mathcal{D} = C_\alpha C_\Omega$) is between 10 and 20 (see BBS01), and hence $C_\alpha = \alpha/(\eta_{\rm T} k_1) \approx 0.25$. Thus, shear dominates strongly over the α effect (C_Ω/C_α is between 150 and 300), which is typical for $\alpha\Omega$-type behavior (i.e. oscillations) rather than α^2-type behavior which would start when C_Ω/C_α is below about 10 (e.g. Roberts and Stix, 1972).

The results shown in Table 1 suggest that the period in this oscillatory dynamo is controlled by the microscopic magnetic diffusivity, because $\omega_{\rm cyc}/(\eta k_1^2)$ is approximately independent of $R_{\rm m}$. Using (77), this means that $\eta_{\rm T}(\text{quenched})/\eta = \mathcal{O}(10)$ for $R_{\rm m}$ between 30 and 200. This result would favor a model where $\eta_{\rm T}$ is still quenched in an $R_{\rm m}$-dependent fashion. In the next section we show that the apparent $R_{\rm m}$-dependent $\eta_{\rm t}$ quenching can easily also be produced when the field possesses a helical component.

Looking at the scaling of the cycle frequency with resistivity may be quite misleading in the present case, because the large-scale magnetic field exceeds the kinetic energy by a large factor (20–30). This would always lead to the usual (non-catastrophic) quenching of α and $\eta_{\rm t}$. Furthermore, such strong magnetic fields will affect the mean shear flow. Most important is perhaps the fact that in the simulation of BBS01 the shear flow varies sinusoidally in the cross stream direction, so the mean field depends on the two coordinate directions perpendicular to the streamwise direction. For this reason BB02 solved the mean-field and dynamical quenching equations in a 2-dimensional model. It turned out to be important to allow for

Table 2. Results from the simulations of BBS01 and BDS02, compared with those of 2-dimensional mean-field models. Model results that are in fair agreement with the simulations are highlighted in bold face. Here, Q is the ratio of toroidal to poloidal rms field

Model	R_m	C_α	C_Ω	$\tilde{g}$	$\dfrac{S}{\eta k_1^2}$	$\dfrac{\langle \boldsymbol{b}^2\rangle}{B_\mathrm{eq}^2}$	$\dfrac{\langle \overline{\boldsymbol{B}}^2\rangle}{B_\mathrm{eq}^2}$	Q^{-1}	ϵ_m	$\dfrac{\omega_\mathrm{cyc}}{S}$	$\dfrac{\lambda}{S}$
BBS01	80	1–2	–	–	2000	6	30	0.014	0.06	0.008	0.015
R1	20	1.0	100	0	2000	0.20	**15**	0.031	0.065	0.016	0.044
AG2	100	0.5	20	3	2000	0.10	**22**	0.011	0.024	**0.006**	**0.021**
BDS02	30	1–2	–	–	1000	4	20	0.018	0.11	0.014	0.006
s3	30	0.35	33	1	1000	0.07	6	0.029	**0.061**	**0.014**	**0.016**
S1	30	0.35	33	3	1000	0.07	**19**	0.009	0.019	0.005	0.016

non-catastrophic quenching of η_t using (76) where the value of $\tilde{g}$ has been varied between 0 and 3. The asymptotic $1/B$ behavior (as opposed to $1/B^2$, for example) was motivated both by simulations (B01) and analytic results (Kitchatinov et al., 1994; Rogachevskii and Kleeorin, 2001).

In order to see whether the models can be made to match the direct simulations, several input parameters were varied. It should be kept in mind, however, that not all input parameters are well known. This has to do with the uncertainty in the correspondence between the magnetic Reynolds number in the model (which measures η_t0/η) and the simulations [where it is defined as $u_\mathrm{rms}/(\eta k_\mathrm{f})$]. Likewise, the dynamo number $C_\alpha = \alpha/(\eta_\mathrm{T} k_1)$ is not well determined. Nevertheless, many of the output parameters are reasonably well reproduced; see Table 2.

6.3 Decay Experiments

Finally, we consider the decay of a magnetic field. This provides a fairly straightforward method of determining η_T from the decay rate λ of a sinusoidal field with wavenumber k_1, so $\lambda = \eta_\mathrm{T} k_1^2$. The result reported by Yousef et al. (2003) suggests that

$$\nu_\mathrm{t} \approx \eta_\mathrm{t} = (0.8 \ldots 0.9) \times u_\mathrm{rms}/k_\mathrm{f} \quad (\text{for } \overline{\boldsymbol{B}}^2 \ll B_\mathrm{eq}^2)\,. \tag{78}$$

Once the mean flow profile has decreased below a certain level ($< 0.1 u_\mathrm{rms}$), it cannot decay further and continues to fluctuate around $0.08 u_\mathrm{rms}$, corresponding to the level of the rms velocity of the (forced!) turbulence at $k = k_1$ (see the dashed line in Fig. 14).

The quenching of the magnetic diffusivity, $\eta_\mathrm{t} = \eta_\mathrm{t}(\overline{\boldsymbol{B}})$, can be obtained from one and the same run by simply determining the decay rate, $\lambda_B(\overline{B})$, at different times, corresponding to different values of $\overline{B} = |\overline{\boldsymbol{B}}|$. To describe departures from purely exponential decay one can adopt a $\overline{\boldsymbol{B}}$-dependent η_t expression of the form (76). It turns out that the value of $\tilde{g}$ is not universal and depends on the field geometry. This is easily demonstrated by comparing the decay of helical and nonhelical initial fields; see Fig. 15.

In the next section we show that the slower decay of $\overline{\boldsymbol{B}}$, and hence the implied stronger quenching of η_t, can also be described by a self-induced magnetic α effect

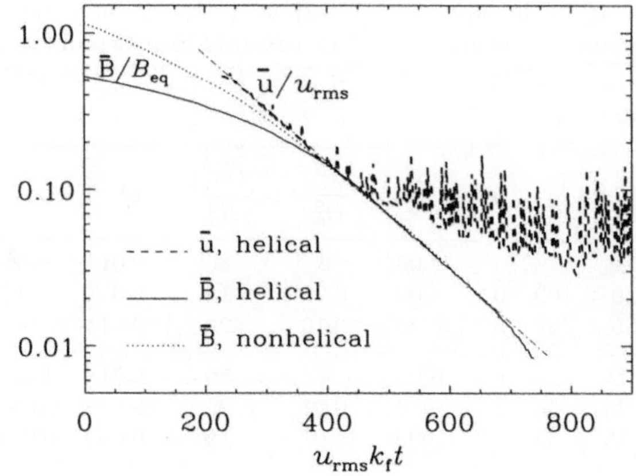

Fig. 14. Decay of large-scale helical velocity and magnetic fields (*dashed and solid lines*, respectively). The graph of $\overline{U}(t)$ has been shifted so that both $\overline{U}(t)$ and $\overline{B}(t)$ share the same tangent (*dash-dotted line*), whose slope corresponds to $\nu_t = \eta_t = 0.86u_{\rm rms}/k_{\rm f}$. The decay of a nonhelical magnetic field is shown for comparison (*dotted line*) [Adapted from Yousef et al. (2003)]

which acts such as to decrease the decay rate. In the case of a helical initial field, we have $\overline{\boldsymbol{J}} \times \overline{\boldsymbol{B}} = 0$, i.e. the large-scale field is force-free and interacts only weakly with the turbulence.

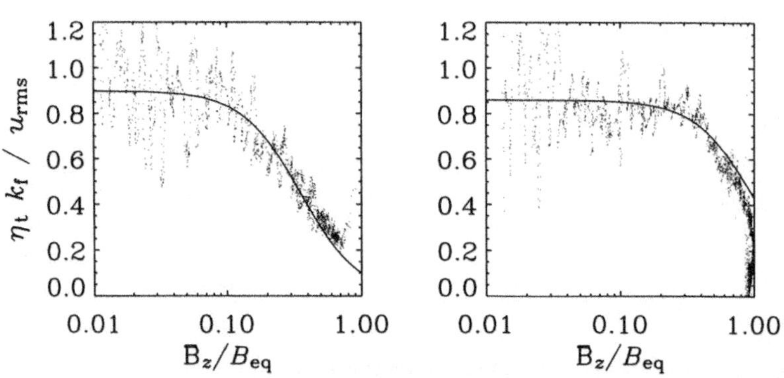

Fig. 15. Dependence of the turbulent diffusion coefficient on the magnitude of the mean field. $R_{\rm m} \approx 20$. *Left*: The initial field is helical and corresponds to data points on the right hand side of the plot. The data are best fitted by $\tilde{g} = 8 = 0.4R_{\rm m}$. *Right*: the same for the nonhelical case. The data are best fitted by $a = 1$, independent of $R_{\rm m}$ [Adapted from Yousef et al. (2003)]

Thus, the indications here are that for non-helical fields, η_t is *not* catastrophically quenched. A resistively slow decay rate occurs however when the magnetic field is helical, but this is *not* to be explained by a catastrophically quenched η_t, but by the magnetic α effect, $\alpha_{\rm M}$, that tries to keep the magnetic field as large as possible, just as enforced by the magnetic helicity constraint. The phenomenon, described in this way, may be more easily described in terms of helicity conservation, because the system has magnetic helicity that can only decay slowly on a resistive time scale, hence lowering the apparent turbulent diffusivity down to the microscopic value η. This will be explained in more detail in the next section.

6.4 Taylor Relaxation or Selective Decay

In the case of a helical field with $\overline{\boldsymbol{B}}^2/B_{\rm eq}^2 \gtrsim R_{\rm m}^{-1}$ the slow decay of $\overline{\boldsymbol{B}}$ is related to the conservation of magnetic helicity. As already discussed by BB02, this behavior is related to the phenomenon of selective decay (e.g. Montgomery et al., 1978) and can be described by the dynamical quenching model. This model applies even to the case where the turbulence is nonhelical and where there is no α effect in the usual sense. However, the magnetic contribution to α is still non-vanishing, because it is driven by the helicity of the large-scale field.

To demonstrate this quantitatively, Yousef et al. (2003) have adopted the one-mode approximation ($\boldsymbol{k} = \boldsymbol{k}_1$) with $\overline{\boldsymbol{B}} = \hat{\boldsymbol{B}}\exp({\rm i}\boldsymbol{k}_1 z)$, the mean-field induction equation

$$\frac{{\rm d}\hat{\boldsymbol{B}}}{{\rm d}t} = {\rm i}\boldsymbol{k}_1 \times \hat{\boldsymbol{\mathcal{E}}} - \eta k_1^2 \hat{\boldsymbol{B}} \,, \tag{79}$$

together with the dynamical α-quenching formula (39),

$$\frac{{\rm d}\alpha}{{\rm d}t} = -2\eta k_{\rm f}^2 \left(\alpha + \tilde{R}_{\rm m} \frac{{\rm Re}(\hat{\boldsymbol{\mathcal{E}}}^* \cdot \hat{\boldsymbol{B}})}{B_{\rm eq}^2} \right) \,, \tag{80}$$

where

$$\hat{\boldsymbol{\mathcal{E}}} = \alpha \hat{\boldsymbol{B}} - \eta_{\rm t} {\rm i}\boldsymbol{k}_1 \times \hat{\boldsymbol{B}} \tag{81}$$

is the electromotive force, and $\tilde{R}_{\rm m}$ is defined as the ratio $\eta_{\rm t0}/\eta$, which is expected to be close to the value of $R_{\rm m}$.

In Fig. 16 we show the evolution of $\overline{\boldsymbol{B}}/B_{\rm eq}$ for helical and nonhelical initial conditions, $\hat{\boldsymbol{B}} \propto (1, {\rm i}, 0)$ and $\hat{\boldsymbol{B}} \propto (1, 0, 0)$, respectively. In the case of a nonhelical field, the decay rate is not quenched at all, but in the helical case quenching sets in for $\overline{\boldsymbol{B}}^2/B_{\rm eq}^2 \gtrsim R_{\rm m}^{-1}$. In the helical case, the onset of quenching at $\overline{\boldsymbol{B}}^2/B_{\rm eq}^2 \approx R_{\rm m}^{-1}$ is well reproduced by the simulations. In the nonhelical case, however, some weaker form of quenching sets in when $\overline{\boldsymbol{B}}^2/B_{\rm eq}^2 \approx 1$ (see the right hand panel of Fig. 15). We refer to this as standard quenching (e.g. (Kitchatinov et al., 1994) which is known to be always present; see (76). BB02 found that, for a range of different values of $R_{\rm m}$, $\tilde{g} = 3$ resulted in a good description of the simulations of cyclic $\alpha\Omega$-type dynamos (BDS02).

Yousef et al. (2003) also showed that the turbulent magnetic Prandtl number is indeed independent of the microscopic magnetic Prandtl number. The resulting values of the flow Reynolds number, ${\rm Re} = u_{\rm rms}/(\nu k_{\rm f})$, varied between 20 and 150, giving $P_{\rm m}$ in the range between 0.1 and 1. Within plot accuracy the three values of λ_B turn out to be identical in the interval where the decay is exponential.

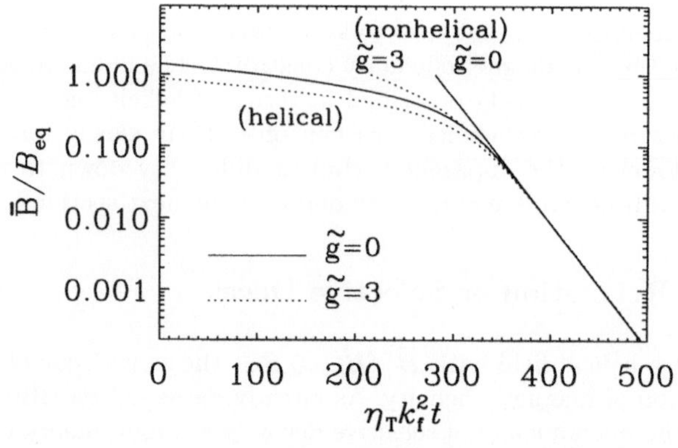

Fig. 16. Dynamical quenching model with helical and nonhelical initial fields. The quenching parameters are $\tilde{g} = 0$ (*solid line*) and 3 (*dotted line*). The graph for the nonhelical cases has been shifted in t so that one sees that the decay rates are asymptotically equal at late times. The value of η_T used to normalize the abscissa is based on the unquenched value [Adapted from Yousef et al. (2003)]

7 Conclusions

In the present review we have tried to highlight some of the recent discoveries that have led to remarkable advances in the theory of mean-field dynamos. Of particular importance are the detailed confirmations of various aspects of mean-field theory using helically forced turbulence simulations. The case of homogeneous turbulence with closed or periodic boundary conditions is now fairly well understood. In all other cases, however, the flux of current helicity becomes important. The closure theory of these fluxes is still a matter of ongoing research (Kleeorin et al., 2000, 2002, 2003), Vishniac and Cho (2001), Subramanian and Brandenburg (2004), and Brandenburg and Subramanian (2005). The helicity flux of Vishniac and Cho (2001) has been independently confirmed (Subramanian and Brandenburg, 2004). A more detailed investigation of current helicity fluxes appears to be quite important when one tries to get qualitative and quantitative agreement between simulations and theory.

The presence of current helicity fluxes is particularly important when there is also shear. This was already recognized by Vishniac and Cho (2001) who applied their calculations to the case of accretion discs where shear is particularly strong. In the near future it should be possible to investigate the emergence of current helicity flux in more detail. This would be particularly interesting in view of the many observations of coronal mass ejections that are known to be associated with significant losses of magnetic helicity and hence also of current helicity (DeVore, 2000; Démoulin et al., 2002; Gibson et al., 2002).

In order to be able to model coronal mass ejections it should be particularly important to relax the restrictions imposed by the vertical field conditions employed in the simulations of Brandenburg and Sandin (2004). A plausible way of doing this would be to include a simplified version of a corona with enhanced temperature and hence decreased density, making this region a low-beta plasma.

In the context of accretion discs the importance of adding a corona is well recognized (Miller and Stone, 2000), although its influence on large-scale dynamo action is still quite open. Regarding hydromagnetic turbulence in galaxies, most simulations to date do not address the question of dynamo action (Korpi et al., 1999; de Avillez and Mac Low, 2002).

This is simply because here the turbulence is driven by supernova explosions which leads to strong shocks. These in turn require large numerical diffusion, so the effective magnetic Reynolds number is probably fairly small and dynamo action may only be marginally possible. In nonhelically driven turbulence has been applied to the galactic medium to argue that it is dominated by small-scale fields (Schekochihin et al., 2002), but the relative importance of small-scale fields remains still an open question (Haugen et al., 2003). Galaxies are however rotating and vertically stratified, so the flows should be helical, but in order to say anything about magnetic helicity evolution, much larger magnetic Reynolds numbers are necessary. At the level of mean-field theory the importance of magnetic helicity fluxes is well recognized. The explicitly time-dependent dynamical α quenching equation with magnetic helicity fluxes has been included in mean-field simulations (Kleeorin et al., 2000, 2002, 2003), but the form of the adopted fluxes is to be clarified in view of the differences with the results of Vishniac and Cho (2001) and Subramanian and Brandenburg (2004). Nevertheless, given that the form of the dynamical quenching equations is likely to be still incomplete, it remains to be demonstrated, using simulations, that magnetic or current helicity fluxes do really allow the dynamo to saturate on a dynamical time scale.

Acknowledgements

The Danish Center for Scientific Computing is acknowledged for granting time on the Horseshoe cluster. This work has been completed while being on sabbatical at the Isaac Newton Institute for Mathematical Sciences in Cambridge.

References

de Avillez, M. A., Mac Low, M.-M.: 2002, ApJ 581, 1047
Balbus, S. A., Terquem, C.: 2001, ApJ 552, 235
Balsara, D., Pouquet, A.: 1999, Phys. Plasmas 6, 89
Beck, R., Brandenburg, A., Moss, D. et al.: 1996, ARA&A 34, 155
Berger, M., Field, G. B.: 1984, JFM 147, 133
Berger, M. A., Ruzmaikin, A.: 2000, JGR 105, 10481
Boldyrev, S., Cattaneo, F.: 2004, Phys. Rev. Lett. 92, 144501
Blackman, E. G.: 2003, MNRAS 344, 707
Blackman, E. G., Field, G. F.: 2000, MNRAS 318, 724
Blackman, E. G., Field, G. B.: 2002, Phys. Rev. Lett. 89, 265007
Blackman, E. G., Brandenburg, A.: 2002, ApJ 579, 359 (BB02)
Blackman, E. G., Brandenburg, A.: 2003, ApJ 584, L99
Brandenburg, A.: 2001, ApJ 550, 824 (B01)

Brandenburg, A.: 2003, in *Simulations of Magnetohydrodynamic Turbulence in Astrophysics*, ed. E. Falgarone, T. Passot (Lecture Notes in Physics, Vol. 614. Berlin: Springer), p. 402
Brandenburg, A., Dobler, W.: 2001, A&A 369, 329
Brandenburg, A., Dobler, W.: 2002, Comp. Phys. Comm. 147, 471
Brandenburg, A., Matthaeus, W. H.: 2004, Phys. Rev. E 69, 056407
Brandenburg, A., Sandin, C.: 2004, A&A 427, 13
Brandenburg, A., Sokoloff, D.: 2002, Geophys. Astrophys. Fluid Dyn. 96, 319
Brandenburg, A., Subramanian, K.: 2005, Phys. Rep. [arXiv:astro-ph/0405052]
Brandenburg, A., Bigazzi, A., Subramanian, K.: 2001, MNRAS 325, 685 (BBS01)
Brandenburg, A., Blackman, E. G., Sarson, G. R.: 2003, Adv. Spa. Sci. 32, 1835
Brandenburg, A., Dobler, W., Subramanian, K.: 2002, AN 323, 99 (BDS02)
Brandenburg, A., Nordlund, Å., Pulkkinen, P. et al.: 1990, A&A 232, 277
Brun, A. S., Miesch, M. S., Toomre, J.: 2004, ApJ 614, 1073
Cattaneo F., Hughes D. W.: 1996, Phys. Rev. E 54, R4532
Chae, J.: 2000, ApJ 540, L115
Cowling, T. G.: 1934, MNRAS 94, 39
Démoulin, P., Mandrini, C. H., van Driel-Gesztelyi, L. et al.: 2002, Sol. Phys. 207, 87
DeVore, C. R.: 2000, ApJ 539, 944
Field, G. B., Blackman, E. G.: 2002, ApJ 572, 685
Gibson, S. E., Fletcher, L., Del Zanna, G. et al.: 2002, ApJ 574, 1021
Gilman, P. A.: 1983, ApJS 53, 243
Glatzmaier, G. A.: 1985, ApJ 291, 300
Gruzinov, A. V., Diamond, P. H.: 1994, Phys. Rev. Lett. 72, 1651
Haugen, N. E. L., Brandenburg, A., Dobler, W.: 2003, ApJ 597, L141
Haugen, N. E. L., Brandenburg, A., Dobler, W.: 2004, Phys. Rev. E 70, 016308
Hollerbach, R., Rüdiger, G.: 2004, MNRAS 347, 1273
Hughes D. W., Cattaneo F., Kim E. J.: 1996, Phys. Lett. 223, 167
Kazantsev, A. P.: 1968, Sov. Phys. JETP 26, 1031
Keinigs, R. K.: 1983, Phys. Fluids 26, 2558
Kitchatinov, L. L., Rüdiger, G., Pipin, V. V.: 1994, AN 315, 157
Kleeorin, N. I., Ruzmaikin, A. A.: 1982, Magnetohydrodynamics 18, 116
Kleeorin, N. I, Moss, D., Rogachevskii, I., Sokoloff, D.: 2000, A&A 361, L5
Kleeorin, N. I, Moss, D., Rogachevskii, I., Sokoloff, D.: 2002, A&A 387, 453
Kleeorin, N. I, Moss, D., Rogachevskii, I., Sokoloff, D.: 2003, A&A 400, 9
Korpi, M. J., Brandenburg, A., Shukurov, A. et al.: 1999, ApJ 514, L99
Krause, F., Rädler, K.-H.: 1980, *Mean-Field Magnetohydrodynamics and Dynamo Theory* (Berlin: Akademie-Verlag; also Oxford: Pergamon Press)
Kulsrud, R. M., Anderson, S. W.: 1992, ApJ 396, 606
Kulsrud, R. M.: 1999, ARA&A 37, 37
Low, B. C.: 2001, JGR 106, 25,141
Maron, J., Blackman, E. G.: 2002, ApJ 566, L41
Meneguzzi, M., Frisch, U., Pouquet, A.: 1981, Phys. Rev. Lett. 47, 1060
Milano, L. J., Matthaeus, W. H., Dmitruk, P.: 2003, Phys. Plasmas 10, 2287
Miller, K. A., Stone, J. M.: 2000, ApJ 534, 398
Mininni, P. D., Gómez, D. O., Mahajan, S. M.: 2003, ApJ 587, 472
Moffatt, H. K.: 1969, JFM 35, 117

Moffatt, H. K.: 1978, *Magnetic Field Generation in Electrically Conducting Fluids* (Cambridge: Cambridge University Press)
Montgomery, D., Turner, L., Vahala, G.: 1978, Phys. Fluids 21, 757
Montgomery, D., Matthaeus, W. H., Milano, L. J., Dmitruk, P.: 2002, Phys. Plasmas 9, 1221
Ossendrijver, M., Stix, M., Brandenburg, A.: 2001, A&A 376, 713
Pouquet, A., Frisch, U., Léorat, J.: 1976, JFM 77, 321
Rädler, K.-H.: 1969, Geod. Geophys. Veröff., Reihe II 13, 131
Roberts, P. H., Stix, M.: 1972, A&A 18, 453
Rogachevskii, I., Kleeorin, N.: 2001, Phys. Rev. E 64, 056307
Rogachevskii, I., Kleeorin, N.: 2003, Phys. Rev. E 68, 036301
Rogachevskii, I., Kleeorin, N.: 2004, Phys. Rev. E 70, 046310
Rogers, N. N., Denton, R. E., Drake, J. E. et al.: 2001, Phys. Rev. Lett. 87, 195004
Rüdiger, G., Kitchatinov, L. L.: 1993, A&A 269, 581
Schekochihin, A. A., Maron, J. L., Cowley, S. C., McWilliams, J. C.: 2002, ApJ 576, 806
Schekochihin, A. A., Cowley, S. C., Maron, J. L., McWilliams, J. C.: 2004, Phys. Rev. Lett. 92, 054502
Shukurov, A.: 2005, in *Mathematical Aspects of Natural Dynamos*, ed. E. Dormy (Dordrecht: Kluwer) (in press) [arXiv:astro-ph/0411739]
Steenbeck, M., Krause, F.: 1969a, AN 291, 49
Steenbeck, M., Krause, F.: 1969b, AN 291, 271
Steenbeck, M., Krause, F., Rädler, K.-H.: 1966, Z. Naturforsch. 21a, 369
Subramanian, K., Brandenburg, A.: 2004, Phys. Rev. Lett. 93, 205001
Vainshtein, S. I., Cattaneo, F.: 1992, ApJ 393, 165
Vainshtein, S. I., Chitre, S. M., Olinto, A.: 2000, Phys. Rev. E 61, 44224430
Vishniac, E. T., Brandenburg, A.: 1997, ApJ 475, 263
Vishniac, E. T., Cho, J.: 2001, ApJ 550, 752
Widrow, L. M.: 2002, Rev. Mod. Phys. 74, 775
Yoshimura, H.: 1975, ApJS 29, 467
Yousef, T. A., Brandenburg, A., Rüdiger, G.: 2003, A&A 411, 321
Zeldovich, Ya. B.: 1957, Sov. Phys. JETP 4, 460

Numerical Magnetohydrodynamics in Astrophysics

Max Camenzind

Landessternwarte Königstuhl, 69117 Heidelberg, Germany
M.Camenzind@lsw.uni-heidelberg.de

Abstract. Newtonian magnetohydrodynamics (MHD) is a very special limit of Maxwell's equations and plasma dynamics. In the last 10 years, powerful numerical algorithms and computational methods have been developed for simulating the time-evolution of magnetic field configurations in the astrophysical environment. The most recent trends go in the direction of fully conservative schemes and adaptive mesh refinement for large-scale supercomputing. The most popular codes in use in astrophysics are briefly discussed, together with their strengths and pitfalls. Progress in understanding fundamental MHD processes have been achieved in the field of flux tube dynamics, magnetic turbulence in accretion and galactic disks, magnetic Herbig–Haro flows from young stellar objects and the evolution of primordial magnetic fields in galaxy clusters.

The true relativistic magnetohydrodynamics includes various terms which are completely neglected in the classical MHD. Some progress has been achieved in the last few years in the numerical modelling of special relativistic MHD (SRMHD), with main applications to the simulation of the propagation of relativistic extragalactic jets. While this field of research is presently under a high dynamical pressure, the development of codes for General Relativistic MHD (GRMHD) is still in its infancy phase. The main application for such schemes is for understanding the extraction of rotational energy of rapidly rotating compact objects, such as neutron stars and Black Holes. In particular, the extraction of rotational energy from Black Holes in microquasars, quasars and radio galaxies and their transformation into collimated jet outflows is one of the main unsolved problem in modern astrophysics.

1 Introduction

With the development of computer technology, numerical simulation has been more and more widely used in many fields of our society. Simulation techniques not only play very important roles in scientific study, but also occupy very important places in education, military, entertainment and almost any fields that we can imagine.

Generally, simulation includes all methods that can reproduce the processes of a system in an analog or digital fashion. It includes numerical simulations, and simulations quite different from numerical simulations. For example, we can simply use a circuit to simulate the generation of Earth's magnetic dynamo (one-disk and two-disk geodynamo models). In this case, we use a circuit to represent the environment of inner earth and by changing the capacitance, resistivity and the combination of the disks and electric wires to achieve different behaviors of the system. In this simulation activity, we only use several oversimplified basic components to simulate the system. Though we can use this simulation to study the geodynamic process qualitatively, it is impractical for us to use the numerical results of such simulation to draw further conclusion. Another example is flight simulation, which is widely used in flight training and entertainment. The concentration of the flight simulation is to give the person who operates the system a visual sensation of flight. In practice, we can use well-organized two-dimensional objects to simulate the visual feeling of 3-dimensions. The result is that, though we are seeing 2-dimensional objects, we have the feeling that what we are seeing is 3-dimensional. Though physically this is totally wrong, it is acceptable, as the goal of flight simulation has been reached. Of course, we also cannot use the result of such simulation to get any physical understanding of the system. In contrast, numerical simulation is the kind of simulation that uses numerical methods to quantitatively represent the evolution of a physical system. It pays much attention to the physical content of the simulation and emphasizes the goal that, from the numerical results of the simulation, knowledge of background processes and physical understanding of the simulation region can be obtained. In practice, numerical simulation uses the values that can best represent the real environment. The evolution of the system also strictly obeys the physical laws that govern the real physical processes in the simulation region. Then the result of such simulation can have a good representation of the real environment. From the result of such simulation we can safely draw proper conclusions and have a better understanding of the system.

2 The Classical MHD Model in Computer Simulation

Classical magnetohydrodynamics (MHD) is the non-relativistic limit of Maxwell's equations combined with dynamical equations for a plasma in the one-component description. We should, however, not forget that originally Maxwell's equations are relativistic and that the non-relativistic version oversimplifies many aspects.

The study of numerical algorithms for MHD simulations is getting a very active field of research, with no one method having become standard. There are two main generic algorithms most widely used in practice: the Method of Characteristics/Constrained Transport (MOC/CT) (Evans and Hawley, 1988; Stone and Norman, 1992), and the shock-capturing (Godunov) methods (Brio and Wu, 1988; Zachary and Colella, 1992; Dai and Woodward, 1994; Balsara and Spicer, 1999). Each of these methods has distinct benefits and drawbacks. Codes based on the MOC/CT algorithms are relatively simple in design, and essentially satisfy the divergence-free condition to machine precision. The ZEUS scheme e.g. (Stone and Norman, 1992) is by construction second-order on Alfvén waves, but does not address the question of the two compressive waves (the slow and fast magnetosonic waves). Codes implementing the shock-capturing methods are more complex (and CPU time-consuming), but give highly acccurate results for strong shocks. They, however, suffer from the drawback that the divergence-free constraint is only satisfied to truncation error. In order to overcome this disease, a number of techniques have been invented in the last years to treat this difficulty.

2.1 MHD in Advective Form

Originally, the basic MHD equations are written in advective form, as used in ZEUS–like codes (e.g. in NIRVANA, Ziegler and Yorke, 1997)

$$\partial_t \rho = -\nabla \cdot (\rho \boldsymbol{v}) \tag{1}$$

$$\partial_t (\rho \boldsymbol{v}) = -\nabla \cdot (\rho \boldsymbol{v} \otimes \boldsymbol{v}) - \nabla P + \frac{1}{8\pi} \nabla \boldsymbol{B}^2 + \frac{1}{4\pi} (\boldsymbol{B} \cdot \nabla)\boldsymbol{B} - \rho \nabla \Phi \tag{2}$$

$$\partial_t e = -\nabla \cdot (e \boldsymbol{v}) - P \nabla \cdot \boldsymbol{v} + \frac{\eta}{16\pi^2} |\nabla \times \boldsymbol{B}|^2 + \nabla \cdot (\kappa \nabla T) + \sigma : \nabla \boldsymbol{v} \tag{3}$$

$$\partial_t \boldsymbol{B} = -\nabla \times \left(\boldsymbol{v} \times \boldsymbol{B} - \frac{\eta}{4\pi} \nabla \times \boldsymbol{B} \right) \tag{4}$$

$$\begin{aligned} \sigma_{\mathrm{ik}} &= -l_{\mathrm{T}}^2 \rho \min(0, \nabla \cdot \boldsymbol{v}) \times \left(\nabla \boldsymbol{v}_{\mathrm{ik}} - \frac{1}{3} \nabla \cdot \boldsymbol{v} \right) \\ &\quad + l_{\mathrm{A}} \rho \left(\delta x_{\mathrm{i}} \times \min(0, \nabla v_{\mathrm{ik}})^2 \right) \delta_{\mathrm{ik}} \end{aligned} \tag{5}$$

The viscous stress tensor σ_{ik} is composed of artificial viscosity, which contains the tensor artificial viscosity and the Neumann–Richtmyer artificial viscosity. l_{T} and l_{A} are the corresponding shock smearing length scales.

In a real physical system, we should have the conservation of momentum, energy and mass. But the above set of equations cannot guarantee strict numerical conservation of momentum and energy, though it can assure the conservation of mass. At the same time, it also has numerical difficulties with convective derivatives. The practical application of this set of equations also shows that non-conservative equations lead to numerical difficulties with strong shocks and to errors in the Rankine–Hugoniot conditions and the shock speed (Falle, 2002).

2.2 MHD in Conservative Form

In conservative form, the MHD equations assume the form

$$\partial_t \rho + \nabla \cdot (\rho \boldsymbol{v}) = 0 \tag{6}$$

$$\partial_t (\rho \boldsymbol{v}) + \nabla \cdot \left[\rho \boldsymbol{v} \otimes \boldsymbol{v} + \left(P + \frac{1}{8\pi} \boldsymbol{B}^2 \right) \mathbf{I} - \frac{1}{4\pi} \boldsymbol{B} \otimes \boldsymbol{B} \right] = 0 \tag{7}$$

$$\partial_t \boldsymbol{B} + \nabla \cdot \left[\boldsymbol{v} \otimes \boldsymbol{B} - \boldsymbol{B} \otimes \boldsymbol{v} \right] = 0 \tag{8}$$

$$\partial_t (\rho E) + \nabla \cdot \left[\left(\rho E + P + \frac{\boldsymbol{B}^2}{8\pi} \right) \boldsymbol{v} - \frac{1}{4\pi} (\boldsymbol{v} \cdot \boldsymbol{B}) \boldsymbol{B} \right] = 0 \,, \tag{9}$$

subject to the constraint $\nabla \cdot \boldsymbol{B} = 0$. The total energy E is given by

$$\rho E = \frac{P}{\Gamma - 1} + \frac{1}{2} \rho v^2 + \frac{1}{8\pi} \boldsymbol{B}^2 \,. \tag{10}$$

This set of equations allows strict numerical conservation of mass, momentum and energy. But practical application of it shows that, in the region of low plasma

β, numerical difficulties will be met. Sometimes pressure becomes negative because P becomes the difference of large numbers (here we can also see that normalization has to be combined with other techniques to avoid the difference of large numbers, though normalization itself can prevent most such cases). The model of ideal MHD is therefore a fully conservative system, except for the induction equation which requires some regulation for maintaining $\nabla \cdot \boldsymbol{B} = 0$ (Dedner et al., 2002).

With the definition of a state vector

$$\mathbf{Q} = \{\rho, \boldsymbol{v}, e, \boldsymbol{B}\}^{\mathrm{T}} \tag{11}$$

these equations can be written in compact form

$$\frac{\partial \mathbf{Q}}{\partial t} + \nabla \cdot \mathcal{F} = 0 \, . \tag{12}$$

The flux vector $\mathcal{F}$ depends in general non-linearly on the state variable $\mathbf{Q}$.

2.3 Time and Space Discretization

In the real world, space and time are continuous, but in an MHD simulation, we have to use discrete space and time to describe the system. The simplest way to discretize space is to divide the whole simulation region into many regions of the same size. For the time discretization, the simplest way is to choose the same time step, Δt, for all the evolution. Though the uniform discretization of space and time is simple, we cannot hope that it can optimize the simulation efficiency and reliability. In order to have a better solution of this problem, we need finer grids in some regions, while at the same time, coarser grids are enough in other regions. A gridding scheme depending on the real physical component value, instead of being predefined, is called adaptive (AMR). The kind of step adjusting that can be done by program itself is called self-adaptive. The same thing can also be performed in time domain.

A package of FORTRAN 90 routines, called PARAMESH, has been developed to provide an application developer with an easy route to extend an existing serial code that uses a logically Cartesian structured mesh into a parallel code with AMR. PARAMESH builds a hierarchy of sub-grids to cover the computational domain, with spatial resolution varying to satisfy the demands of the application (Fig. 1). These sub-grid blocks form the nodes of a tree data-structure (quad-tree in 2D or oct-tree in 3D). Each grid block has a logically Cartesian mesh, and the index ranges are the same for every block. Thus, in 2D, if we begin with a 10×20 grid on one block covering the entire domain, the first refinement step would produce 4 child blocks, each with its own 10×20 mesh, but now with mesh spacing 1/2 that of its parent. Any or all of these children can themselves be refined in the same manner. This process continues until the domain is covered with a quilt-like pattern of blocks with the desired spatial resolution everywhere. A similar package called Chombo provides a set of tools for implementing finite difference methods for the solution of partial differential equations on block-structured adaptively refined rectangular grids. Both elliptic and time-dependent modules are included. Support for parallel platforms and standardized self-describing file formats are included. Chombo provides a distributed infrastructure for parallel calculations over block-structured, adaptively refined grids.

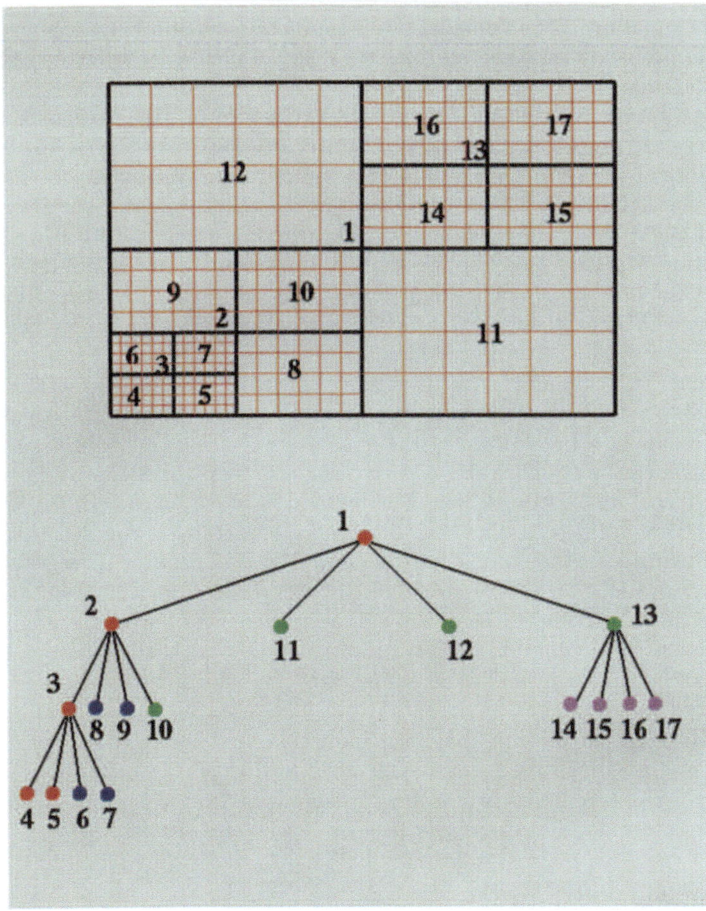

Fig. 1. A 2D 6 × 4 grid is created on each block. The numbers assigned to each block designate the blocks location in the quad-tree below. The colors assigned to the nodes of the tree indicate one possible distribution of the blocks during a 4 processor calculation [Source: PARAMESH]

2.4 Error Handling and Divergence Cleaning

The errors of analytical results come from the approximation of the theoretical model and the improper presentation of the physical environments and procedures. When a simulation model comes from a theoretical model, it also has errors generated by the analytical results. In addition, a numerical method has its own errors during the simulation process due to computation accuracy and imperfect representation of those continuous processes, such as differentiations in time and space. Simple error analysis can be conducted on the basis of each of the simulation equations, as shown in the error terms (Dedner et al., 2002). Further error analysis of the whole system becomes much more prohibitive because of the complexity of the MHD equations. The comparison of simulation results with analytical results for simple problems can give us some very useful ideas of the errors of the MHD equation system.

Diffusive errors come from the diffusive property of MHD equations. In some cases, there are some sharp boundaries between two parts of the simulation region. But because of the error of the numerical computation, such sharp boundaries can be smoothed. Generally, diffusive errors smooth the parameter variation in space and make the spatial configuration more flat. In some cases, the error in a numerical computation can feed back on the simulation system. If the computation scheme cannot erase this error, some wiggles may appear. Sometimes these wiggles can be greatly enhanced through the feedback mechanism to an extent that the real physical process can be totally hidden.

In numerical simulations the divergence constraint on the magnetic field causes severe stability problems. Accumulating errors can lead to an unphysical situation and can result in a breakdown of the simulation. Various authors have developed an approach to the stabilization of numerical schemes which can be easily used as an extension of an existing solver. The method is based on a modified formulation of the MHD equations in which the divergence constraint is coupled to the system by introducing a further unknown function. The evolution of divergence errors is strongly dependent on the type of the equation chosen for this function. For the one-dimensional setting, these errors can be transported out of the computational domain by a wave equation or can be dissipated by a heat equation. In Dedner et al. (2002) a mixed formulation is suggested, by which the divergence errors are transported and dissipated at the same time. The resulting system is still hyperbolic and the density, momentum, magnetic induction, and the total energy density are still conserved. Numerical examples demonstrate that this method decreases the divergence errors by up to two orders of magnitude even compared with the often used source term stabilization approach by Powell and coworkers.

2.5 Boundary Conditions

In a numerical simulation, it is impossible and unnecessary to simulate the whole universe. Generally we choose a region of interest in which we conduct a simulation. The interesting region has a certain boundary with the surrounding environment. Numerical simulations also have to consider the physical processes in the boundary region. In most cases, the boundary conditions are very important for the simulation region's physical processes. Different boundary conditions may cause quite different simulation results. Improper sets of boundary conditions may introduce nonphysical influences on the simulation system, while a proper set of boundary conditions can avoid that. So arranging the boundary conditions for different problems becomes very important. While at the same time, different variables in the environment may have different boundary conditions according to certain physical problems.

For fixed boundary condition fits for those environment values that do not change with time and physical processes well interior to the simulation region. If the physical process concentrates in the center of the simulation region and causes very little influence on the boundary, and at the same time, the surrounding environment of the simulation region is stable, then we can set the boundary to fixed boundary conditions.

Linear Boundary Conditions

If the influence of the physical processes in the simulation region is large enough to reach boundary, then we have to consider the interaction of this influence and outer environment. At this time, the boundary will change according to the result of this interaction. A simple treatment of this case is to look on the boundary as a linear continuous boundary. Assuming that $f(x_{\mathrm{b}})$ is the boundary value, the value of the inner point adjacent to x_{b}, $f(x_{\mathrm{b}}+\Delta x)$ the value of the outer point adjacent to x_{b}. Then for a linear boundary condition we can have

$$f(x_{\mathrm{b}}+\Delta x) = 2f(x_{\mathrm{b}}) - f(x_{\mathrm{b}}-\Delta x) \tag{13}$$

Using the newly found $f(x_{\mathrm{b}}+\Delta x)$, we can easily continue our simulation on the boundary.

Symmetric Boundary Conditions

In some of the simulations, we can assume a symmetric state exists on the boundary. This treatment of the boundary condition corresponds to the physical assumption that, on the two sides of boundary, the same physical processes exist. The variable values at the same distance from the boundary at the two sides are the same. The function of such a boundary is that of a mirror that can reflect all the fluctuations generated by the simulation region.

2.6 Initial Conditions

For any numerical simulation, we need to have an initial state to begin the simulation. So we need to set the initial environment values at the beginning of simulation. The initial condition of a numerical simulation thus gives an initial state for the simulation region. The system evolution will start from this initial condition. The initial condition should be able to reflect the real physical environment, at least it should be an acceptable simplification of the real environment.

In many cases, we can decide the initial values of the simulation region directly. But in some other cases, the initial states are so complex that it is very difficult to give the initial condition directly. So we need some other methods to derive the initial condition. For example, in global MHD simulation, generally we need to run a specific program first to determine the general configuration of magnetosphere. Only after that can we use the resulting configuration as the initial condition for the further simulation.

2.7 Time-Stepping

To have sufficient accuracy in computation, we should have the proper time step Δt. Theoretical study has shown that, if a simulation is stable, it must abide by Courant–Friedrichs–Lewy (CFL or Courant) condition

$$\alpha = \frac{v\,\Delta t}{\Delta x} \tag{14}$$

where α is called Courant number and v is the sum of all characteristic speeds and plasma speed. From the CFL condition we can see that, the smaller the space step, the smaller the time step to keep the program stable. After having defined a certain $\alpha < 1$ and having determined the spatial step, we can decide on the Δt to use.

2.8 Public Domain Prominent Codes

There are a few classical MHD codes which are publicly available and are well tested:

- **ZEUS3D:** ZEUS is several different numerical codes for astrophysical gas dynamics in two- and three-dimensions. The basic numerical algorithms employed are simple, but accurate and robust. A great deal of physics has been added to the codes, making them useful tools for investigation of a wide variety of problems. The codes and their tests are well documented in the refereed literature, and each version is freely available from the corresponding websites.

- **NIRVANA2.0:** The software NIRVANA2.0 contains the packages NIRVANA and NIR2VIZ. NIRVANA is a general-purpose numerical code for non-relativistic, visco-resistive-conductive, compressible, time-dependent magnetohydrodynamics in two or three space dimensions using Cartesian, cylindrical or spherical coordinates. NIRVANA implements a Poisson solver to handle selfgravitating flows. NIRVANA2.0 allows adaptive mesh refinement based on gradients in the solution (Ziegler and Yorke, 1997; Ziegler, 2002). The tool NIR2VIZ gives you a hand in converting flow data from NIRVANA simulations into visualization data suited for IDL or the IBM Data Explorer. In the version NIRVANA_CP it runs on shared memory machines and includes atomic networks for handling non-equlibrium cooling in astrophysical environments (Thiele, 2000; Camenzind et al., 2003).

- **FLASH**: The FLASH code developed at the University of Chicago Flash Center is a modular, general-purpose, adaptive, parallel simulation code capable of handling compressible flow problems in various physical, in particular astrophysical, environments. FLASH is designed to allow users to configure initial and boundary conditions, change algorithms, and add new physics modules with minimal effort. It uses the PARAMESH library to manage a block-structured adaptive grid, placing resolution elements only where they are needed most. FLASH uses the Message–Passing Interface (MPI) library to achieve portability and scalability on a variety of different parallel computers. Starting with version release 2.0, the FLASH code now also supports the equations of magnetohydrodynamics. The FLASH code is available for public distribution. Both information regarding code licensing and code distribution can be found at the Flash Center web site: `http://flash.uchicago.edu`.

Most of the numerical methods used in astrophysical fluid dynamics rely on explicit time–stepping schemes, whereas the higher robustness of implicit methods which constitute the core of modern computational fluid dynamics is rarely explored. Hujeirat and Rannacher (2002) survey some modern implicit solvers which are specially adapted to multi–dimensional problems and discuss their potential and range of application in comparison to common explicit methods. Special emphasis is put on the aspect of efficiency and robustness. The reference set of equations are those corresponding to radiative magnetohydrodynamics modelling self-gravitating and partially and/or fully ionized flows. The authors particularly present a new three–stages implicit numerical method for searching strongly time–dependent, quasi–stationary and steady–state solutions for the above equations. Preconditioned

Krylov–space and multilevel techniques are employed for enhancing the efficiency and robustness of the computation.

3 Progress in Understanding Fundamental MHD Processes

MHD simulations have widespread application in astrophysics ranging from solar physics to extragalactic scales and magnetic fields in the cosmological environment. In the following we discuss two examples where new insights have been gained in the last years.

3.1 MRI Driven Turbulence in Disks

Accretion Disks

Recent increases in supercomputer performance have significantly improved the ability to evolve the basic equations of accretion disk structure and evolution. These developments, along with continuing progress in understanding the most important physical processes that occur within accretion disks, suggest that predictive disk simulations are a realistic goal. Such disk simulations will be global, fully three–dimensional, and incorporate physical processes such as magnetohydrodynamics (MHD) and radiation transport. At present, we are still some ways from this goal; global simulations are still rather idealized in terms of disk structure, energetics, and dynamical range. However, because almost any three dimensional disk simulation is relatively novel, there remain many significant questions to be investigated even with such simplified models.

Much of this work has focused on thick accretion disks. With a pressure scale height H comparable to the disk radius R, the thick disk, or accretion torus, is more easily resolved in a numerical simulation than disks for which $H/R \ll 1$. Matsumoto (1999) followed the evolution of a thick torus embedded in an external vertical field, and found significant outflow collimated along the global vertical field lines. Hawley (2000) considered tori containing toroidal fields and poloidal field loops. In these studies the initial field was entirely contained within the disk and the resulting outflows were confined to the creation of a magnetized corona. A generic feature of all these thick disk simulations is the presence of large amplitude fluctuations in accretion rate, density, and other variables, in both space and time.

At a minimum these efforts have established that the magnetorotational instability, or MRI, (Balbus and Hawley, 1991) is just as efficacious in thick disks as in local simulations to produce MHD turbulence and angular momentum transport. Thick accretion tori with initially non-Keplerian angular momentum distributions are highly unstable (Balbus, 2005). MHD turbulence develops rapidly and is sustained by a self-consistent dynamo process within the disk. The constant or near-constant specific angular momentum distribution of the initial torus rapidly evolves to one that is near Keplerian (Fig. 3). The main focus for dynamical studies would therefore seem to be Keplerian disks, both hot (high internal sound speed), and thin and cold (low internal sound speed).

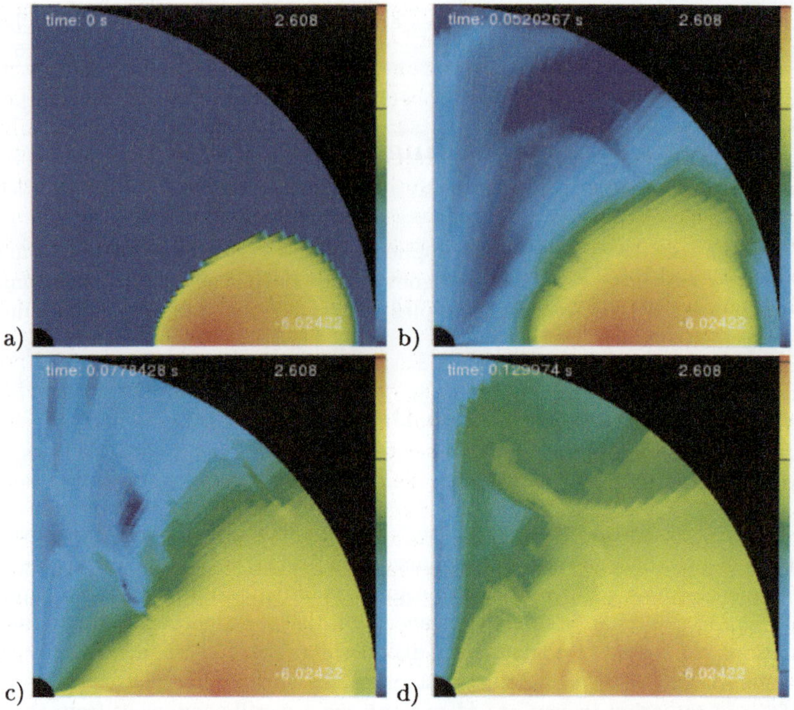

Fig. 2. Evolution of a torus including weak magnetic fields (plasma $\beta = 10^{-3}$) for various time steps (Brinkmann, 2004). The Black Hole of 10 solar masses (*dark central region*) is treated in the pseudo–Newtonian approach. The density is represented in logarithmic scale. Shown is the original torus configuration (**a**) and the density distribution at one (**b**), 1.5 (**c**) and 2.5 (**d**) revolution periods at density maximum. White numbers (rhs) indicate the values of the density maximum and minimum

The simulation shown in Fig. 2 assumes a toroidal configuration located at 16 Schwarzschild radii, having initially a constant specific angular momentum. The computational domain is in spherical coordinates (r, θ, ϕ) running from 2 to 30 Schwarzschild radii in r, $0 \leq \theta \leq \pi/2$. An outflow boundary condition is used at both the outer and inner radial boundaries, and periodic boundary conditions are used for ϕ.

Already after one revolution at the torus center, the MRI sets in, field energy is amplified, and soon the characteristic radial streaming structures (referred to as the channel solution) of the vertical field instability appear, much as they do in the local shearing box models. In the present simulation, these structures develop first at the inner part of the torus where the rotation frequency is the highest. The amplitude of the MRI becomes nonlinear by 3 orbits at the center of the grid, and filaments of strong magnetic field are carried inward and outward by fluid elements well out of Keplerian balance. These reach the outer part of the disk even before the local MRI in that region becomes fully nonlinear. Thus there are two immediate

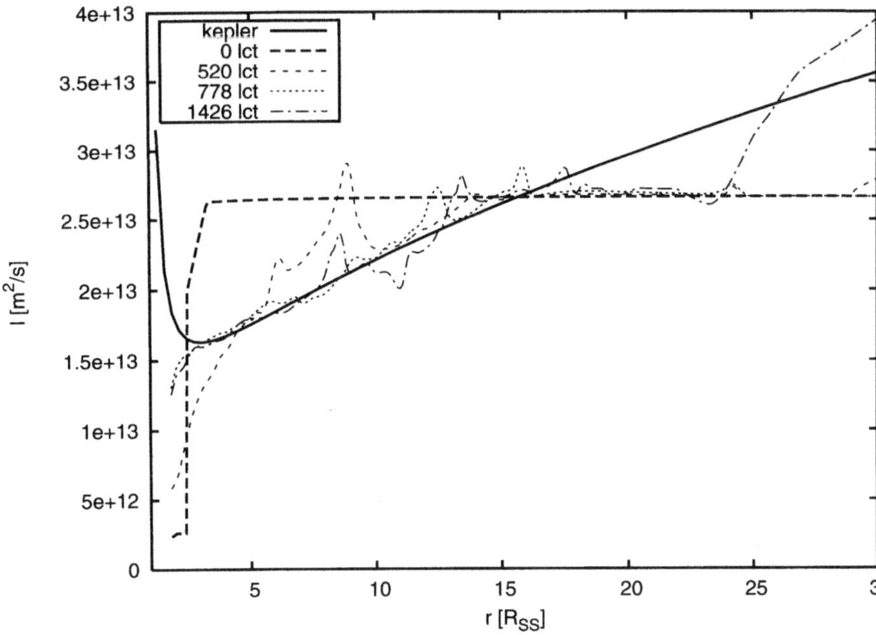

Fig. 3. Evolution of the angular momentum distribution in the accreting torus (Brinkmann, 2004). Radii are given in units of Schwarzschild radii. The initial constant specific angular momentum (*dashed line*) quickly evolves towards a quasi–Keplerian distribution (*solid line*). Compare with Fig. 2

global effects not seen in local simulations: linear growth rates that vary strongly with radius ($\omega_{MRI} \simeq \Omega \simeq R^{-3/2}$), and extended radial motions of significantly non–Keplerian plasma.

Galactic Disks

The Galactic intersstellar medium (ISM) is characterized by a complex distribution of density, temperature and magnetic fields, as well as a turbulent velocity field. Similar to accretion disks, the structure and dynamics of diffuse gas in the Milky Way and other disk galaxies may be strongly influenced by thermal and magnetorotational instabilities (TI and MRI) on scales of about 1–100 pc. Piontek and Ostriker (2004) have initiated a study of these processes, using two–dimensional numerical hydrodynamic and magnetohydrodynamic (MHD) simulations with conditions appropriate for the atomic interstellar medium (ISM). They demonstrate, consistent with previous work, that nonlinear development of *pure TI* produces a network of filaments that condense into cold clouds at their intersections, yielding a distinct two–phase warm/cold medium within about 20 Myr. TI–driven turbulent motions of the clouds saturate at subsonic amplitudes for uniform initial $P/k = 2000$ K cm^{-3}. MRI has previously been studied in near–uniform media; their simulations include both TI+MRI models, which begin from uniform–density conditions, and cloud+MRI models, which begin with a two–phase cloudy medium.

Both the TI+MRI and cloud+MRI models show that MRI develops within a few galactic orbital times, just as for a uniform medium. The mean separation between clouds can affect which MRI mode dominates the evolution. Provided intercloud separations do not exceed half the MRI wavelength, they find the MRI growth rates are similar to those for the corresponding uniform medium. This opens the possibility, if low cloud volume filling factors increase MRI dissipation times compared to those in a uniform medium, that MRI–driven motions in the ISM could reach amplitudes comparable to observed H I turbulent linewidths.

3.2 Magnetized Jets in Astrophysics

The evidence for highly collimated jets in astrophysics goes back to the early radio observations of twin lobes in extended radio galaxies, of which the prototype is Cygnus A. After associating them with optical galaxies at cosmological distances, it was clear that they had gigantic dimensions (up to megaparsec scales) and astonishing powers (up to 10^{40} Watt) emitted as nonthermal radio continua of synchrotron type. These facts made a single ejection event from the nucleus of the parent galaxy unlikely and, in general, posed a serious energetic problem. In addition, the short synchrotron lifetimes of relativistic electrons do not allow radio emission for more than a few million years, unless reacceleration is introduced to the picture, and the situation is obviously much worse for higher frequencies. Again, this phenomenology could be explained more economically in terms of fluid jets continuously transferring energy, momentum and magnetic fields from the galactic nuclei into the lobes and maintaining in situ particle reacceleration. Very long baseline interferometry (VLBI) observations traced the outflow collimation down to subparsec scales and allowed measurements in several cases of superluminal proper motions. This fact, together with a statistically significant presence of one–sided jets in strong sources, was considered evidence that jets may, at least in some cases, be relativistic.

In this framework, modeling of supersonic, relativistic, collimated outflows from AGNs has been one of the most challenging problems in astrophysics in recent years. The early development of the numerical study of supersonic hydrodynamic and magnetohydrodynamic flows has been connected with the observations of the solar and stellar winds and plasma motions in solar magnetic loops. Although the global and specific energetics of stellar and galactic phenomena differ by orders of magnitude, most of the dynamical events and the underlying physical processes may not be conceptually far apart.

The original phenomenological model was proposed by Rees and Scheuer. A pictorial scheme is illustrated in Fig. 4. Twin opposite jets are produced and collimated in the innermost cores of AGNs (sizes of milliparsec) by some powerful engine that most likely derives its energy from accretion onto a gravitational well and thrusts continuously supersonic and/or super-Alfvénic magnetized plasma along the angular momentum axis. The twin jets plough their way through the ambient intergalactic gas, transferring energy and momentum far away from the parent core. Jets are structurally affected by the interaction with the external medium originating shocks, filaments, and wiggles (Krause, 2002). Local electron acceleration to relativistic energies supports synchrotron emission. The "head" where the jet pushes against the external medium is a turbulent working surface producing a bow shock and a cocoon around the entire source.

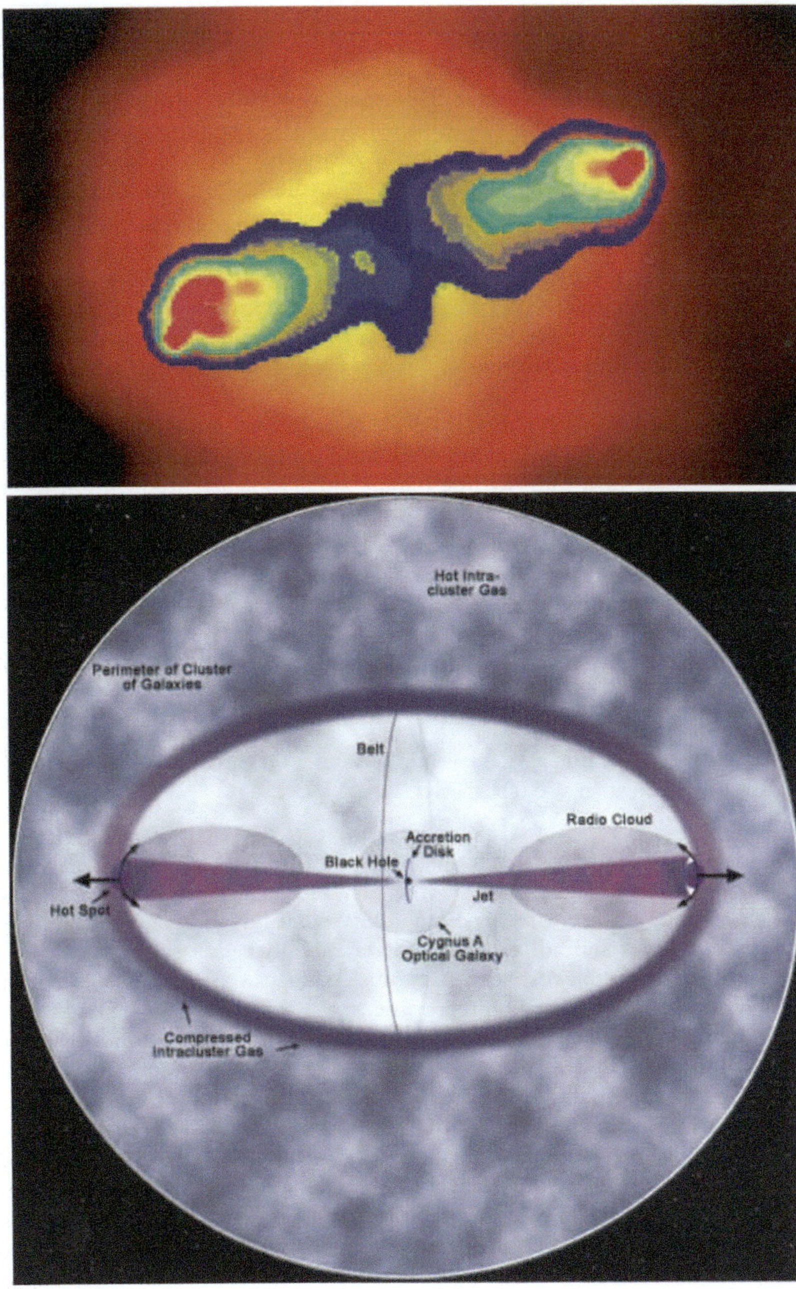

Fig. 4. The jets of quasars and radio galaxies propagate in a complex environment. *Top*: Cygnus A in X–rays (Chandra) superposed with low–frequency radio emission (VLA 330 MHz). The radio emission forms a cylindrical cocoon around the beam which is only visible in high frequency radio emission. *Bottom*: A schematic drawing for the confinement of radio jets by the ambient cluster medium. Initially, the jet drives an elliptical bow shock into the cluster gas

The Parameter Space

Jets of young low–mass stellar objects, as well as jets of quasars and radio galaxies are thought to be accelerated and collimated by rapidly rotating magnetospheres of the underlying central object (Camenzind, 1997). The propagation of these collimated outflows not only depends on the beam–velocity and Mach–number, but mainly on the structure of the surrounding medium. In clusters of galaxies, jets propagate through the hot cluster gas whose density profile is now known from X–ray observations. Bow–shocks have been found in the bright radio galaxy Cygnus A, while usually only the beam and shocked cocoon plasma is visible in synchrotron radiation. Despite velocity, the biggest difference between jets of low–mass stars and extragalactic jets in Radio Galaxies and Quasars is their environment. Both enevironments are highly turbulent, but the cluster medium has a definite density profile.

Once outflows are collimated to jets, their beams have to work against the ambient medium. Since the first simulations of jet propagation 20 years ago, it has become clear that in a pure hydrodynamic simulation jet propagation essentially depends on four parameters: the overall beam speed V_b, the density constrast $\eta = \rho_b \Gamma_b^2 h_b / \rho_M$, the internal Mach–number M and the pressure adjustment at inlet (Fig. 5). Γ_b is the Lorentz factor of the beam with specific enthalpy h_b. The question whether the jet starts in pressure equilibrium or not is a key issue. For $M \gg 1$ the advance speed of the bow shock follows from momentum conservation (see e.g. Camenzind, 1997)

$$V_{\rm head} = V_b \, \frac{\sqrt{\eta\epsilon}}{1 + \sqrt{\eta\epsilon}} \, . \tag{15}$$

Since in a cluster medium $\eta \ll 1$, we get a handle on the required beam speeds for bright radio galaxies ($\eta \leq 0.001$ in Cyg A)

$$V_b = 100 \, \frac{\rm kpc}{\rm Myr} \, \frac{0.0001}{\sqrt{\eta\epsilon}} \, \frac{V_{\rm head}}{\rm kpc/Myr} \, . \tag{16}$$

The sound speed in the cluster medium is somewhat less than one kpc/Myr, so that the beam speed is required to be at least marginally relativistic, $V_b > 0.3c$. [1] The Mach–number of the bow–shock in Cygnus A is less than two due to constraints from the jumps in the X–ray temperature.

Originally, only jets with high density contrast, $\eta > 0.01$ could be simulated on long time–scales. Due to enhanced computer power, low–density contrast simulations are now achievable downto $\eta > 10^{-5}$, typically encountered in galaxy cluster environments (Fig. 5). We have carried out simulations involving bipolar jets, removing the artificial boundary condition at the symmetry plane (Krause and Camenzind, 2003). We use a very low jet density (IGM/jet $\simeq 10^4$) and take into account a decreasing gas density profile in the cluster medium, as observed by Chandra and XMM. The jet bow shock undergoes two phases: First a nearly spherical one and second the well–known cigar–shaped structure. Cygnus A is obviously in a transition phase, showing clear signs from both phases. Due to inward

[1] In these units, the speed of light is 307 kpc/Myr, indicating that bow shocks are indeed propagating very slowly through the medium in view of lifetimes for the sources of 10 – 100 million years.

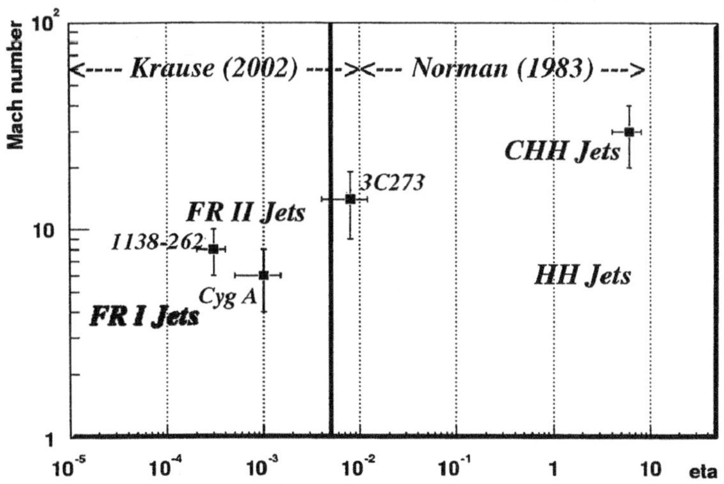

Fig. 5. Jet parameters for extragalactic jets compared to Herbig–Haro flows (HH jets). The sonic Mach–number is not a free parameter, but adjusts to values given by inner shocks in the beam. HH jets have high sonic Mach–numbers due to efficient cooling by line emission (CHH jets). The jet of the quasar 3C 273 is an extreme example due to high beam Lorentz factors. In the early Universe, the cluster gas has a higher density, the density contrast is even smaller and cooling of the shocked cluster gas leads to Lyα haloes (PKS 1138–262 with redshift 2.15). The density contrast for radio jets is normalized at a distance of one kpc from the center of the galaxy. Inclusion of magnetic fields at least doubles the dimension of the parameter space

growing of Kelvin–Helmholtz instabilities between cocoon and shocked IGM, mass entrainment is observed predominantly in the symmetry plane. This mechanism could explain some of the so far enigmatic X– ray features in the symmetry plane of Cygnus A.

While magnetic fields are probably not crucial for the understanding of the interaction between advancing jets and the galaxy cluster medium, the internal structure, and in particular the synchrotron emission from high energetic electrons, require a detailed information about the structure of the magnetic field. These are by far the most expensive simulations, since the plasma state has to be covered correctly. In the case of Herbig–Haro flows, the shock heated plasma will cool down by standard atomic processes. Fig. 6 shows the internal magnetic field distribution in cocoon and beam for a jet propagating with density contrast $\eta = 10$. The magnetic field, which is originally helical and only confined to the beam, undergoes pinching and kinking when it is dragged along the beam. In these knots, plasma is heated and cools by strong emission lines (Thiele and Camenzind, 2002).

Fig. 6. A 3D simulation of the propagation of magnetized Herbig–Haro flows simulated with NIRVANA_CP (Camenzind et al., 2003). The jet propagates from top–right to bottom–left. Helical magnetic fields are injected by the beam plasma, the ambient medium is not magnetized. Turbulence in the molecular cloud excites internal pinch and kink modes which appear in the observations as propagating knots (Thiele and Camenzind, 2002). The magnetic fields are shown here in the spaghetti representation

4 Special Relativistic MHD (SRMHD) Limits of the Classical MHD

As we have discussed in the beginning, classical MHD is a particular limit of relativistic magnetohydrodynamics which is the ab initio marriage of Maxwell's equations with plasma motion. There are many situations in astrophysical plasmas where this limit is no longer guaranteed.

4.1 SRMHD in Conservative Form

For this we write the SR MHD in fully conservative form (Komissarov, 1999)

$$\partial_\mu J^\mu = 0 \tag{17}$$

$$\partial_\mu T^{\mu\nu} = 0 \tag{18}$$

$$\partial_\mu (*F)^{\mu\nu} = 0\,. \tag{19}$$

The complete energy–momentum tensor $T^{\mu\nu}$ is given by a plasma part and a part generated by electromagnetic fields

$$T^{\mu\nu} = \left(\rho + P + \frac{b^2}{4\pi}\right) U^\mu U^\nu + \left(P + \frac{b^2}{8\pi}\right) g^{\mu\nu} - \frac{1}{4\pi} b^\mu b^\nu \,. \qquad (20)$$

The antisymmetric Faraday tensor $F_{\mu\nu}$ follows from the magnetic field b^μ and the 4–velocity U^μ (in its dual form)

$$*F^{\mu\nu} = b^\mu U^\nu - U^\mu b^\nu \,. \qquad (21)$$

From this we get the magnetic field strengths $\boldsymbol{B}$ in the lab system ($i = 1, 2, 3$)

$$B^i = *F^{it} = b^i U^t - U^i b^t \,. \qquad (22)$$

The homogeneous Maxwell equation is nothing else than the induction equation in the lab system

$$\partial_t \boldsymbol{B} = \nabla \times (\boldsymbol{V} \times \boldsymbol{B}) = \nabla \cdot (\boldsymbol{V} \otimes \boldsymbol{B} - \boldsymbol{B} \otimes \boldsymbol{V}) \,. \qquad (23)$$

In the framework of ideal MHD, the inhomogeneous Maxwell equations are merely used to calculate the current densities, but they are not used as dynamical equations.

In this form, ideal SRMHD is expressed in conservative form for the 8–dimensional state vector Q^a (Komissarov, 1999)

$$\frac{\partial Q^a}{\partial t} + \frac{\partial F^{ai}[\boldsymbol{Q}]}{\partial x^i} = 0 \,, \quad a = 1, \ldots, 8 \,. \qquad (24)$$

This vector contains the plasma velocities $\boldsymbol{V}$, the Lorentz factor $W = 1/\sqrt{1 - \boldsymbol{V}^2}$, the total energy τ, as well as the induction field b^μ

$$\boldsymbol{Q} = \begin{bmatrix} \rho_0 W \\ (\rho_0 h + b^2/4\pi) W^2 V^x - b^t b^x/4\pi \\ (\rho_0 h + b^2/4\pi) W^2 V^y - b^t b^y/4\pi \\ (\rho_0 h + b^2/4\pi) W^2 V^z - b^t b^z/4\pi \\ (\rho_0 h + b^2/4\pi) W^2 - (P_g + b^2/4\pi) - (b^t)^2/4\pi \\ B^x \\ B^y \\ B^z \end{bmatrix} \qquad (25)$$

This state vector can be expressed completely in terms of the general density D, the momentum vectors $\boldsymbol{S}$, the lab field $\boldsymbol{B}$ and the energy τ

$$\boldsymbol{Q} = \begin{bmatrix} D \equiv \rho_0 W \\ S_x \equiv \rho_0 \bar{h} W^2 V^x - (\boldsymbol{B} \cdot \boldsymbol{V}) B^x/4\pi \\ S_y \equiv \rho_0 \bar{h} W^2 V^y - (\boldsymbol{B} \cdot \boldsymbol{V}) B^y/4\pi \\ S_z \equiv \rho_0 \bar{h} W^2 V^z - (\boldsymbol{B} \cdot \boldsymbol{V}) B^z/4\pi \\ \tau \equiv \rho_0 h W^2 - P + \boldsymbol{B}^2/8\pi + \boldsymbol{E}^2/8\pi - D \\ B^x \\ B^y \\ B^z \end{bmatrix} \qquad (26)$$

This system generalizes the conservative scheme of special relativistic hydrodynamics (Marti and Müller, 2003). The corresponding fluxes $\boldsymbol{F}^a$ have the form

$$\boldsymbol{F}^i = \begin{bmatrix} DV^i \\ S_x V^i - B^i[B_x/4\pi W^2 + V_x(\boldsymbol{B}\cdot\boldsymbol{V})/4\pi] + P_T\delta^i_x \\ S_y V^i - B^i[B_y/4\pi W^2 + V_y(\boldsymbol{B}\cdot\boldsymbol{V})/4\pi] + P_T\delta^i_x \\ S_z V^i - B^i[B_z/4\pi W^2 + V_z(\boldsymbol{B}\cdot\boldsymbol{V})/4\pi] + P_T\delta^i_x \\ EV^i \\ B^iV^x - B^xV^i \\ B^iV^y - B^yV^i \\ B^iV^z - B^zV^i \end{bmatrix} \tag{27}$$

P_T represents the total pressure consisting of plasma pressure and pressure due to magnetic fields. In this formulation, it is obvious that Poynting flux not only contributes to the total energy flux, but also to momentum fluxes. This is the secret behind magnetically driven jets: momentum originally stored in magnetic fields can be converted into directed plasma motion. And this effect can be very efficient when the energy density in the magnetic fields dominates the plasma inertia, as is the case in Quasar jets. As a consequence, any attempt to simulate magnetic jet generation within classical MHD is waste of time, since the appropriate physics is not included in the classical MHD model.

4.2 Primitive Variables

The simulation of relativistic flows has one severe drawback. In contrast to the classical MHD, the primitive variables, such as ρ_0, $\boldsymbol{V}$ and temperature T are now in a nonlinear relation to the state vector $\mathbf{Q}$. The electric field e.g. follows from the MHD condition

$$\boldsymbol{E}^2 = \boldsymbol{B}^2\boldsymbol{V}^2 - (\boldsymbol{B}\cdot\boldsymbol{V})^2 \,, \tag{28}$$

and the momentum vector has various contributions

$$\boldsymbol{S} = \rho_0 \bar{h} W^2 c^2 \boldsymbol{V} - (\boldsymbol{B}\cdot\boldsymbol{V})\boldsymbol{B}/4\pi \tag{29}$$

with the modified enthalpy of the form

$$\bar{h} \equiv 1 + \epsilon + \frac{P}{\rho_0 c^2} + \frac{\boldsymbol{B}^2}{4\pi W^2 \rho_0 c^2} \,. \tag{30}$$

The inversion of these relations can only be done numerically and is in general quite CPU–time consuming (for the hydro case, see e.g. Marti and Müller, 2003). Conservative schemes are the basis of codes developed with a Godunov–type scheme by Komissarov (1999) and Koldoba et al. (2002), a third–order shock–capturing code avoiding the use of a Riemann solver by Del Zanna et al. (2002, 2003). Similar to classical MHD, such schemes will not preserve any numerical representation of $\nabla\cdot\boldsymbol{B} = 0$, except when an Evans–Balbus and Hawley (1998) type constrained transport scheme is used (i.e. a staggered mesh, where the magnetic fields are zone–face centered). Also here divergence–cleaning algorithms can be introduced.

4.3 Beyond SRMHD for Simulations of Extragalactic Jets

SRMHD has been applied mainly in the field of the propagation of relativistic jets and pulsar winds. But exactly in this field, SRMHD is probably not the correct vehicle to model jets of quasars and radio galaxies. In a relativistic jet, the plasma is heated to extreme temperatures (of the order of at least 10^{12} K), so that electrons readily achieve relativistic Lorentz factors and are now prone to acceleration to even higher energies in reconnection sheets and shock fronts. As a consequence, the one–component description of the jet plasma is a very sad approach and must be replaced by a true two–component plasma modelling in the future, before definite conclusions can be drawn about the plasma and magnetic field distribution in extragalactic jets.

5 Relativistic MHD for Rotating Black Holes (GRMHD)

The discovery by Balbus and Hawley (1998) that the weak magnetic fields in accretion disks provide the neccessary angular momentum transport by turbulence motivated the development of general relativistic MHD schemes (in general called GRMHD). The first numerical scheme for GRMHD was in fact written by Wilson already in 1977. Only twenty years later, modern schemes appeared slowly on the horizon which now can be used in extensive modelling. A second issue for the development of GRMHD codes has its roots in the theory of relativistic jet production. The presently–favored mechanism is an electrodynamic one, in which the hot plasma is accelerated by magnetic fields that are generated by strong differential rotation (Camenzind, 2005). The most pressing issues of current interest are understanding what factors control the jet power and its speed. These will have a direct bearing on understanding the origins of radio and blazar activity in active galactic nuclei and on the fundamental difference between radio loud and radio quiet objects. Clues to the answers to these questions may lie in related galactic sources – the microquasars in close binary stellar systems – and in their progenitors, the supernovae and gamma–ray bursters.

5.1 Kerr Magnetohydrodynamics

As already indicated in the last section, the extension of SRMHD to include gravitational effects is quite straightforward and follows the general covariance principle. No new elements appear in GRMHD, except that the conservation laws for momentum now have source terms given by the connection coefficients of the underlying metric field, which are well known in the case of general relativistic hydrodynamics (Font, 2003)

$$\frac{1}{\sqrt{\gamma}} \frac{\partial(\sqrt{\gamma} Q^a)}{\alpha\, \partial t} + \frac{1}{\alpha\sqrt{\gamma}} \frac{\partial(\alpha\sqrt{\gamma} F^{ai}[\boldsymbol{Q}])}{\partial x^i} = \mathcal{S}^a[g] \quad , \quad a = 1, \ldots, 8\,. \tag{31}$$

This particular formulation is adapted to the gravitaional field of rotating compact objects (such as neutron stars and Black Holes, see e.g. Camenzind, 2003), given by the line element

$$ds^2 = -\alpha^2\, dt^2 + R^2(d\phi - \omega\, dt)^2 + \exp(2\mu_r)\, dr^2 + \exp(2\mu_\theta)\, d\theta^2 . \qquad (32)$$

α is the redshift factor of the metric and ω the frame–dragging potential generated by the angular momentum of the compact object. γ in the above formula is the determinant of the absolute 3–space. This 3+1–split is nowadays the standard way to formulate relativistic MHD (Camenzind, 1998). In particular, the stationary limit of these equations is well known and can be used to formulate e.g. suitable initial conditions (Camenzind, 1996, 1998; Fendt and Memola, 2001).

5.2 Wave Speeds

The dispersion relation $D(k) = 0$ for MHD waves has only a simple form in a comoving frame. It can be formulated in terms of the relativistic sound speed $c_S^2 \equiv [\partial(\rho + u)/\partial p]_s^{-1} = \Gamma p/w$ (if $p = (\Gamma - 1)u$) and in terms of the relativistic Alfvén speed $\boldsymbol{V}_A \equiv \boldsymbol{B}/\sqrt{\mathcal{E}}$, where $\mathcal{E} \equiv b^2 + w$ and $w = \rho + u + p$ is the specific enthalpy (Appl and Camenzind, 1988)

$$\omega^2[\omega^2 - (\boldsymbol{k} \cdot \boldsymbol{V}_A)^2] \times \qquad (33)$$
$$\left[\omega^4 - \omega^2(k^2[\boldsymbol{V}_A^2 + c_S^2(1 - \boldsymbol{V}_A^2/c^2)] + c_S^2(\boldsymbol{k} \cdot \boldsymbol{V}_A^2/c^2) + k^2 c_S^2(\boldsymbol{k} \cdot \boldsymbol{V}_A^2)\right] = 0 .$$

The first frequency is the zero frequency entropy mode, the second is the Alfvén mode, and the third term contains the fast and slow modes.

In modelling the GRMHD equations, also here the first choice is between conservative and non–conservative schemes (ZEUS–type schemes as implemented by De Villiers and Hawley (2003). Conservative schemes are the basis of codes developed originally by Koide et al. (1999), and recently by Gammie et al. (2003).

The relativistic sound speed asymptotes to $\sqrt{\Gamma - 1}c = c/\sqrt{3}$ for $\Gamma = 4/3$. The Alfven speed asymptotes to the speed of light. In the limit $B^2/4\pi\rho \gg 1$ [2] and $p/\rho \ll 1$, the GRMHD equations degenerate to the force–free electrodynamic equations recently discussed by Komissarov (2002). In this limit, there is no slow mode, we only find fast modes and Alfvén modes that move at the speed of light. In a way, they are indistinguishable from vacuum electromagnetic modes.

Numerical implementation can be made second order in time by taking half–step from t^n to $t^{n+1/2}$, evaluating $\mathcal{F}(\mathcal{P}(t^{n+1/2}))$ and then update $\mathbf{U}(t^n)$ to $\mathbf{U}(t^{n+1})$. As always in relativistic calculations, it is appropriate to modify the total energy by subtracting out the particle number conservation – magnetic and internal energy densities can be order of magnitudes smaller than the rest mass energy density.

5.3 MRI near Rotating Black Holes

In Fig. 7 we show the time evolution of the density distribution in a weakly magnetized torus around a rotating Black Hole with Kerr parameter $a/M_H = 0.5$, simulated with the GRMHD code HARMS (Gammie et al., 2003). The pressure maximum is at 12 gravitational radii. Superposed on this equilibrium is a purely

[2] In this limit magnetic energy completely dominates inertia, which would correspond to superluminal Alfvén speeds in classical MHD !

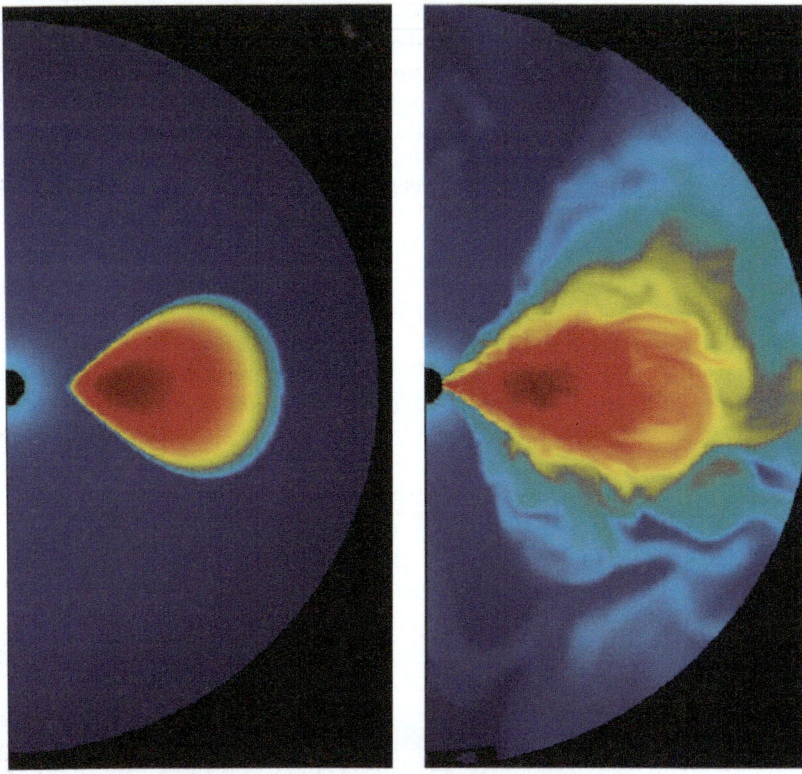

Fig. 7. Disk evolution near a rotating Kerr Black Hole with spin parameter $a/M_H = 0.5$ at $t = 0$ (*left*) and after 2000 light crossing times (*right*) (Gammie et al., 2003). The resolution is 300×300 grid points. After a few orbital periods at the pressure maximum the accretion becomes fully turbulent, as in the newtonian calculations

poloidal magnetic field with plasma beta of 0.01. The orbital period at the pressure maximum is 264 light crossing times. The simulation runs to 2000 light crossing times, corresponding to 7.6 orbital periods at pressure maximum. It will be interesting to investigate the coupling between magnetic fields and the gravitomagnetic field of the Black Hole occuring near the ergosphere of the Black Hole, i.e. typically in a region smaller than 3 gravitational radii (Krolik et al., 2005). This is also the region where the velocities in pseudo–Newtonian calculations are to be the most inaccurate. In relativity, speeds are limited to the speed of light, while in pseudo–Newtonian simulations they can easily exceed this causal limit. For this purpose, simulations with extremely high resolution are needed and the singular Boyer–Lindquist coordinates must be avoided.

5.4 Jets as Outflows from the Ergospheric Region

There are theoretical reasons for postulating that the AGN jets are relativistic outflows (Lorentz gamma > 5) initiated on scales of order of a few gravitational radii $r_g = GM/c^2$. From models of central engines (massive black holes) the jet energy may be extracted electromagnetically from the black holes and the jet plasma is believed to be either normal plasma consisting of thermal ions and relativistic electrons, or to be electron–positron pairs. One of the most efficient process to accelerate the plasma to relativistic speeds is by means of transformation of Poynting flux into kinetic energy along collimated flux tubes (Camenzind, 1998). This process is understood in the stationary limit of GRMHD, but could not be simulated successfully in a time–dependent fashion. When plasma is embedded into a strong magnetic field within the ergosphere of the Kerr Black Hole, it can attain a total negative energy and angular momentum (Fig. 8). As a consequence, the frame–dragging effect produces a positive Poynting–flux flowing away from the ergosphere and channelled along the flux tubes. Within the lightcylinder, this Poynting–flux is converted into kinetic energy of the outflowing disk wind (Camenzind, 2005). It is one of the big challenges of GRMHD to simulate this process self–consistently with codes based on conservative schemes.

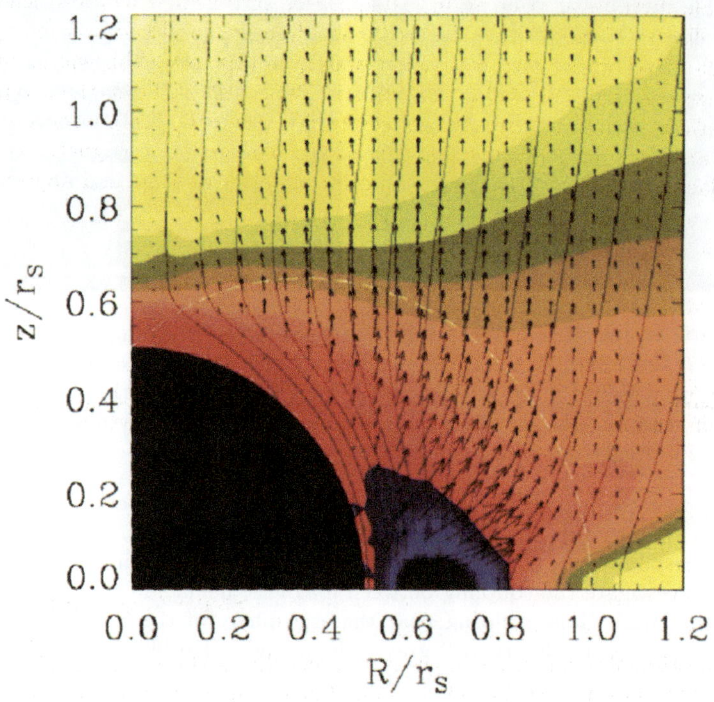

Fig. 8. The ergospheric region near a very rapidly rotating Kerr Black Hole. Large–scale magnetic fields are swamped inwards through the disk and embed the Black Hole into a rotating magnetosphere (*solid lines*). In the blue region, the plasma has negative total energy and angular momentum. Positive energy flows away as Poynting flux (*arrows*)

6 Future Prospects

The inclusion of magnetic fields is of great importance for many astrophysical systems. The formation and collimation process of (relativistic) jets (powering powerful extragalactic radio sources, galactic microquasars, and GRBs) most likely involves dynamically important magnetic fields and occurs in strong gravitational fields. The same is likely to be true for accretion discs around black holes. Magneto–relativistic effects even play a non–negligible role in the formation of proto-stellar jets in regions close to the light cylinder (Camenzind, 1997). Thus, relativistic MHD codes are a very desirable tool in astrophysics. The non–trivial task of developing such a kind of code is considerably simplified by the fact that because of the high conductivity of astrophysical plasmas one must only consider ideal SRMHD in most applications. Despite the fact that the field of modeling MHD processes in Astrophysics has made considerable progress in the last years, there are still many questions open. In particular, ideal MHD is not the appropriate vehicle in many applications, it has to be extended towards radiative MHD and two–component theories, especially in the case of extragalactic jet propagation.

The purpose of any (Newtonian or relativistic) MHD code is to evolve the induction equation to obtain the magnetic fields from which to calculate the Lorentz force. Magnetic fields are divergence free. Hence, numerical schemes are required to maintain this constraint (if fulfilled for the initial data) during the evolution. The question of divergence cleaning algorithms in the relativistic case is still somewhat obscure and has to be formulated in a conformity with causality.

The understanding of the formation of relativistic collimated outflows from Black Holes is still not yet settled. However, despite their success, present simulations only cover a tiny fraction of dynamical time scales (typicall a few rotational periods of the accretion disk) and jets are formed with very small terminal speeds (Lorentz factors less than 2). Hence, the quest for robust codes able to follow the formation of steady relativistic jets is still open. In this field, conservative codes are probably the adequate vehicle to model these issues.

The future in understanding Black Hole physics will be mainly based on heavy numerical simulations of accretion processes, magnetic coupling near rotating Black Holes and time–dependent formation of collimated outflows. For all these purposes, simple ideal MHD is not sufficient. The above discussed frameworks have to be extended to include radiative processes and propagation of photons in the background of Black Holes in order to understand the observed spectra (Müller and Camenzind, 2004). In particular, the magnetorotational instability in radiation pressure dominated accretion disks is mainly unexplored.

Acknowledgement

Part of this research has been supported by the Sonderforschungsbereich 439 of the Deutsche Forschungsgemeinschaft *Galaxies in the Early Universe* in Heidelberg, and by the *Star Formation* Schwerpunkt of the Deutsche Forschungsgemeinschaft (M. Thiele).

References

Appl, S., Camenzind, M.: 1988, A&A 206, 258

Balbus, S. A.: 2005, in *The Astrophysics of Cataclysmic Variables and Related Objects*, ed. J. M. Hameury, J. P. Lasota (San Francisco: ASP), ASP Conf. Proc. 330, p. 185

Balbus, S. A., Hawley, J. F.: 1991, ApJ 376, 214

Balbus, S. A., Hawley, J. F.: 1998, Rev. Mod. Phys. 70, 1

Balsara, D. S., Spicer, D. S.: 1999, J. Comp. Phys. 149, 270

Brinkmann, S.: 2004: *MHD–Instabilities in Accretion Disks*, Diploma Thesis, University of Heidelberg

Brio, M., Wu, C. C.: 1988, J. Comp. Phys. 75, 400

Camenzind, M.: 1996, in: *Solar and Astrophysical Magnetohydrodynamic Flows*, ed. K. Tsinganos (Dordrecht: Kluwer), p. 699

Camenzind, M.: 1997, in IAU 118 *Herbig–Haro Flows and the Birth of Low Mass Stars*, ed. B. Reipurth, C. Bertout (Dordrecht: Kluwer) p. 241

Camenzind, M.: 1998, in *Relativistic Astrophysics*, ed. H. Riffert, H. Ruder, H.-P. Nollert, F. W. Hehl (Braunschweig: Vieweg), p. 82

Camenzind, M., Krause, M., Thiele, M.: 2003, in *High Performance Supercomputing in Science and Engineering*, ed. E. Krause, W. Jäger (Berlin: Springer), p. 62

Camenzind, M.: 2003, in *Accretion Discs, Jets and High Energy Phenomena in Astrophysics*, Les Houches Lecture, Session LXXVIII, ed. V. Beskin, G. Henri, F. Menard, G. Pelletier, J. Dalibard (Paris: EDP Sciences), p. 405

Camenzind, M.: 2005, Mem. Soc. Astr. Ital. 76, 98

Chombo: `http://seesar.lbl.gov/anag/chombo/`

Crockett, R. K., Colella, P., Fisher, R. T., Klein, R. I., McKee, C. F.: 2005, J. Comp. Phy. 203, 422

Dai, W., Woodward, P. R.: 1994, J. Comp. Phys. 115, 485

Dedner, A., Kemm, F., Kröner, D., Munz, C.-D., Schnitzer, T., Wesenberg, M.: 2002, J. Comp. Phys. 175, 645

Del Zanna, L., Bucciantini, N.: 2002, A&A 390, 1177

Del Zanna, L., Bucciantini, N., Londrillo, P.: 2003, A&A 400, 397

De Villiers, J.-P., Hawley, J. F.: 2003, ApJ 592, 1060

De Villiers, J.-P., Hawley, J. F., Krolik, J. H.: 2003a, ApJ 589, 458

De Villiers, J.-P., Hawley, J. F., Krolik, J. H.: 2003b, ApJ 599, 1238

Evans, C. R., Hawley, J. F.: 1988, ApJ 332, 695

Falle, S. A.: 2002, ApJ 577, L123

Fendt, C., Memola, E.: 2001, A&A 365, 631

Font, J. A.: 2003, Living Rev. Rel. 3, 2003–4

Gammie, C. F., McKinney, J. C., Toth, G.: 2003, ApJ 589, 444

Godunov, S. K.: 1972, Num. Methods Mech. Cont. Media 1, 26

Hawley, J. F.: 2000, ApJ 528, 462

Hawley, J. H., Krolik, J. H.: 2002, ApJ 566, 164

Hujeirat, A., Rannacher, R.: 2002, New Astron. Rev. 45, 425

Koide, S., Shibata, K., Kudoh, T.: 1999, ApJ 522, 727

Koldoba, A. V., Kuznetsov, O. A., Ustyugova, G. V.: 2002, MNRAS 333, 932

Krause, M.: 2002, *On the Interaction of Jets with the Dense Medium of the Early Universe*, PhD Thesis, University Heidelberg

Krause, M., Camenzind, M.: 2003, New Astron. Rev. 47, 573
Komissarov, S.: 1999, MNRAS 303, 343
Komissarov, S.: 2002, MNRAS 336, 759
Krolik, J. H., Hawley, J. F., Hirose, S.: 2005, ApJ 622, 1008
Marti, J.-M., Müller, E.: 2003, Living Rev. Rel. 2003-7
Matsumoto, R.: 1999: in *Numerical Astrophysics*, ed. Miyama, S., Tomisaka, K., Hanawa, T. (Dordrecht: Kluwer), p. 195
Müller, A., Camenzind, M.: 2004, A&A 413, 861
PARAMESH: http://ct.gsfc.nasa.gov/paramesh/Users_manual/amr.html
Piontek, R. A., Ostriker, E. C.: 2004, ApJ 601, 905
Stone, J. M., Norman, M. L.: 1992, ApJS 80, 791
Thiele, M.: 2000: *Numerische Simulationen Protostellarer Jets*, PhD Thesis, University Heidelberg
Thiele, M., Camenzind, M.: 2002, A&A 381, L53
Zachary, A. L., Colella, P.: 1992, J. Comp. Phys. 99, 341
Ziegler, U., Yorke, H.: 1997, Comp. Phys. Comm. 101, 54
Ziegler, U.: 2002, http://www.aip.de/groups/MHD/projects/NIRVANA/nirvana.html

Lecture Notes in Physics

For information about earlier volumes please contact your bookseller or Springer LNP Online archive: springerlink.com

Vol.618: M.D. Esposti, S. Graffi (Eds.), The Mathematical Aspects of Quantum Maps

Vol.619: H.M. Antia, A. Bhatnagar, P. Ulmschneider (Eds.), Lectures on Solar Physics

Vol.620: C. Fiolhais, F. Nogueira, M. Marques (Eds.), A Primer in Density Functional Theory

Vol.621: G. Rangarajan, M. Ding (Eds.), Processes with Long-Range Correlations

Vol.622: F. Benatti, R. Floreanini (Eds.), Irreversible Quantum Dynamics

Vol.623: M. Falcke, D. Malchow (Eds.), Understanding Calcium Dynamics, Experiments and Theory

Vol.624: T. Pöschel (Ed.), Granular Gas Dynamics

Vol.625: R. Pastor-Satorras, M. Rubi, A. Diaz-Guilera (Eds.), Statistical Mechanics of Complex Networks

Vol.626: G. Contopoulos, N. Voglis (Eds.), Galaxies and Chaos

Vol.627: S.G. Karshenboim, V.B. Smirnov (Eds.), Precision Physics of Simple Atomic Systems

Vol.628: R. Narayanan, D. Schwabe (Eds.), Interfacial Fluid Dynamics and Transport Processes

Vol.629: U.-G. Meißner, W. Plessas (Eds.), Lectures on Flavor Physics

Vol.630: T. Brandes, S. Kettemann (Eds.), Anderson Localization and Its Ramifications

Vol.631: D. J. W. Giulini, C. Kiefer, C. Lämmerzahl (Eds.), Quantum Gravity, From Theory to Experimental Search

Vol.632: A. M. Greco (Ed.), Direct and Inverse Methods in Nonlinear Evolution Equations

Vol.633: H.-T. Elze (Ed.), Decoherence and Entropy in Complex Systems, Based on Selected Lectures from DICE 2002

Vol.634: R. Haberlandt, D. Michel, A. Pöppl, R. Stannarius (Eds.), Molecules in Interaction with Surfaces and Interfaces

Vol.635: D. Alloin, W. Gieren (Eds.), Stellar Candles for the Extragalactic Distance Scale

Vol.636: R. Livi, A. Vulpiani (Eds.), The Kolmogorov Legacy in Physics, A Century of Turbulence and Complexity

Vol.637: I. Müller, P. Strehlow, Rubber and Rubber Balloons, Paradigms of Thermodynamics

Vol.638: Y. Kosmann-Schwarzbach, B. Grammaticos, K.M. Tamizhmani (Eds.), Integrability of Nonlinear Systems

Vol.639: G. Ripka, Dual Superconductor Models of Color Confinement

Vol.640: M. Karttunen, I. Vattulainen, A. Lukkarinen (Eds.), Novel Methods in Soft Matter Simulations

Vol.641: A. Lalazissis, P. Ring, D. Vretenar (Eds.), Extended Density Functionals in Nuclear Structure Physics

Vol.642: W. Hergert, A. Ernst, M. Däne (Eds.), Computational Materials Science

Vol.643: F. Strocchi, Symmetry Breaking

Vol.644: B. Grammaticos, Y. Kosmann-Schwarzbach, T. Tamizhmani (Eds.) Discrete Integrable Systems

Vol.645: U. Schollwöck, J. Richter, D.J.J. Farnell, R.F. Bishop (Eds.), Quantum Magnetism

Vol.646: N. Bretón, J. L. Cervantes-Cota, M. Salgado (Eds.), The Early Universe and Observational Cosmology

Vol.647: D. Blaschke, M. A. Ivanov, T. Mannel (Eds.), Heavy Quark Physics

Vol.648: S. G. Karshenboim, E. Peik (Eds.), Astrophysics, Clocks and Fundamental Constants

Vol.649: M. Paris, J. Rehacek (Eds.), Quantum State Estimation

Vol.650: E. Ben-Naim, H. Frauenfelder, Z. Toroczkai (Eds.), Complex Networks

Vol.651: J.S. Al-Khalili, E. Roeckl (Eds.), The Euroschool Lectures of Physics with Exotic Beams, Vol.I

Vol.652: J. Arias, M. Lozano (Eds.), Exotic Nuclear Physics

Vol.653: E. Papantonoupoulos (Ed.), The Physics of the Early Universe

Vol.654: G. Cassinelli, A. Levrero, E. de Vito, P. J. Lahti (Eds.), Theory and Appplication to the Galileo Group

Vol.655: M. Shillor, M. Sofonea, J.J. Telega, Models and Analysis of Quasistatic Contact

Vol.656: K. Scherer, H. Fichtner, B. Heber, U. Mall (Eds.), Space Weather

Vol.657: J. Gemmer, M. Michel, G. Mahler (Eds.), Quantum Thermodynamics

Vol.658: K. Busch, A. Powell, C. Röthig, G. Schön, J. Weissmüller (Eds.), Functional Nanostructures

Vol.659: E. Bick, F.D. Steffen (Eds.), Topology and Geometry in Physics

Vol.660: A.N. Gorban, I.V. Karlin, Invariant Manifolds for Physical and Chemical Kinetics

Vol.661: N. Akhmediev, A. Ankiewicz (Eds.) Dissipative Solitons

Vol.662: U. Carow-Watamura, Y. Maeda, S. Watamura (Eds.), Quantum Field Theory and Noncommutative Geometry

Vol.663: A. Kalloniatis, D. Leinweber, A. Williams (Eds.), Lattice Hadron Physics

Vol.664: R. Wielebinski, R. Beck (Eds.), Cosmic Magnetic Fields

GPSR Compliance

The European Union's (EU) General Product Safety Regulation (GPSR) is a set of rules that requires consumer products to be safe and our obligations to ensure this.

If you have any concerns about our products, you can contact us on ProductSafety@springernature.com

In case Publisher is established outside the EU, the EU authorized representative is:

Springer Nature Customer Service Center GmbH
Europaplatz 3
69115 Heidelberg, Germany

Batch number: 10682471

Printed by Printforce, the Netherlands